LLYFRGELL
LIBRARY
ABERYSTWYTH

D0239229

The Practical Approach Series

SERIES EDITORS

D. RICKWOOD
Department of Biology, University of Essex,
Wivenhoe Park, Colchester, Essex CO4 3SQ, UK

B. D. HAMES
Department of Biochemistry and Molecular Biology
University of Leeds, Leeds LS2 9JT, UK

★ indicates new and forthcoming titles

Affinity Chromatography
Anaerobic Microbiology
Animal Cell Culture (2nd edition)
Animal Virus Pathogenesis
Antibodies I and II
★ Antibody Engineering
Basic Cell Culture
Behavioural Neuroscience
Biochemical Toxicology
Bioenergetics
Biological Data Analysis
Biological Membranes
Biomechanics—Materials
Biomechanics—Structures and Systems
Biosensors
Carbohydrate Analysis (2nd edition)
Cell–Cell Interactions
The Cell Cycle
Cell Growth and Apoptosis
Cellular Calcium
Cellular Interactions in Development
Cellular Neurobiology
Clinical Immunology
Crystallization of Nucleic Acids and Proteins
★ Cytokines (2nd edition)
The Cytoskeleton
Diagnostic Molecular Pathology I and II
Directed Mutagenesis
★ DNA and Protein Sequence Analysis
DNA Cloning 1: Core Techniques (2nd edition)
DNA Cloning 2: Expression Systems (2nd edition)
★ DNA Cloning 3: Complex Genomes (2nd edition)
★ DNA Cloning 4: Mammalian Systems (2nd edition)
Electron Microscopy in Biology

Electron Microscopy in Molecular Biology
Electrophysiology
Enzyme Assays
★ Epithelial Cell Culture
Essential Developmental Biology
Essential Molecular Biology I and II
Experimental Neuroanatomy
★ Extracellular Matrix
Flow Cytometry (2nd edition)
★ Free Radicals
Gas Chromatography
Gel Electrophoresis of Nucleic Acids (2nd edition)
Gel Electrophoresis of Proteins (2nd edition)
Gene Probes 1 and 2
Gene Targeting
Gene Transcription
Glycobiology
Growth Factors
Haemopoiesis
Histocompatibility Testing
★ HIV Volumes 1 and 2
Human Cytogenetics I and II (2nd edition)
Human Genetic Disease Analysis
Immunocytochemistry
In Situ Hybridization
Iodinated Density Gradient Media
★ Ion Channels
Lipid Analysis
Lipid Modification of Proteins
Lipoprotein Analysis
Liposomes
Mammalian Cell Biotechnology
Medical Bacteriology
Medical Mycology
Medical Parasitology
Medical Virology
Microcomputers in Biology
Molecular Genetic Analysis of Populations
Molecular Genetics of Yeast
Molecular Imaging in Neuroscience
Molecular Neurobiology
Molecular Plant Pathology I and II
Molecular Virology
Monitoring Neuronal Activity
Mutagenicity Testing
★ Neural Cell Culture
Neural Transplantation
★ Neurochemistry (2nd edition)
Neuronal Cell Lines
NMR of Biological Macromolecules
Non-isotopic Methods in Molecular Biology
Nucleic Acid Hybridization
Nucleic Acid and Protein Sequence Analysis
Oligonucleotides and Analogues
Oligonucleotide Synthesis
PCR 1
PCR 2
Peptide Antigens

348 NOV 99

Photosynthesis: Energy Transduction
Plant Cell Biology
Plant Cell Culture (2nd edition)
Plant Molecular Biology
Plasmids (2nd edition)
★ Platelets
Pollination Ecology
Postimplantation Mammalian Embryos
Preparative Centrifugation
Prostaglandins and Related Substances
Protein Blotting
Protein Engineering
Protein Function
Protein Phosphorylation
Protein Purification Applications
Protein Purification Methods
Protein Sequencing
Protein Structure
★ Protein Structure Prediction
Protein Targeting
Proteolytic Enzymes
★ Pulsed Field Gel Electrophoresis
Radioisotopes in Biology
Receptor Biochemistry
Receptor–Ligand Interactions
RNA Processing I and II
★ Subcellular Fractionation
Signal Transduction
Solid Phase Peptide Synthesis
Transcription Factors
Transcription and Translation
Tumour Immunobiology
Virology
Yeast

Subcellular Fractionation

A Practical Approach

Edited by

J. M. GRAHAM
John Graham Research Consultancy,
School of Biomolecular Sciences,
Liverpool John Moores University,
Liverpool L3 3AF, UK

and

D. RICKWOOD
Department of Biology, University of Essex,
Colchester CO4 3SQ, UK

at
OXFORD UNIVERSITY PRESS
Oxford New York Tokyo

Oxford University Press, Great Clarendon Street, Oxford OX2 6DP

Oxford New York
Athens Auckland Bangkok Bogota Bombay Buenos Aires
Calcutta Cape Town Dar es Salaam Delhi Florence Hong Kong
Istanbul Karachi Kuala Lumpur Madras Madrid Melbourne
Mexico City Nairobi Paris Singapore Taipei Tokyo Toronto

and associated companies in
Berlin Ibadan

Oxford is a trade mark of Oxford University Press

Published in the United States
by Oxford University Press Inc., New York

A catalogue record for this book is available from the British Library

Library of Congress Cataloging in Publication Data
(Data available)

ISBN 0 19 963495 5 (Hbk)
ISBN 0 19 963494 7 (Pbk)

Typeset by Footnote Graphics, Warminster, Wilts
Printed in Great Britain by Information Press, Ltd, Eynsham, Oxon.

Preface

Subcellular fractionation is a core procedure in cell biology and it relies very heavily on the use of centrifugation as the primary technique. The editors hope therefore that this latest addition to the Practical Approach series will complement earlier volumes of the series. While any text on subcellular fractionation is naturally concerned with the isolation of subcellular particles, this book attempts to present these techniques in the context of a number of related functional topics that are particularly relevant to modern research in cell and molecular biology. Thus the chapter on endocytosis is concerned not only with the isolation of various endocytic compartments but also covers for example the labelling of ligands, the kinetics of their internalization and analysis of data. The isolation methods for rough and smooth endoplasmic reticulum, Golgi membranes, and plasma membrane are not only concerned with these membranes as individual compartments but also as part of the complex secretory process. Methods for the purification of specific organelles (nuclei, mitochondria, peroxisomes, chloroplasts, and ribosomes) are accompanied by important analytical procedures and protocols concerned with assessing their function and their structure. The seven chapters devoted to the isolation and analysis of these different subcellular compartments are prefaced by two chapters: one on homogenization which presents the available strategies and enables the reader to choose a technique which is best suited to a particular tissue and subcellular fraction, and a second which reviews the choice of separation techniques and some of the problems that might be encountered. The final chapter is devoted to a discussion of the principal approaches to the determination of membrane structure by electron microscopy.

Colchester and Liverpool D. R.
September 1996 J. M. G.

Contents

Contributors

E. BAUMGART
Ruprecht-Karls Universität Heidelberg, Institut für Anatomie und Zellbiologie (II), Im Neuenheimer Feld 307, D-69120 Heidelberg, Germany.

TROND BERG
Department of Biology, Faculty of Mathematics and Natural Sciences, University of Oslo, PO Box 1050, Blindern, N-0316 Oslo, Norway.

TROND OLAV BERG
Department of Biology, Faculty of Mathematics and Natural Sciences, University of Oslo, PO Box 1050, Blindern, N-0316 Oslo, Norway.

ULRICH A. BOMMER
Division of Biochemistry, Department of Cellular and Molecular Sciences, St. George's Hospital Medical School, Cranmer Terrace, London SW17 0RE, UK.

NILS BURKHARDT
Max-Planck-Institut für Molekulare Genetik, Ihnestr. 73, 14195 Berlin, Germany.

H. D. FAHIMI
Ruprecht-Karls Universität Heidelberg, Institut für Anatomie und Zellbiologie (II), Im Neuenheimer Feld 307, D-69120 Heidelberg, Germany.

TOR GJØEN
Department of Biology, Faculty of Mathematics and Natural Sciences, University of Oslo, PO Box 1050, Blindern, N-0316 Oslo, Norway.

JOHN M. GRAHAM
John Graham Research Consultancy, School of Molecular Sciences, Liverpool John Moores University, Liverpool L3 3AF, UK.

RICHARD H. HINTON
School of Biological Sciences, University of Surrey, Guildford, Surrey GU2 5XH, UK.

NIGEL JAMES
Department of Biomedical Science, University of Sheffield, PO Box 601, Sheffield S10 2TN, UK.

RALF JÜNEMANN
Max-Planck-Institut für Molekulare Genetik, Ihnestr. 73, 14195 Berlin, Germany.

J. G. LINDSAY
Department of Biochemistry, University of Glasgow, Glasgow G12 8QQ, Scotland, UK.

ANTHEA MESSENT
Department of Biology, University of Essex, Colchester CO4 3SQ, UK.

ANTHONY L. MOORE
Biochemistry Department, University of Sussex, Falmer, Brighton BN1 9QG, UK.

BARBARA M. MULLOCK
Department of Clinical Biochemistry, University of Cambridge, Addenbrookes Hospital, Cambridge CB2 2QR UK.

KNUD H. NIERHAUS
Max-Planck-Institut für Molekulare Genetik, Ihnestr. 73, 14195 Berlin, Germany.

DIPAK PATEL
Department of Biology, University of Essex, Colchester CO4 3SQ, UK.

J. E. RICE
Department of Biochemistry, University of Glasgow, Glasgow G12 8QQ, Scotland, UK.

DAVID RICKWOOD
Department of Biology, University of Essex, Colchester CO4 3SQ, UK.

CHRISTIAN M. T. SPAHN
Max-Planck-Institut für Molekulare Genetik, Ihnestr. 73, 14195 Berlin, Germany.

FRANCISCO J. TRIANA-ALONSO
Max-Planck-Institut für Molekulare Genetik, Ihnestr. 73, 14195 Berlin, Germany.

A. VÖLKL
Ruprecht-Karls Universität Heidelberg, Institut für Anatomie und Zellbiologie (II), Im Neuenheimer Feld 307, D-69120 Heidelberg, Germany.

DAVID G. WHITEHOUSE
Biochemistry Department, University of Sussex, Falmer, Brighton BN1 9QG, UK.

Abbreviations

AOM	asialoorosomucoid
β-AGA	β-acetylglucosaminidase
APS	ammonium persulfate
BSA	bovine serum albumin
CFE	continuous flow electrophoresis
DAB	diaminobenzidine
DEPC	diethylpyrocarbonate
DMSO	dimethyl sulfoxide
DNA	deoxyribonucleic acid
DNase	deoxyribonuclease
DNP	2,4-dinitrophenol
DTT	dithiothreitol
EDTA	ethylenediaminetetraacetic acid
EGTA	ethylene glycol-bis(β-aminoethyl ether) *N,N,N′, N′*-tetraacetic acid
EM	electron microscopy
EOP	effective osmotic pressure
ER	endoplasmic reticulum
FRP	final reaction product
G-6-P	glucose-6-phosphate
GA	glutaraldehyde
GB	gradient buffer
HM	homogenization medium
hnRNP	heterogeneous ribonucleoprotein particle(s)
HRP	horse-radish peroxidase
IM	inner membrane
IMS	intermembrane space
LDL	low density lipoprotein
MPR	mannose 6-phosphate receptor
OM	outer membrane
PD	percentage distribution
PMP	peroxisomal membrane proteins
PMS	phenazine methosulfate
PMSF	phenylmethylsulfonyl fluoride
PO	peroxisomes
PRP	primary reaction product
PVP	polyvinylpyrrolidone
PVS	polyvinyl sulfate
RCF	relative centrifugal force
RNA	ribonucleic acid

RNase	ribonuclease
RNP	ribonucleoprotein particle(s)
RSA	relative specific activity
RSW	ribosomal salt wash
SC	secretory component
SDS	sodium dodecyl sulfate
SEM	scanning electron microscope
SMP	submitochondrial particle
snRNP	small nuclear ribonucleoprotein particle(s)
TC	tyramine cellobiose
TCA	trichloroacetic acid
TEM	transmission electron microscope
TEMED	*N*,*N*,*N*′,*N*′-tetramethylethylenediamine
TGN	*trans*-Golgi network
Tris	tris(hydroxymethyl)-aminomethane
VSV	vesicular stomatitis virus

1

Homogenization of tissues and cells

JOHN M. GRAHAM

1. Introduction

Many investigations into the structure and function of cells and tissues require, at some stage, the isolation of a particular membrane or subcellular organelle. The successful isolation and fractionation of subcellular membranes from any source, from animal cells, plant cells, or microbial organisms depends critically on the ability to disrupt the cell in such a way that the subcellular particle of interest can be isolated in high yield and as free from contaminating structures as possible. Moreover, the isolated biological particle should retain as much of its normal structural and functional integrity as possible and be in a form which is amenable to the analytical techniques to be used subsequently.

In Sections 2 and 3, the aims and selection of homogenization procedures appropriate to the particular tissue or cell will be discussed. Sections 4 and 5 will describe the media used and some of the commercially available homogenization apparatus. Section 6 describes the application of homogenization techniques to some common tissues and cell types and gives a few typical protocols.

2. Aims of the homogenization procedure

Any homogenization technique must stress the cells sufficiently to cause the disruption of the surface membrane and so release the cytosol and internal organelles (if present). The principal aim must be to achieve, reproducibly, the highest degree of cell breakage using the minimum of disruptive forces, but at the same time cause no damage to any of the organelles of interest, and to avoid protein denaturation. It is important that the imposition of the disruptive force be for the shortest possible time, otherwise organelles and membranes released early in the process will also be exposed to these forces; while those released towards the end of the process will be relatively little affected. A target cell lysis of at least 90% is generally acceptable. Many methods involve continuous or repetitive exposure of the entire sample to the disruptive force(s): these homogenization procedures tend to produce membrane organelles and membranes in various stages of fragmentation. This is not only detrimental in itself but causes great difficulty in any subsequent separations which use differential centrifugation or rate-zonal gradient centrifugation, since the existing problem of size heterogeneity of organelles will be exacerbated.

A parallel aim is that the homogenization procedure should not cause serious breakdown of the membrane vesicles which may already exist in the cytoplasm of eukaryotic cells, that is, those involved in membrane synthesis and the transport of macromolecules to and from the surface. There is evidence, for example, that during endocytosis the size of endosomes increases as the ligand is processed (1, 2), if the homogenization procedure caused the larger vesicles to fragment and reseal, then a possible means of fractionation would be lost.

An obvious requirement of homogenization is to maintain, as far as is feasible, the functional integrity of the subcellular components. During the isolation of a particular subcellular organelle or membrane, the homogenization procedure is the first of many assaults which are imposed on the cell and its contents: it should be recognized that as soon as the cell is broken, the environment of all its components is changed drastically so as to facilitate the subsequent fractionation steps. In the cytosol of eukaryotic cells, organelles are closely juxtaposed to their neighbours and are bathed in a medium rich in proteins, ions, and metabolites, often at 20–37 °C. Upon homogenization the organelles are substantially diluted into an ice-cold medium largely devoid of these solutes and containing instead organic molecules (e.g. sucrose and Tris) never encountered within the cell. It would therefore be surprising if the functional competence of organelles *in vitro* was identical to that *in vivo*.

Functional integrity of subcellular components is also compromised by foaming: this can lead to protein denaturation and it is a particular problem with mechanical shear methods (Section 5.2.2) and sonication (Section 5.2.3). The foaming which occurs with nitrogen cavitation (Section 5.1), although posing a handling problem, is less deleterious because the bubbles are of an oxygen-free gas rather than air.

A major problem is the redistribution of macromolecules which occurs on the cytoplasmic surface of membranes and organelles to the soluble cytosol fraction or to other organelles when the cell is broken and its components exposed to the dual assault of the homogenization medium and the disruptive force. Work from immunohistochemistry has shown for example that the type III isoenzyme of hexokinase which is clearly localized to the nuclear periphery in kidney and liver, is found in the soluble fraction of homogenates of these tissues (3). A similar problem exists with DNA polymerase which may be found in the cytosol after release from damaged nuclei. Moreover, any protein in the soluble fraction, whether native or dissociated from the surface of an organelle during the homogenization procedure can be trapped adventitiously by vesicles formed during the process.

An attendant problem with nuclei is that any damage to the organelle which results in the release of DNA can have serious consequences for any subsequent fractionation since it tends to adhere to other particles. Not only does this cause aggregation of these particles, studies of the DNA in mitochondria or chloroplasts are going to be compromised.

3. Influence of sample type

The selection of an appropriate homogenization procedure depends on the following important factors: the presence or absence of intracellular organelles, the presence or absence of a strong cell wall, and in the case of an organized tissue, its structural organization.

All eukaryotic cells with the exception of the human erythrocyte, contain, in addition to a surface (plasma) membrane (sometimes called the cell envelope, limiting membrane, cytoplasmic membrane, or plasmalemma) a number of intracellular organelles: the homogenization process must be sufficiently severe to break the surface membrane efficiently but cause minimal damage to the organelles released. Bacteria and the human erythrocyte on the other hand contain no other internal membranes; homogenization is concerned only with rupture of the surface membrane and the form (sheets, vesicles, or inverted vesicles) in which it is isolated.

Virtually all cells possess some macromolecular structure peripheral to the surface or limiting membrane. In most animal cells this cell coat (glycocalyx), whilst the site of many important macromolecules involved in adhesion, movement, development, and recognition, has no significant effect on the homogenization process; only in the case of the cells lining the lumen of the mammalian digestive tract, in which this superficial layer is elaborated into an extensive mucous layer, does it play a determining role as far as homogenization is concerned.

The presence and the organization of a cytoskeleton also has a significant bearing on the homogenization process and its products. In many cultured cells, particularly suspension culture cells, the cytoskeleton tends to be rather diffuse and subcellular particles may adhere to it when it is released during homogenization—this is particularly noticeable if the cells are swollen in a hypotonic buffer, although in such cases it is often difficult to ascertain whether the aggregation of particles is to the cytoskeleton or to the DNA released by nuclei. In the case of an organized tissue such as the intestinal mucosa the extensive cytoskeleton under the microvilli remains attached to the membrane and is a positive benefit in isolating such a membrane. The presence of structures such as tight junctions, desmosomes, and a domain-localized cytoskeleton in organized tissues appear to facilitate the fragmentation of the plasma membrane at points away from such structures, while their absence in cultured cells implies a lack of 'weak points' on the surface.

In many bacteria, in fungi, algae, and in all plants, the presence of a thick and resilient cell wall also has important consequences for the homogenization of these cells. By its very nature it resists all but the most severe disruptive forces. The molecular details of this outer layer vary with the organism: in Gram-positive bacteria for example it is a thick peptidoglycan layer up to 80 nm thick; while in Gram-negative bacteria it is more complex comprising a much thinner peptidoglycan layer together with an outer layer of lipopolysaccharide. In plants a complex and rigid wall comprising cellulose, hemicellulose, and pectins are present. In these cases severe mechanical shearing forces must be applied unless the outer wall material is enzymically digested. The latter technique is now the method of choice for many workers as once the outer wall has been digested, the released cell (called a protoplast from plants and a spheroplast from organisms such as yeast) can be homogenized by much more mild techniques (see Sections 6.5 and 6.6). This is particularly important with plants, as their organelles appear especially fragile.

Mammalian tissues can be divided into soft, such as the liver and intestinal mucosa, or hard such as skeletal and heart muscle. The former can be effectively homogenized using relatively mild liquid shear techniques (e.g. Potter–Elvehjem or Dounce homogenization) while the latter need more severe mechanical shear methods (mechanical blenders and mixers). Another more subtle consideration with tissues relates to the differential breakage of different domains of the plasma membrane: certain domains may be stabilized such that they can be isolated as large intact sheets, rather than as vesicles—this, together with a more detailed discussion of the problems posed by specific tissues and cells is considered in more detail in Section 6.1.

4. Homogenization media

The standard isotonic medium used for the isolation of all organelles (other than nuclei) from mammalian cells is 0.25 M sucrose containing 1 mM EDTA and buffered with a suitable organic buffer: Tris, Hepes, or Tricine are commonly used at pH 7.0–7.6. Triethanolamine, particularly when used in conjunction with acetic acid, appears to promote a high per cent cell lysis, especially for cultured cells, although the reasons for this are not at all clear. If the principal aim of the homogenization is to recover nuclei however, EDTA should be replaced by low concentrations of divalent cations (as $MgCl_2$ or $CaCl_2$) and KCl. While $CaCl_2$ is a common component of plant media, $MgCl_2$ is generally preferred in media for animal cells because Ca^{2+} can activate certain phospholipases and proteases and inhibit RNA polymerase. Specific media tailored to the requirements of specific organelles are given in *Table 1*. The osmolarity of the media for the isolation of plant organelles tends to be slightly higher than that of mammalian tissue and concentrations of up to 0.33 M sucrose, sorbitol, or mannitol are used routinely. The release of phenolic compounds and quinones from the central vacuole can be a major problem unless 2-mercaptobenzothiazole or polyvinylpyrrolidone (PVP) are added (see Section 6.5); defatted bovine serum albumin (BSA) is also often included (*Table 2*). For micro-organisms (see Section 6.9) the composition of the media tends to be more varied though often comprising a simple 50 mM Tris or phosphate buffer, containing low concentrations of $MgCl_2$ (1–5 mM) and/or EDTA (1–2 mM).

To protect organelles from the potentially damaging effects of proteases which may be released, for example, from lysosomes during homogenization, it is common to include a cocktail of protease inhibitors in the homogenization medium. A typical mixture for mammalian cells is 1 mM phenylmethylsulfonyl fluoride (PMSF), 2 μg/ml each of leupeptin, antipain, and aprotinin. These are normally stored at 100 × concentration solutions, PMSF in ethanol (or isopropanol), leupeptin and antipain in 10% dimethyl sulfoxide, and aprotinin in water.

Table 1. Homogenization media for liver organelles

Liver organelle	Homogenization medium
Mitochondria	200 mM mannitol, 50 mM sucrose, 1 mM EDTA, 10 mM Hepes–NaOH pH 7.4
Peroxisomes	250 mM sucrose, 1 mM EDTA, 0.1% ethanol, 10 mM Tris–HCl pH 7.5 (4)
Nuclei	250 mM sucrose, 25 mM KCl, 5 mM $MgCl_2$, 20 mM Tris–HCl pH 7.4 (5)

Table 2. Homogenization media for plant organelles

Plant organelle	Medium
Chloroplast[a]	0.33 M sorbitol, 5 mM ascorbate, 50 mM Hepes pH 7.5, 2 mM EDTA, 1 mM $MnCl_2$, 1 mM $MgCl_2$ (6)
	0.33 M mannitol, 25 mM pyrophosphate, 0.1% BSA pH 7.8 (7)
	0.33 M sorbitol, 10 mM pyrophosphate, 5 mM $MgCl_2$, 2 mM isoascorbate pH 6.5 (8).
Mitochondria	250 mM sucrose, 10 mM Hepes, 2 mM EGTA, 3 mM cysteine, 1 mg/ml BSA pH 8.0 (9)
	450 mM mannitol, 30 mM MOPS, 2 mM EDTA, 4 mM cysteine, 0.6% (w/v) PVP, 2 g/litre BSA pH 7.4 (10)
	0.3 M mannitol, 7 mM cysteine, 2 mM EDTA, 0.1% (w/v) BSA, 0.6% (w/v) PVP-40, 40 mM MOPS–KOH pH 7.5 (11)
Microbodies	0.5 M sucrose, 150 mM Tricine, 10 mM KCl, 1 mM EDTA, 1 mM $MgCl_2$, 0.1% DTT pH 7.5 (12)

[a] Pyrophosphate-containing media are not suitable for peas (8).

Although some cultured cells can be homogenized in one of the standard iso-osmotic media, very often it is necessary to stress the cells osmotically by using a hypo-osmotic medium, 1 mM sodium bicarbonate pH 8.0 is a commonly used medium, but organic buffers (Tris, Hepes, triethanolamine, etc.) can be used at approximately 10 mM concentration. If EDTA is included in such media, the nuclei will be particularly fragile and prone to leak DNA. Although it is very much a 'trial and error' procedure to optimize the medium for a particular cell type, inclusion of $MgCl_2$ may be beneficial to nuclear integrity. Inclusion of DNase I (10 μg/ml) can guard against aggregation due to the release of DNA.

Mitochondria are very stringent in their requirements for an isolation medium and it is worthwhile noting some of the important factors. The homogenization medium which is optimal for the isolation of mitochondria from a variety of sources has been studied by Nedergaard and Cannon (13). Although sucrose is often used as the main osmotic component, mannitol or sorbitol are often preferred because of reduced binding to glycogen or (in the case of plant tissues) starch grains. In the case of tissues which may lead to the formation of gelatinous homogenates (brain or muscle), salts such as KCl (100–150 mM) or potassium aspartate (160 mM) may be preferred. The disadvantage of any salt medium is the tendency for ionically-bound membrane peripheral proteins to be lost. As an alternative to the use of salt solutions for brain or muscle tissue, anticoagulants such as heparin or polyethylene sulfonate may be included.

Since mitochondria are uncoupled by Ca^{2+}, a chelator is essential, EGTA or EDTA; for the same reason 0.1–1.0% (w/v) defatted bovine serum albumin is often included to remove free fatty acids, acyl CoA esters, and lysophospholipids. A sulfhydryl reagent, 2-mercaptoethanol or dithiothreitol (0.1–0.5 mM) may be included, along with protease inhibitors (see above). In some methods ATP is also added: the widely used Chappel–Perry medium (14) for muscle mitochondria contains 100 mM KCl, 20 mM MOPS, 5 mM $MgSO_4$, 1 mM ATP, 5 mM EDTA, and 0.2% BSA.

5. Methods of homogenization

Methods which use mechanical or liquid shear are by far the most widely applied, and the nature and severity of shearing forces required to disrupt the tissue dictate what sort of device is applicable. Techniques which rely on the properties of cells can be used in some cases. Since all cells are surrounded by an osmotically-active surface membrane, the latter can be stressed by suspending cells in a hypo-osmotic medium. This may be sufficient in itself to cause lysis or it may be used in combination with a mechanical or liquid shear method. Its only uncontroversial use is for lysing cells which do not contain cytoplasmic organelles: bacterial protoplasts (15) and the human erythrocyte

(16). However, it is often found necessary to stress tissue culture cells osmotically before they can be disrupted by other methods (Section 6.4). Chemical methods are mainly restricted to the use of hydrolytic enzymes for the degradation of the cell wall of plant tissues, bacteria, algae, fungi, etc., thereby producing a protoplast or spheroplast whose surface membrane is then susceptible to more modest disruptive forces. Proteolytic enzymes (e.g. Nagarse) may also be used to make hard animal tissues, such as skeletal muscle, more amenable to disruption. In a few instances low concentrations (less than 0.2%, w/v) of a mild detergent such as Nonidet NP-40 may be used to weaken the surface of cultured cells, especially if the biological particle of interest is an internal organelle such as the nucleus, which is rather less susceptible to detergents than the plasma membrane.

Homogenizers fall broadly into two categories: in Type 1 homogenizers the material is subjected to the disruptive force once, while in Type 2 devices it is repeatedly or continuously exposed to the disruptive force. Within these two categories different homogenizers achieve the disruption of cells and tissues by a variety of means. In terms of homogeneity of products and reproducibility the former type is certainly to be preferred, but is generally only applicable to cell suspensions and not to tissues. Type 1 homogenizers rely on either liquid shear or cavitation as a means of cell disruption.

5.1 Type 1 homogenizers

In those Type 1 homogenizers which rely on liquid shear to disrupt cells, a suspension is propelled by a piston through a narrow orifice, often the annulus around a ball bearing. Most frequently the pressure is applied to the piston by compressed air, but it may be electrically driven. The French pressure cell is commonly used in the laboratory, for the homogenization of microorganisms, particularly those with a tough outer wall (15), but it has been used by Higgins (17) to break membrane vesicles. It operates at high hydraulic pressures of 100–150 MPa (15000–20000 psi). The 'Stansted' cell disruptor (*Figure 1*) which operates at lower pressures has been used to homogenize lymphocytes, in addition to yeast and bacteria (18). The severity of the shearing forces is controlled by the size of the orifice and the imposed pressure. The method is highly reproducible.

The pressure vessel used in nitrogen cavitation consists of a robust stainless steel casing with an inlet port for delivery of the gas from a cylinder and an outlet tube with a needle valve (*Figure 2*). Nitrogen dissolves in the suspending medium and in the cytosol of the cells under high pressure and when the needle valve is opened, the internal pressure forces the suspension through the outlet tube and as it becomes exposed to atmospheric pressure, it experiences a rapid decompression. Cell disruption occurs due to the sudden formation of bubbles of nitrogen gas in the cytosol and in the medium (gaseous shear). In nitrogen cavitation, a stirred cell suspension is equilibrated with oxygen-free nitrogen at pressures of about 5500 kPa (800 psi) for between 10

Figure 1. The Stansted cell disruptor (Energy Service Co). The cell suspension from the reservoir (E) is fed, via the pump (A) to the valve unit (B) where disruption takes place. The homogenate is delivered through a tube (C) to a beaker (D). Compressed air is supplied to the pump through the regulator (F) which controls the pumping rate and to the adjustable valve regulator (G) which controls the pressure applied to the cell suspension. The air outlet (J) allows exhaust air to be delivered into a sterilizing solution in a flask (H) when biohazardous material is used (reproduced with permission from ref. 20).

and 30 minutes. The method has many advantages: it is highly reproducible, it can be applied to a wide range of volumes (10 ml–1 litre), the results are little influenced by cell concentration, and it is successful with any tissue culture cell. However, it also has some very significant problems: all the membranes of the endosomal compartment and the plasma membrane are converted to small vesicles of similar size, ribosomes tend to be stripped off the rough endoplasmic reticulum, and the nuclei tend to be very fragile. A variety of chamber sizes are available (for sample sizes down to about 10 ml) and large ones can be adapted down to accommodate smaller containers (using a plastic collar) but normally 1–2 ml of the sample remains in the vessel after collection of the homogenate.

(b)

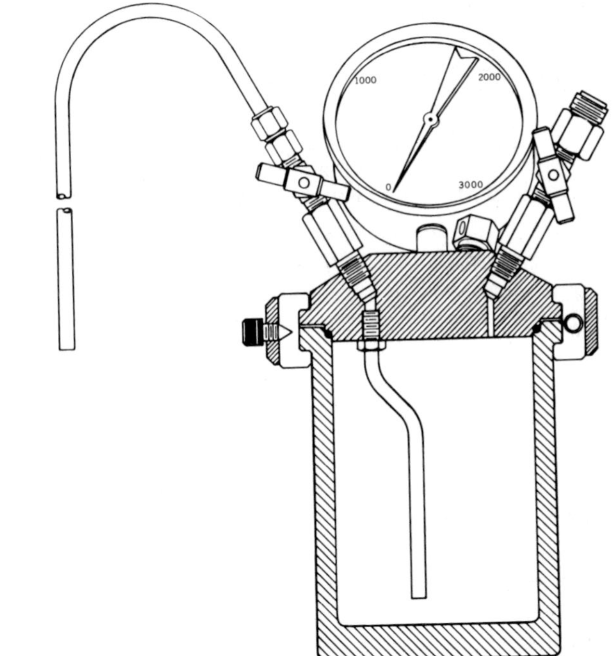

Figure 2. Nitrogen cavitation vessel. (a) Disassembled parts of the homogenizer, (b) diagrammatic cross-section (reproduced with permission from ref. 20). The version shown in (a) is manufactured by Baskervilles, Manchester, UK. Artisan Industries, Waltham, Mass, USA make a similar version which uses the pressure gauge on the gas cylinder to monitor the pressure.

5.2 Type 2 homogenizers

Type 2 homogenizers cover a wide range of shearing methods and equipment; the methods are broadly, grinding, mincing or blending, and liquid shear. The selection of the appropriate form of homogenizer depends very much on the type of tissue or cell.

5.2.1 Liquid shear

The most widely used homogenizers for soft tissues and cells without a tough outer coat rely on liquid shear for the disruptive force. The Potter–Elvehjem (sometimes called Teflon-and-glass) homogenizer consists of a glass containing vessel and a pestle of machined Teflon which is normally attached to an overhead electrical motor (*Figure 3*). Typically the pestle is rotated at 500–1000 r.p.m. In the Dounce (sometimes called all-glass) homogenizer, the pestle which is hand operated, is normally a smooth glass ball rather than a shaped Teflon cylinder (*Figure 3*), although some small volume variants have a ground glass cylinder.

The magnitude of the liquid shear force is controlled by:

- the clearance between the pestle and the containing vessel
- the thrust of the pestle
- the speed of rotation of the pestle (in the case of the Potter–Elvehjem type only)

The clearances of Dounce homogenizers can be smaller than those of Potter–Elvehjem ones: a tight-fitting Dounce homogenizer normally has a clearance of 0.05–0.08 mm, while in a loose-fitting one the clearance is 0.1–0.3 mm. The clearance of a Teflon pestle is normally in the range 0.08–0.3 mm. Obviously the smaller the clearance, the greater the shearing force. Generally, for soft tissues such as liver, a Potter–Elvehjem homogenizer, with a clearance of about 0.09 mm is used for routine applications. Smaller clearances can lead to damage to released organelles, particularly the nuclei, and cause difficulty in moving the glass vessel relative to the pestle in the early stages of homogenization.

The thrust of the pestle is impossible to quantify as it is solely operator-dependent: in published methods it is described nebulously as ‘gentle’ or ‘vigorous’. The speed of rotation can be well defined, normally 500–1500 r.p.m. In practice it is difficult to achieve a constant speed because of the tendency of the rotating pestle to slow down as it reaches the bottom of the vessel. As to the number of up-and-down strokes (passes) of the pestle, it is obviously important to restrict this to the minimum, for the reasons given in Section 2: normally six to ten up-and-down strokes are sufficient for good disruption of liver.

Quantification of Dounce homogenization suffers from the same problems. Generally, Dounce homogenization is only used on tissues if it is very import-

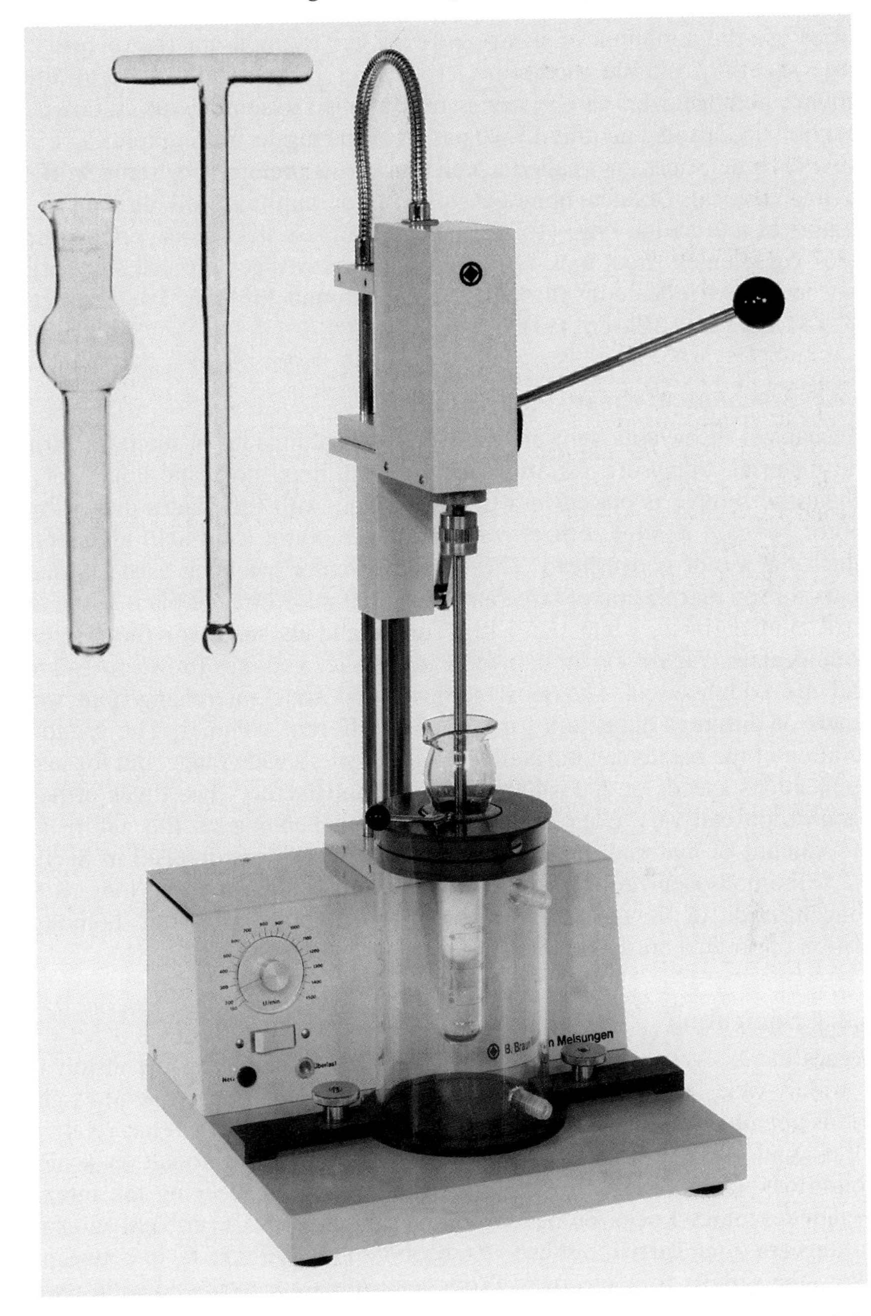

Figure 3. A Potter–Elvehjem homogenizer (inset Dounce homogenizer). The device holding the Potter–Elvehjem homogenizer allows the glass vessel to be cooled in an ice–water mixture and the pestle can be advanced using a remote handle (reproduced with permission from ref. 20).

ant to use the minimum of shearing forces, for example for the recovery of large sheets of plasma membrane from liver (Section 6.1.2). Tight-fitting Dounce homogenizers can be successfully applied to some tissue culture cells (Section 6.4.2) and generally 15–20 passes of the pestle are required.

A machine, sometimes called a 'cell cracker' is preferred by some workers to the classical Dounce homogenizer (8) for cultured cell disruption. It comprises a precision bore (1.270 cm) in a stainless steel block, containing a 1.267 cm stainless steel ball. Using a system of syringes the cell suspension can be forced repeatedly through the gap around the ball. For most cells 10–12 passes are sufficient (19).

5.2.2 Mechanical shear

Mechanical shear homogenizers use rotating metal blades or teeth to disrupt the material. Many are variations of the traditional domestic liquidizer, in which the sample is placed in a glass reservoir with the blades driven by a motor beneath it, while others resemble the modern hand-held blenders in which the motor is overhead. The Waring blender has been used for many years for the maceration of large amounts (100 ml–1 litre) of plant tissue and hard animal tissue, while the Ultra-Turrax and its successor the Polytron homogenizer (*Figure 4*) can be used with smaller volumes (down to 1–2 ml) and are widely used. The most sophisticated have interchangeable work heads of different diameters for use with different volumes. The speed of rotation of the blades can normally be varied over a wide range, but for many applications speeds of 2000–4000 r.p.m. are satisfactory. The times of homogenization can vary from 15 to 60 seconds depending on the nature and the amount of material. Unlike the liquid shear devices detailed in Section 5.2.1, the homogenization conditions are easily standardized so long as the concentration of tissue is kept constant. An application in the homogenization of skeletal muscle is given in Section 6.3.

5.2.3 Sonication

Sonication is a very effective means of breaking open cells in suspension and is widely used in the preparation of bacterial membranes. Sometimes glass beads are added to the cell suspension for improved efficacy. It has relatively little application to animal or plant cells as it is very difficult to achieve conditions which disrupt the surface membrane while keeping the internal organelles intact. Localized heating in the sample is also a problem and sonications are often carried out in 5–10 second pulses, with 'rests' in between to allow the sample to cool down. Probe sonicators are available with a wide variety of work heads and those with a maximum power of 150 W are normally adequate for most tasks. Because of the variability of work heads, current, and amplitude it is strongly advised that any sonication process is closely monitored by microscopy to determine the degree of breakage.

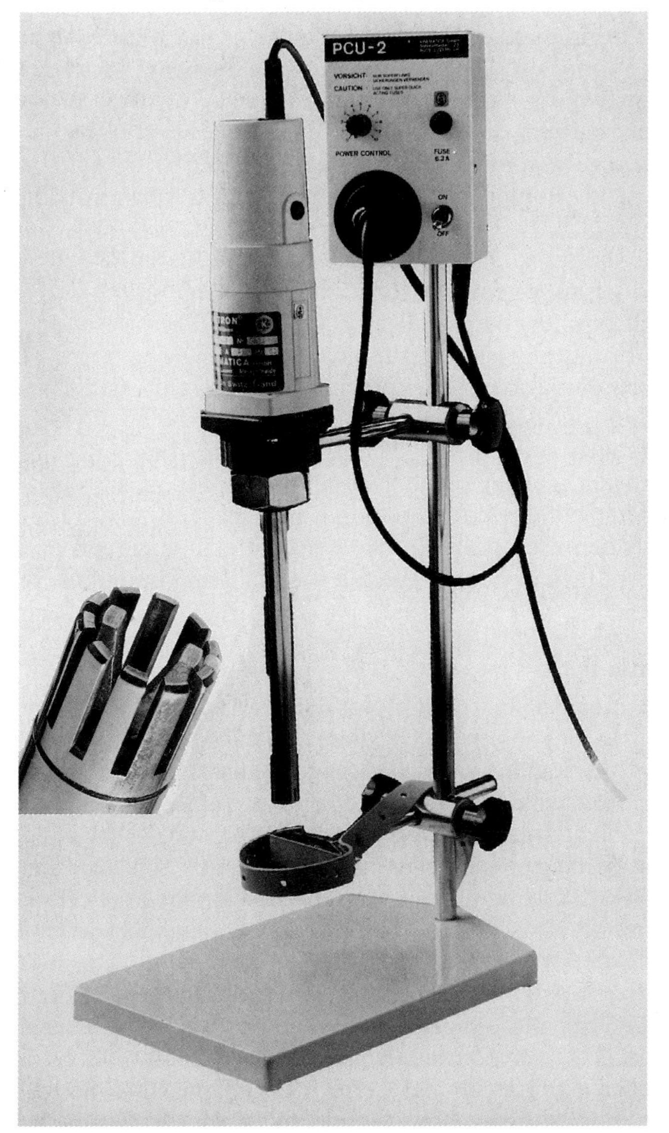

Figure 4. A Polytron homogenizer. Inset shows (inverted) the rotating teeth of the work head (reproduced with permission from ref. 20).

5.2.4 Other abrasive methods

The use of small pore metal or nylon screens can be used as very simple means of breaking some cells. Tissue grinders use a glass pestle to force cells through a stainless steel screen (pore size down to 0.7 mm). They have been used with cells from lymphoid tissue (20) and Avruch and Wallach (21) reported that the method of choice for adipocytes was to force them through a metal screen (pore size 100 μm) held within a Swinney™ filter holder from a syringe; liquid shear methods resulted in a poor recovery of membranes and nitrogen cavitation led to nuclear damage. The protoplasts prepared from enzymic digestion of plant tissue (see Section 6.5) are quite fragile and can be very gently homogenized by forcing them through a nylon mesh of pore size 20 μm (22).

Plant tissue, bacteria, algae, yeast, and fungi are susceptible to grinding in a mortar and pestle, often with some abrasive agent such as carborundum powder, silica powder, fine sand, or glass beads. The size of the abrasive particles can be quite important, the smaller ones causing fragmentation not only of the cell, but also of some organelles (if present). Glass beads which are available in specified size ranges may therefore be an advantage. Yeast cells are effectively broken using 0.5–0.75 mm glass beads (23). Sometimes the abrasive agent is shaken with a suspension of the yeast; sometimes they are used in combination with a rotating blades macerator. Abrasion of plant leaf tissue is often used prior to treatment with cellulase and pectinase, to permit access of the enzymes to the cell wall material, during the production of protoplasts.

5.2.5 Osmotic lysis

Exposure of mammalian cells, protoplasts, or spheroplasts (from plants, bacteria, etc.) to a hypo-osmotic medium will cause them to swell and then, depending on the fragility of the structure, burst. The method is relatively little used for eukaryotic cells, except to facilitate subsequent rupture of tissue culture cells by liquid shear (see Section 6.4.2). Although such a medium will also cause the swelling of organelles in eukaryotic cells, some cultured cells are so resistant to breakage in their native state that swelling the cells is the only answer. Osmotic lysis is used however to produce membrane vesicles from bacterial spheroplasts (see Section 6.7) and human erythrocyte ghosts (16).

An interesting variation was used by Barber and Jamieson (24) for homogenizing platelets. They allowed platelets to take up glycerol by incubating them in 4.3 M glycerol. The hyperosmotic platelets then lysed when transferred to iso-osmotic sucrose. Graham and Coffey (25) adapted the method to some tissue culture cells. Normally however only hypo-osmotic swelling is used.

6. Homogenization of tissues and cells

6.1 Mammalian liver

6.1.1 Organelles other than the plasma membrane

Certainly the most widely investigated of any tissue type, mammalian liver is relatively easy to homogenize and its subcellular particles have been well characterized. Potter–Elvehjem homogenization using 0.25 M sucrose containing 1 mM EDTA (see *Protocol 1*) will release the major organelles; nuclei, lysosomes, peroxisomes, and mitochondria in an intact form, while the endoplasmic reticulum (both the smooth and rough) will vesiculate. Golgi membranes may be recovered as intact stacks if the homogenization conditions are mild and if the Golgi is well developed and discrete, as in liver. If, on the other hand, it is rather diffuse and the homogenization conditions more severe, then it too will vesiculate. Nuclei are very delicate in this medium; it is therefore very important to use the minimum amount of liquid shear to produce a satisfactory lysis. If nuclei are to be isolated specifically then the homogenization medium should contain 25 mM KCl and 5 mM $MgCl_2$, instead of the EDTA; other media are given in *Table 1*.

The animal is usually starved overnight to reduce the amount of glycogen granules which might otherwise interfere in the subsequent subcellular fractionation. The following protocol is for a single rat liver, about 10 g wet weight. Carry out all operations and have all solutions and glassware at 0–4°C.

Use an homogenizer of about 50 ml capacity. If only one of a smaller capacity is available (e.g. 30 ml) carry out the homogenization in two batches; never overfill the homogenizer. It is often both easier and preferable in any case, to carry out the process using two or more batches of liver to minimize the problems associated with any 'continuous' homogenization process (i.e. some tissue will be disrupted and its organelles released during the first pass of the pestle, while some tissue will only be disrupted several passes later). Use a Teflon pestle with a clearance of about 0.09 mm. A wall-mounted thyristor-controlled, high torque electric motor is essential for facilitating the homogenization process and for reproducible results; a variable speed electric drill is ideal for this purpose.

Protocol 1. Homogenization of rat liver in iso-osmotic sucrose

Equipment and reagents

- Dissection equipment
- Potter–Elvehjem homogenizer
- Cheesecloth
- Homogenization medium (see *Table 1*)

Method

1. After the animal is sacrificed, normally by cervical dislocation, open the abdominal cavity and remove the liver as quickly as possible, to a 100 ml beaker containing a small amount of ice-cold homogenization medium.
2. Decant the medium and, on ice, chop up the liver into small pieces (no more than about 30 mm^3) with fine scissors. Add about 50 ml medium, swirl the pieces around, and decant again to remove some of the blood if the tissue has not been perfused prior to excision.

Protocol 1. *Continued*

3. Transfer the minced liver to the Potter–Elvehjem glass vessel and with the pestle rotating at about 500 r.p.m. homogenize the liver mince using six to eight up-and-down passes of the pestle. Do not hold the glass homogenizer in the unprotected hand—wear a protective glove, this also reduces heat transfer to the sample.
4. If the nuclear pellet is going to be used for further processing, then filter the homogenate through three layers of cheesecloth, muslin, surgical gauze, or a single layer of nylon mesh (75 μm pore size), to remove residual clumps of cells and connective tissue. Do not force the homogenate through the filter by squeezing—gentle agitation with a glass rod should suffice.

6.1.2 The plasma membrane

Generally speaking a membrane such as the sinusoidal domain of the hepatocyte plasma membrane (of intact liver) will form vesicles under any homogenization conditions, but the contiguous domain, which is stabilized by various junctional complexes and by desmosomes can remain as an intact sheet if the homogenization conditions are relatively mild. For example, if Dounce homogenization is used rather than Potter–Elvehjem homogenization, if the cells are swollen in a hypotonic medium so as to reduce the shearing forces needed to rupture the cells, and if stabilizing divalent cations are included in the homogenization medium—all these factors favour the retention of the contiguous membranes as intact structures. As such they are relatively easy to purify, since they sediment at low relative centrifugal forces (RCFs) along with the nuclei and the heaviest mitochondria. Plasma membrane vesicles on the other hand will behave in differential and gradient centrifugation much like any other vesicle, unless they are specifically, relatively dense (e.g. coated vesicles). Another membrane which tends to retain its planar sheet-like structure is the brush-border membrane of intestinal epithelium, due to the extensive underlying terminal web structure.

Dounce homogenization of liver is used if the aim is to isolate the contiguous membranes as large sheets. It is essential to perfuse the liver with an isotonic saline medium to remove as much of the blood as possible, since the presence of erythrocytes in the nuclear fraction may seriously compromise any subsequent attempts to resolve this fraction by gradient centrifugation.

Protocol 2. Perfusion of rat liver and Dounce homogenization

Equipment and reagents

- Peristaltic pump, tubing, and cannula
- Dounce homogenizer
- Cheesecloth
- 0.9% NaCl
- Homogenization medium (see below)

Method

1. Anaesthetize the animal. This procedure requires the operator to be in possession of the required licence and it should be carried out by a trained operative.
2. Set-up the perfusion line using a reservoir of ice-cold 0.9% NaCl, or if preferred, the homogenization medium, a peristaltic pump and the appropriate tubing for connection to a cannula, and fill the line with medium.
3. Open the abdominal cavity, insert the cannula into the hepatic portal vessel, and secure it with a ligature.
4. Attach the perfusion line to the cannula. A syringe (20 or 50 ml) containing ice-cold saline could be used as an alternative.
5. Open the thoracic cavity and cut the vessels leading from the liver and perfuse with medium until all the lobes are cleared of blood. This should take no more than 1 min.
6. Mince the liver as in *Protocol 1* and homogenize in about 30 ml of homogenization medium, using 10–15 up-and-down passes of the pestle of a loose-fitting Dounce homogenizer (clearance about 0.08 mm).
7. Filter the homogenate as described in *Protocol 1*.

There are a number of variations on *Protocol 2*; Scott *et al.* (26) decapitated the anaesthetized animal to drain most of the blood prior to perfusion, but in the author's experience, cannulation of the portal vein is then less easy. The homogenization medium may be 1 mM $NaHCO_3$ pH 7.4 (27) or 0.25 M sucrose, 1 mM $MgCl_2$, 10 mM Tris–HCl pH 7.4 (26). The hypo-osmotic bicarbonate medium may make the disruption process easier by causing cell swelling. To stabilize both the nuclei and the membrane sheets the $MgCl_2$-containing medium may be beneficial. The use of $CaCl_2$ is generally avoided because of its role in the activation of certain phospholipases and proteases (20) and inhibition of RNA polymerase.

6.2 Brain

The only notable difference between homogenization methods for liver and brain is the tendency to use 0.32 M sucrose rather than 0.25 M. Otherwise, the approaches are broadly similar: the tissue is first finely chopped with scissors and then disrupted. Commonly, the homogenization schedule calls for up to 12 strokes of the pestle (500–600 r.p.m.) of a standard Potter–Elvehjem homogenizer although Cebrian-Perez (28) specified 20 strokes of a loose-fitting Teflon pestle at 100 r.p.m. Sequential use of a loose-fitting (clearance 0.12 mm) and a tight-fitting (clearance 0.05 mm) Dounce homogenizer has also been used (28). Organelles such as mitochondria (29), synaptosomes (30), and synaptic vesicles (31) are isolated in a buffered sucrose medium containing 1 mM EDTA (or EGTA) while for synaptic plasma membranes the EDTA should be replaced by 1 mM $MgCl_2$ (32).

Rat brain is a common source of coated vesicles, for these the homogenization medium usually omits sucrose and includes instead a high concentration of buffer or salt. Maycox *et al.* (33) used 100 mM Mes–NaOH pH 6.5 containing 1 mM EGTA and 0.5 mM $MgCl_2$, while Jackson (34) replaced the 100 mM Mes–NaOH with 100 mM KCl, 10 mM Mes–NaOH and included 0.2 mM dithiothreitol (DTT), PMSF (50 μg/ml), and azide (0.2%, w/v). Safety note: azide is extremely toxic.

6.3 Muscle

The more fibrous nature of muscle, both striated and cardiac, renders the tissue more difficult to homogenize than liver, requiring some form of rotating blades homogenizer either alone or as a prelude to the use of a Dounce or Potter–Elvehjem homogenizer. Tissue from a young animal may only require fine chopping with scissors prior to liquid shear homogenization. Sometimes a meat grinder is used as a preliminary step to produce a mince before homogenization using either an Ultra-Turrax or Polytron blender. For example, Scarpa *et al.* (35) used a meat grinder to produce a mince of striated muscle and then 5 g of minced tissue was homogenized in a Polytron homogenizer in 25 ml of Chappel–Perry medium (see Section 4), at setting 5 for 15 seconds.

To overcome these problems, proteinases have been used to soften the tissue prior to the use of a blender or Potter–Elvehjem—a commonly used one for muscle mitochondria is the commercially available Nagarse™. The concentrations of enzyme are normally 0.1–0.2 mg/ml and the medium is either the Chappel–Perry medium (14, 36) or contains a polyhydric alcohol, e.g. 250 mM sucrose, 1 mM EDTA, 20 mM Hepes–NaOH pH 7.4 (37), or 0.1 M sucrose, 5 mM $MgCl_2$, 10 mM EDTA, 180 mM KCl, 0.05 M Tris–HCl pH 7.4 (38) at 4 °C for 5–10 min.

According to Bhattacharya *et al.* (39) the high shear forces used by many workers may contribute to the impairment of mitochondrial function and these authors have produced a method which may be a useful one. *Protocol 3* is adapted from Bhattacharya *et al.* (39) and Thakar and Ashmore (38).

Protocol 3. Homogenization of skeletal muscle

Equipment and reagents

- Scalpels
- Potter–Elvehjem homogenizer
- Wash solutions: 0.2 M mannitol, 70 mM sucrose, 0.1 mM EDTA, 10 mM Tris–HCl pH 7.4
- Homogenization medium: 0.1 M sucrose, 10 mM EDTA, 46 mM KCl, 100 mM Tris–HCl pH 7.4
- BSA
- Nagarse

Method

1. Wash the striated muscle in 0.2 M mannitol, 70 mM sucrose, 0.1 mM EDTA, 10 mM Tris–HCl pH 7.4.
2. Mince the muscle finely using scalpels to produce pieces approx. 30 mm^3.
3. Add 4–5 g tissue to 40 ml 0.1 M sucrose, 10 mM EDTA, 46 mM KCl, 0.5% BSA, 100 mM Tris–HCl pH 7.4, containing 20 mg Nagarse per 100 ml, and incubate for 5 min at 4°C.
4. Homogenize in a Potter–Elvehjem homogenizer (0.66 mm clearance with the pestle rotating at 700–1000 r.p.m.) using seven up-and-down strokes, and then incubate with stirring at 0°C for 5 min.
5. Dilute with an equal volume of medium (minus BSA and Nagarse) and then rehomogenize.

Bullock *et al.* (40) compared Nagarse, trypsin, and Pronase and chose the trypsin method without further homogenization for rat leg muscle. They stirred minced rat leg muscle (10 g) in 100 ml 210 mM mannitol, 70 mM sucrose, 10 mM EDTA, 10 mM Tris–HCl pH 7.4 containing 2.4 mg trypsin, which they fractionated without further homogenization. Makinen and Lee (41) worked out the incubation conditions for human (2 min), dog (5 min), and rat (7 min) using *B. subtilis* proteinase (5 mg/g tissue) in a Chappel–Perry medium. They carried out a single incubation, diluted with medium, and homogenized with a hand-held Ultra-Turrax TP N/2 disintegrator for 12–15 seconds.

6.4 Mammalian tissue culture cells

6.4.1 Nitrogen cavitation

Tissue culture cells are very effectively disrupted by nitrogen cavitation; routinely a suspension of cells is equilibrated with oxygen-free nitrogen at pressures of approximately 5500 kPa (800 psi). Cell concentrations of up to 5×10^7/ml can be used, but slightly lower values are recommended. The commercially available vessels (for manufacturers of nitrogen pressure vessels see Appendix 1) have capacities ranging from approximately 10 ml to 1 litre or more. Small volumes (as low as 25–50 ml) can be used in the larger vessels if the suspension is accommodated in a plastic beaker, centred within the vessel by a plastic collar. Volumes smaller than 25 ml are not advisable in these larger vessels as they are difficult to recover. Make sure that the delivery tube up which the cell suspension is ejected, does not quite reach the bottom of the beaker, otherwise sealing the vessel may be difficult (*Figure 2*).

It may be necessary to include low concentrations of a divalent cation (1–2 mM $MgSO_4$) to protect the nuclei during the disruptive process. The cation can be chelated with EDTA before fractionation, if required.

Protocol 4. Homogenization of tissue culture cells by nitrogen cavitation

Equipment and reagents

- Nitrogen pressure vessel
- Oxygen-free nitrogen
- Magnetic stirrer
- Phosphate-buffered saline
- Homogenization medium: 0.25 M sucrose, 20 mM Tris–HCl pH 7.4

Method

1. Pre-cool the nitrogen pressure vessel in a bath of ice on a magnetic stirrer.
2. Wash the cells two or three times with phosphate-buffered saline, and once with the chosen homogenization medium (normally 0.25 M sucrose, 20 mM Tricine–HCl pH 7.4, but see above), and finally re-suspend in 25–100 ml of this medium.
3. Transfer the suspension to the plastic beaker (containing a small magnetic flea).
4. Seal the vessel according to the manufacturer's instructions and start the stirrer. Connect the vessel to a cylinder of oxygen-free nitrogen. Close the exit and venting valves (*Figure 2*). If you are inexperienced in the handling of high pressure gases always seek expert help before using gas cylinders.
5. Open the inlet valve and pressurize the vessel to about 5500 kPa.
6. Close the inlet valve and equilibrate for up to 15 min.
7. Turn off the stirrer, open the delivery valve very slowly, and collect the homogenate in a beaker. The homogenate is expelled with great force and this operation needs to be carried out very carefully, particularly if you are using material which is hazardous. Because of the potential for creating an aerosol, this operation should not be carried out in the open laboratory.
8. When all the homogenate has been expelled, close the delivery valve and vent the remaining gas.
9. It is very important that the foam is allowed to subside by gentle stirring before any centrifugation is carried out.

One of the main problems of nitrogen cavitation is that the organelles are rendered rather fragile and great care is required during subsequent manipulations to prevent organelle breakage. Another is the rather homogeneous size of the population of vesicles which is produced from the plasma membrane, endosomes, Golgi membranes, and endoplasmic reticulum, making their resolution by rate-zonal density gradient centrifugation difficult.

6.4.2 Liquid shear

Attempts to maintain the plasma membrane of tissue culture cells as sheets meet with little success. Generally the disruptive forces required to break open tissue culture cells are rather more severe and without some stabilizing structure, the plasma membrane tends to vesiculate rather than to remain as a sheet. Moreover, only under conditions which allow the plasma membrane to be recovered as a cell 'ghost' are such membranes stable, resisting further fragmentation during subsequent manipulations. Small fragments of the plasma membrane are inherently unstable because the exposure of their hydrophobic cores, at the periphery of the sheet, is thermodynamically unfavourable and the tendency is for them to vesiculate spontaneously.

Different tissue culture cells respond individually to liquid shear techniques, both in their susceptibility to the shearing conditions and their results. It is recommended that the investigator determine the optimum conditions required for the particular cell type used and also to subject the homogenate to differential centrifugation (e.g. 1000 *g* for 10 min, 3000 *g* for 10 min, 10 000 *g* for 10 min, 20 000 *g* for 15 min, and 100 000 *g* for 30 min) to discover the size of the vesicles or fragments of plasma membrane produced. The following protocol is probably satisfactory for most confluent fibroblastic cells.

Protocol 5. Homogenization of monolayer tissue culture cells by liquid shear

The volumes are for a single 10 cm dish of just confluent monolayer cells and the method is adapted from Marsh *et al.* (42).

Equipment and reagents

- Rubber policeman
- Dounce homogenizer
- Phase-contrast microscope
- Homogenization medium: 0.25 M sucrose, 10 mM triethanolamine–acetic acid pH 7.6, 1 mM EDTA
- Wash medium: as above minus EDTA

Method

1. Remove the growth medium by aspiration and wash the monolayer at least twice with 0.25 M sucrose, 10 mM triethanolamine–acetic acid pH 7.6. If it is necessary to adjust the pH of the wash medium add either triethanolamine or acetic acid, not NaOH or HCl.
2. Add 2 ml of the medium containing 1 mM EDTA and scrape off the cell monolayer with a rubber policeman.
3. Transfer the crudely suspended monolayer to a tight-fitting 5 ml Dounce homogenizer. Do not attempt to obtain a single cell suspension—cell disruption is best carried out with small sheets of cells.

Protocol 5. *Continued*

4. Use approx. ten vigorous up-and-down strokes of the pestle to homogenize the cells.
5. Monitor the disruption by phase-contrast microscopy and repeat the homogenization if necessary.
6. Repeat the process with other dishes of cells. If you have just two or three dishes it is best to process each individually. Larger numbers of dishes can be treated in batches, using a larger homogenizer.

If the cells fail to homogenize adequately after 20 strokes of the pestle, it is best not to continue the process as those organelles and membranes which have been released from disrupted cells will increasingly be stressed by the continued imposition of shearing forces and such procedures tend to be difficult to reproduce accurately. This method seems to rely on the use of the triethanolamine–acetic acid buffer—if other organic buffers are substituted the method is ineffective.

Suspension culture cells do not seem to break very easily using these iso-osmotic conditions and it may be necessary to render the cells more susceptible to liquid shear forces by swelling them in a hypo-osmotic medium, for example 1 mM $NaHCO_3$ or a low concentration (5–10 mM) organic buffer. If this approach is adopted it is particularly important to use as high a cell concentration as possible, so that the released organelles are to some extent protected from osmotic stress by the released soluble proteins of the cytosol.

6.5 Plant organelles

There are two basic approaches to the isolation of plant organelles. The intact tissue can be macerated using a blender or by grinding with silica or glass beads in a mortar and pestle; alternatively, the thick cell wall can be digested with enzymes and the released protoplasts can be disrupted by very gentle shearing conditions. The latter avoids exposing the organelles to the harsh shearing conditions of the blender or grinding processes. See Chapter 8 for details regarding suitable homogenization for the preparation of chloroplasts and mitochondria.

Generally the osmolality of plant cells is higher than that of animal cells, consequently higher amounts of osmotic balancers such as sucrose, sorbitol, or mannitol are required in homogenization media. A few of the different types of homogenization media are given in *Table 2*. Iso-osmotic (approx. 330 mM) sorbitol (or glucose) media are preferred over sucrose, but the fine detail of the media depends very much on the type of tissue. For plant mitochondria, sulfhydryl reagents are often included, 2-mercaptoethanol (43), dithiothreitol, or cysteine. Polyvinylpyrrolidone (PVP) at about 1% (w/v) is a common additive to counteract polyphenols in the tissue (43), and 20 μm leupeptin and 0.5 mM PMSF have been used as protease inhibitors (44). For more information see Chapter 8.

For the processing of green leaves it is common to use a rotating blades-type homogenizer, for example, a Waring blender, Ultra-Turrax, or Polytron for very short periods of time up to ten seconds. Sometimes a preliminary partial maceration is carried out using a hand-held domestic blender, to reduce the time required in the Ultra-Turrax or Polytron to a minimum—this is particularly important for the isolation of plant mitochondria (43). The addition of bovine serum albumin (0.25 g/litre) helps to alleviate the tendency of leaves to float in the homogenization medium thus rendering the homogenization ineffective. Some membrane and organelle preparations rely purely on manual methods of maceration. For the isolation of Golgi and plasma membranes from pea seedlings Dhugga *et al.* (45) used a mortar and pestle or gentle mincing with razor blades (45, 46), microbodies too have been prepared using this type of maceration (47). Special equipment has also been constructed whereby a rotating paddle wheel sweeps the plant tissue over a set of blades, while any released material is collected beneath.

Protoplasts are produced by enzymic digestion of the cell wall, normally a mixture of cellulase and pectinase, or cellulase and Macerozyme (48); e.g. 2% cellulase and 0.3% Macerozyme in a 500 mM sorbitol medium at pH 5.5 (containing 1 mM $CaCl_2$ and 5 mM Mes) is allowed to act on the leaf tissue: access of the enzymes to their substrates is enhanced by cutting the leaf with a razor blade or abrading with carborundum. See Chapter 8 for more details on the production of plant protoplasts. Homogenization of the protoplasts can be achieved by gentle homogenization in a glass–Teflon homogenizer (49), or by forcing them from a syringe through a fine nylon mesh (20–25 μm) (48), or by osmotic lysis in 200 mM K_2HPO_4–HCl pH 8.0 (50). The isolation of intact vacuoles requires specially gentle treatment and osmotic lysis of protoplasts can be used.

6.6 Yeast

Because of its tough outer coat, mechanical methods, for example shaking with glass beads (51), have been used for the disruption of intact yeast cells. Guerin *et al.* (52) have reviewed the many mechanical shear techniques which involve the application of homogenizers and ball mills. These are still used widely for large scale homogenization of yeast. However, in most of the more recent applications of genetically engineered yeast cells, for example in the study of membrane synthesis and the secretory process, spheroplast formation is the most common method. The yeast is commonly grown in YPD medium, 1% yeast extract, 2% bactopeptone, and 2% glucose (53).

There are a number of variations on this basic technique, including the use of a pre-enzyme incubation step. Rothblatt and Meyer (53) washed the yeast cells twice in distilled water and pre-incubated them in 100 mM Tris–SO_4, 10 mM DTT at 30°C for 30 min before washing and suspending the spheroplast medium, which contained 1.2 M rather than 1.4 M sorbitol. A similar approach was used by Daum *et al.* (55). Goud *et al.* (54) included a 10 mM azide wash after a pre-incubation in YP medium containing 0.2% glucose.

Protocol 6. Homogenization of yeast by spheroplast formation

This protocol is adapted from Goud *et al.* (54).

Equipment and reagents

- High-speed centrifuge and rotor
- Spheroplast medium: 1.4 M sorbitol, 50 mM potassium phosphate buffer pH 7.5, 10 mM azide, 40 mM 2-mercaptoethanol, 0.25 mg/ml zymolase-100T
- Dounce homogenizer
- Lysis buffer: 0.8 M sorbitol, 1 mM EDTA, 10 mM triethanolamine pH 7.2
- Protease inhibitors

Method

1. Suspend the cells in spheroplast medium: 1.4 M sorbitol, 50 mM potassium phosphate buffer pH 7.5, 10 mM azide, 40 mM 2-mercaptoethanol, and zymolase-100T (0.25 mg/ml). Caution: azide is extremely toxic and should not be used for the preparation of sensitive organelles such as mitochondria.
2. Incubate at 37 °C for 45 min.
3. Cool the spheroplast suspension to 4 °C.
4. Harvest the spheroplasts by centrifugation, at 5000 *g* for 10 min.
5. Resuspend the pellet in lysis buffer, 0.8 M sorbitol, 1 mM EDTA, 10 mM triethanolamine, adjusted to pH 7.2, containing any cocktail of protease inhibitors. For spheroplasts from 100 OD_{599} U use 1.5 ml of buffer.
6. Homogenize using 20 strokes for the pestle in a Dounce homogenizer.

To prevent any damage to the spheroplasts prior to homogenization they can be pelleted on to a cushion of 1.7 M sorbitol, 20 mM phosphate pH 7.5 (56). Alternatively, to remove the spheroplasts more effectively from the enzyme digesting solution, the spheroplast incubation mixture can be layered over 1.5% (w/v) Ficoll 400, 20 mM Hepes–KOH pH 7.4 and centrifuged at 5000 *g* for 15 min at 4 °C (53). This could be combined with a cushion of 1.7 M sorbitol if necessary.

6.7 Other fungi and algae

The mode of homogenization depends very much on the particular organism. *Polytomella caeca* can be disrupted by gentle manual homogenization in an all-glass homogenizer (47), although only about 50% of the total cells were disrupted. Rather more commonly used methods involve grinding in a mortar and pestle with glass beads for 45 sec to 5 min (57). The media are the standard iso-osmotic sucrose or sorbitol media buffered with Tris or Hepes, usually with 1 or 2 mM EDTA. The only notable exception was the use of 1.0 M sucrose for *Prototheca* (57). *Chlamydomonas* has been effectively homogenized using a French pressure cell (4000 psi, 27.5 MPa) at 2×10^8 cells/ml in a medium containing: 0.3 M sucrose, 2 mM Hepes (58), supplemented with 5 mM $MgCl_2$, 5 mM ε-aminocaproic acid, 1 mM benzamidine, and 1 mM PMSF. The walls of *Neurospora crassa* can be weakened sufficiently by treatment with β-glucuronidase, so that the cells can subsequently be disrupted in a Potter–Elvehjem homogenizer (59).

6.8 Trypanosomes

Zymolase is also used as a pre-treatment for trypanosomes such as *Clathridia fasiculata*, in order to render the cells more susceptible to osmotic lysis which is a commonly used homogenization technique for such organisms. The DNA of the kinetoplast of trypanosomes is a research topic of considerable interest and the method given in *Protocol 7* is aimed specifically at studying this organelle and is adapted from ref. 60.

Protocol 7. Homogenization of trypanosomes

Carry out all operations at 0–4°C, except where stated.

Equipment and reagents

- High-speed centrifuge and rotor
- Suspension medium: 43.5 mM EDTA, 43.5 mM Tris–HCl, 0.1 M 2-mercaptoethanol pH 7.9
- 4.2 mg/ml zymolase
- Sorbitol medium: 1 M sorbitol, 0.5 mM $CaCl_2$, 20 mM phosphate buffer pH 7.5
- Lysis buffer: 1 mM Tris–HCl, 1 mM EDTA pH 7.5
- 2.25 M sucrose

Method

1. Suspend the harvested cells (approx. 10^{11}) in 100 ml of 43.5 mM EDTA, 43.5 mM Tris–HCl, 0.1 M 2-mercaptoethanol pH 7.9, and keep at 0°C for 30 min, with occasional agitation.
2. Pellet the cells at 3000 g_{av} for 5 min, and resuspend in 100 ml of 20 mM phosphate buffer pH 7.5, containing 1 M sorbitol and 0.5 mM $CaCl_2$.
3. Repellet the cells, resuspend in 40 ml of the same medium and 4 ml of zymolase (4.2 mg/ml), and incubate at 28°C for 45 min with agitation.
4. Centrifuge the suspension at 3000 g_{av} for 5 min and suspend the cells in the sorbitol medium (100 ml).
5. Centrifuge at 4000 g_{av} for 5 min and remove ALL of the supernatant.
6. Resuspend the cells in 400 ml of lysis buffer (1 mM Tris–HCl, 1 mM EDTA pH 7.5).
7. After 10 min add 50 ml 2.25 M sucrose to return the medium to iso-osmoticity.

Any DNA in the medium will cause any high speed pellet to take on a gelatinous consistency. Birkenmeyer and Ray (60) resuspended such pellets in 0.25 M sucrose, 20 mM Tris–HCl pH 7.5 using a manual Teflon–glass homogenizer, and adjusted the resulting suspension to 3.0 and 0.3 mM with respect to Mg^{2+} and Ca^{2+}, respectively, and 10 μg/ml DNase I. After 35 min at 0 °C the suspension was diluted with an equal volume of 0.25 M sucrose, 20 mM Tris–HCl pH 7.5, 2 mM EDTA and recentrifuged.

6.9 Bacteria

The method depends very much on the type of bacterium, whether it is Gram-positive or Gram-negative. In Gram-positive bacteria the inner cytoplasmic membrane is surrounded by a layer of murein (composed predominantly of peptidoglycans) which may be up to 80 nm thick. In Gram-negative bacteria this layer is much thinner (2 nm) but it has in addition an outer membrane (8 nm thick). This outer membrane has a bilayer construction not unlike that of the cytoplasmic membrane, but the outer leaflet is composed of a complex lipopolysaccharide rather than phospholipid.

The most common methodology involves the digestion of the outer layers by enzymes in a hypotonic medium which will lyse the released protoplasts (or spheroplasts); if the digestion is carried out in a hypertonic medium the protoplasts remain relatively intact. As an alternative, sonication or the French press (normally two passages at 100–150 MPa) may be used to produce cytoplasmic membrane vesicles which are largely 'inside out', while those produced by osmotic lysis retain their normal orientation (15). Freeze-pressed (Hughes press) material tend to produce large fragments (sheets) of membrane rather than vesicles (15).

The enzyme incubation conditions depend on the type of bacteria. In the case of Gram-positive bacteria the spheroplasts that are exposed by the enzyme digestion are lysed immediately, while with Gram-negative bacteria the method produces spheroplasts which are isolated and subsequently lysed. *Protocols 8–10* are adapted from Poole (15).

Protocol 8. Preparation of membrane vesicles from Gram-positive bacteria

Reagents

- 50 mM phosphate buffer pH 8.0
- Lysozyme
- DNase I
- Ribonuclease
- K–EDTA
- $MgSO_4$

Method

1. Suspend the bacteria in 50 mM phosphate buffer pH 8.0.
2. Incubate at 37 °C with 300 μg/ml lysozyme, and 10 μg/ml each of deoxyribonuclease I and ribonuclease for 30 min.

3. Add K–EDTA to 15 mM.
4. After 1 min add $MgSO_4$ to 10 mM.

Protocol 9. Preparation of membrane vesicles from Gram-negative bacteria

Equipment and reagents

- High-speed centrifuge and rotor
- 20% (w/v) sucrose, 30 mM Tris–HCl pH 8.0
- Chloramphenicol
- Lysozyme
- K–EDTA
- 10 mM phosphate buffer pH 6.6
- $MgSO_4$
- DNase I
- Ribonuclease
- Syringe

Method

1. Suspend bacteria in 20% (w/v) sucrose, 50 μg/ml chloramphenicol, 30 mM Tris–HCl pH 8.0.
2. Add K–EDTA to 10 mM and lysozyme to 0.5 mg/ml while stirring, and incubate at 20 °C for 30 min.
3. Harvest the spheroplasts at 13 000 *g* for 30 min.
4. Use a syringe to resuspend the spheroplast pellet in 10 mM phosphate buffer pH 6.6, containing 2 mM $MgSO_4$, ribonuclease and deoxyribonuclease (as in *Protocol 8*).
5. Incubate at 37 °C for 30 min with shaking. If the suspension remains viscous after 15 min, the $MgSO_4$ and enzyme concentrations can be doubled.

Protocol 10. Preparation of membrane vesicles by sonication

Equipment and reagents

- Probe sonicator
- DNase I
- 2 mM $MgCl_2$, 1 mM EGTA, 50 mM Tris–HCl pH 7.4

Method

1. Wash and suspend the bacteria in 2 mM $MgCl_2$, 1 mM EGTA, 50 mM Tris–HCl pH 7.4, and add DNase (10 μg/ml).
2. Transfer the suspension to a suitable glass tube and immerse in an ice–water mixture.
3. Sonicate using a probe sonicator, using five 30 sec periods at 150 W, separated by 15–30 sec 'rests'.

References

1. Berg, T., Kindberg, G. M., Ford, T. C., and Blomhoff, R. (1985). *Exp. Cell Res.*, **161**, 285.
2. Ford, T. C. (1993). In *Methods in molecular biology* (ed. J. M. Graham and J. A. Higgins), Vol. 19, p. 71. Humana Press, Totowa, NJ, USA.
3. Preller, A. and Wilson, J. E. (1992). *Arch. Biochem. Biophys.*, **294**, 482.
4. Völkl, A. and Fahimi, H. D. (1985). *Eur. J. Biochem.*, **149**, 257.
5. Blobel, G. and Potter, V. R. (1966). *Science*, **154**, 1662.
6. Cline, K., Werner-Washbourne, M., Lubben, T. H., and Keegstra, K. (1985). *J. Biol. Chem.*, **260**, 3691.
7. Malherbe, A., Block, M. A., Joyard, J., and Douce, R. (1992). *J. Biol. Chem.*, **267**, 23546.
8. Whitehouse, D. G. and Moore, A. L. (1993). In *Methods in molecular biology* (ed. J. M. Graham and J. A. Higgins), Vol. 19, p. 123. Humana Press, Totowa, NJ, USA.
9. Beavis, A. D. and Vercesi, A. E. (1992). *J. Biol. Chem.*, **267**, 3079.
10. Rustin, P. and Lance, C. (1991). *Biochem. J.*, **274**, 249.
11. Moore, A. L., Fricaud, A.-C., Walters, A. J., and Whitehouse, D. G. (1993). In *Methods in molecular biology* (ed. J. M. Graham and J. A. Higgins), Vol. 19, p. 133. Humana Press, Totowa, NJ, USA.
12. Vigil, E. L., Wanner, G., and Theimer, R. R. (1979). In *Plant organelles* (ed. E. Reid), p. 89. Ellis Horwood, Ltd., UK.
13. Nedergaard, J. and Cannon, B. (1979). In *Methods in enzymology* (ed. S. Fleischer and L. Packer), Vol. 55, p. 3.
14. Chappel, J. B. and Perry, S. V. (1954). *Nature*, **173**, 1094.
15. Poole, R. K. (1993). In *Methods in molecular biology* (ed. J. M. Graham and J. A. Higgins), Vol. 19, p. 109. Humana Press, Totowa, NJ, USA.
16. Dodge, J. T., Mitchell, C., and Hanahan, D. J. (1963). *Arch. Biochem. Biophys.*, **100**, 119.
17. Higgins, J. A. and Fieldsend, J. K. (1987). *J. Lipid Res.*, **28**, 268.
18. Wright, B. M., Edwards, A. J., and Joue, V. E. (1974). *J. Immunol. Methods*, **4**, 281.
19. Balch, W. E., Dunphy, W. G., Braell, W. A., and Rothman, J. E. (1984). *Cell*, **39**, 405.
20. Evans, W. H. (1992). In *Preparative centrifugation: a practical approach* (ed. D. Rickwood), p. 233. Oxford University Press, Oxford, UK.
21. Avruch, J. and Wallach, D. F. H. (1971). *Biochim. Biophys. Acta*, **223**, 334.
22. Markwell, J., Bruce, B. D., and Keegstra, K. (1992). *J. Biol. Chem.*, **267**, 13933.
23. Chen, Y.-S. and Beattie, D. S. (1981). *Biochemistry*, **20**, 7557.
24. Barber, A. J. and Jamieson, G. A. (1970). *J. Biol. Chem.*, **245**, 6357.
25. Graham, J. M. and Coffey, K. H. M. (1979). *Biochem. J.*, **182**, 165.
26. Scott, L., Schell, M. J., and Hubbard, A. L. (1993). In *Methods in molecular biology* (ed. J. M. Graham and J. A. Higgins), Vol. 19, p. 59. Humana Press, Totowa, NJ, USA.
27. Meier, P. J., Sztul, E. S., Reuben, A., and Boyer, J. L. (1984). *J. Cell Biol.*, **98**, 991.
28. Cebrian-Perez, J. A., Muino-Blanco, M. T., and Johansson, G. (1991). *Int. J. Biochem.*, **23**, 1491.
29. Sims, N. R. (1990). *J. Neurochem.*, **55**, 698.
30. Luzio, J. P. and Richardson, P. J. (1993). In *Methods in molecular biology* (ed. J. M. Graham and J. A. Higgins), Vol. 19, p. 141. Humana Press, Totowa, NJ, USA.

31. Bennett, M. K., Calakos, N., Kreiner, T., and Scheller, R. H. (1992). *J. Cell Biol.*, **116**, 761.
32. Cotman, C. W. and Taylor, D. (1972). *J. Cell Biol.*, **55**, 711.
33. Maycox, P. R., Link, E., Reetz, A., Morris, S. A., and Jahn, R. (1992). *J. Cell Biol.*, **118**, 1379.
34. Jackson, A. P. (1993). In *Methods in molecular biology* (ed. J. M. Graham and J. A. Higgins), Vol. 19, P. 83. Humana Press, Totowa, NJ, USA.
35. Scarpa, A., Vallieres, J., Sloane, B., and Somlyo, A. P. (1979). In *Methods in enzymology* (ed. S. Fleischer and L. Packer), Vol. 55, p. 60.
36. Ernster, L. and Nordenbrand, K. (1967). In *Methods in enzymology* (ed. R. W. Estabrook and M. E. Pullman), Vol. 10, p. 88.
37. Rodnick, K. J., Slot, J. W., Studelska, D. R., Hanpeter, D. E., Robinson, L. J., Geuze, H. J. *et al.* (1992). *J. Biol. Chem.*, **267**, 6278.
38. Thakar, J. H. and Ashmore, C. R. (1975). *Anal. Biochem.*, **69**, 545.
39. Bhattacharya, S. K., Thakar, J. H., Johnson, P. L., and Shanklin, D. R. (1991). *Anal. Biochem.*, **192**, 344.
40. Bullock, G., Carter, E. E., and White, A. M. (1970). *FEBS Lett.*, **8**, 109.
41. Makinen, M. W. and Lee, C. P. (1968). *Arch. Biochem. Biophys.*, **126**, 75.
42. Marsh, M., Schmid, S., Kern, H., Harms, E., Male, P., Mellman, I., *et al.* (1987). *J. Cell Biol.*, **104**, 875.
43. Jackson, J. and Moore, A. L. (1979). In *Plant organelles* (ed. E. Reid), p. 1. Ellis Horwood, Ltd., UK.
44. Beers, E. P., Moreno, T. N., and Callis, J. (1992). *J. Biol. Chem.*, **267**, 15432.
45. Dhugga, K. S., Ulvskov, P., Gallagher, S. R., and Ray, P. M. (1991). *J. Biol. Chem.*, **266**, 21977.
46. Morré, D. J. and Buckhout, T. J. (1979). In *Plant organelles* (ed. E. Reid), p. 117. Ellis Horwood, Ltd., UK.
47. Vigil, E. L., Wanner, G., and Theimer, R. R. (1979). In *Plant organelles* (ed. E. Reid), p. 89. Ellis Horwood, Ltd., UK.
48. Robinson, S. P., Edwards, G. E., and Walker, D. A. (1979). In *Plant organelles* (ed. E. Reid), p. 13. Ellis Horwood, Ltd., UK.
49. Hall, J. L. and Taylor, A. R. D. (1979). In *Plant organelles* (ed. E. Reid), p. 103. Ellis Horwood, Ltd., UK.
50. Leigh, R. A., Branton, D., and Marty, F. (1979). In *Plant organelles* (ed. E. Reid), p. 69. Ellis Horwood, Ltd., UK.
51. Vitols, E. V. and Linnane, A. (1961) *J. Biophys. Biochem. Cytol.*, **9**, 701.
52. Guerin, B., Labbe, P., and Somlo, M. (1979). In *Methods in enzymology* (ed. S. Fleischer and L. Packer), Vol. 55, p. 149.
53. Rothblatt, J. A. and Meyer, D. I. (1986). *Cell*, **44**, 619.
54. Goud, B., Salminen, A., Walworth, N. C., and Novick, P. J. (1988). *Cell*, **53**, 753.
55. Daum, G., Bohni, P. C., and Schatz, G. (1982). *J. Biol. Chem.*, **257**, 13028.
56. Ruohola, H. and Ferro-Novick, S. (1987). *Proc. Natl. Acad. Sci. USA*, **84**, 8468.
57. Lloyd, D. (1979). In *Methods in enzymology* (ed. S. Fleischer and L. Packer), Vol. 55, p. 135.
58. Chua, N. H. and Bennoin, P. (1975). *Proc. Natl. Acad. Sci. USA*, **72**, 2175.
59. Bowman, E. J. and Bowman, B. J. (1988). In *Methods in enzymology* (ed. S. Fleischer and L. Packer), Vol. 157, p. 574.
60. Birkenmeyer, L. and Ray, S. R. (1986). *J. Biol. Chem.*, **261**, 2362.

2

Isolation of subcellular fractions

RICHARD H. HINTON and BARBARA M. MULLOCK

1. Introduction

The aim of subcellular fractionation is to separate cell organelles with as little damage as possible. The first thing to remember is that it will never be possible to separate organelles completely undamaged. In the living cell, most of the cell organelles are attached to cytoskeletal elements and are surrounded by cytosol which is a 20% solution of protein in which most of the water is bound to the hydration sphere of the proteins. Breakage of the linkage to the cytoskeleton and alterations in the environment will thus inevitably result in alterations to the cell organelles and these alterations will undoubtedly be exacerbated by damage occurring during homogenization, as discussed in Chapter 1, and damage occurring during the course of separation, as discussed in this and subsequent chapters. Thus it will never be possible to separate cell organelles in a completely natural state and, in fact, there is never a single best way to fractionate a tissue. For example, the method used to determine the distribution of some material between cell organelles may differ markedly from the method used to isolate a particular organelle with the minimum of damage. The first aim of this chapter will be to discuss the general principles of both analytical and preparative methods for cell fractionation.

Probably the most common error made by newcomers in the field of cell fractionation is to assume that methods developed for the separation of organelles from one tissue can be applied directly to another. A moment's thought shows that this is not the case. For example mitochondria and lysosomes in hepatocytes and the proximal tubule of the kidney differ markedly in their relative sizes. Both these cell types contain large numbers of formed structures in their cytoplasm (*Figure 1a,b*) but in other cell types, such as the small lymphocytes in the thymus, few formed structures are seen in the cytoplasm (*Figure 1c*), while in specialized cells, such as the acinar cells of the pancreas, the cytoplasm may be filled with specialized structures, such as secretion granules, which are completely absent from other cell types (*Figure 1d*). It is therefore clear that methods for the separation of cell organelles will differ from tissue to tissue. It must also be remembered that tissues are not

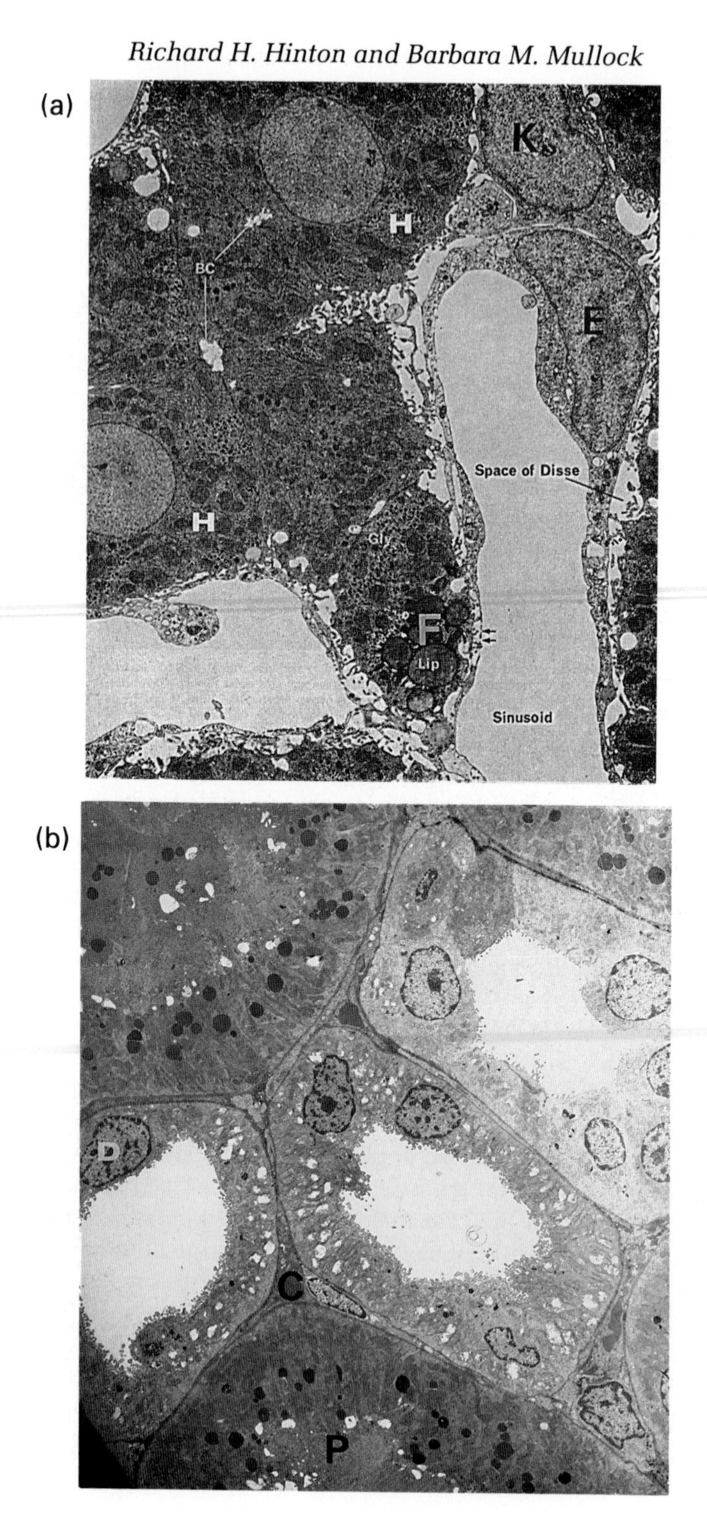
(a)
K
H
BC
E
Space of Disse
H
Gly
F
Lip
Sinusoid
(b)
D
C
P

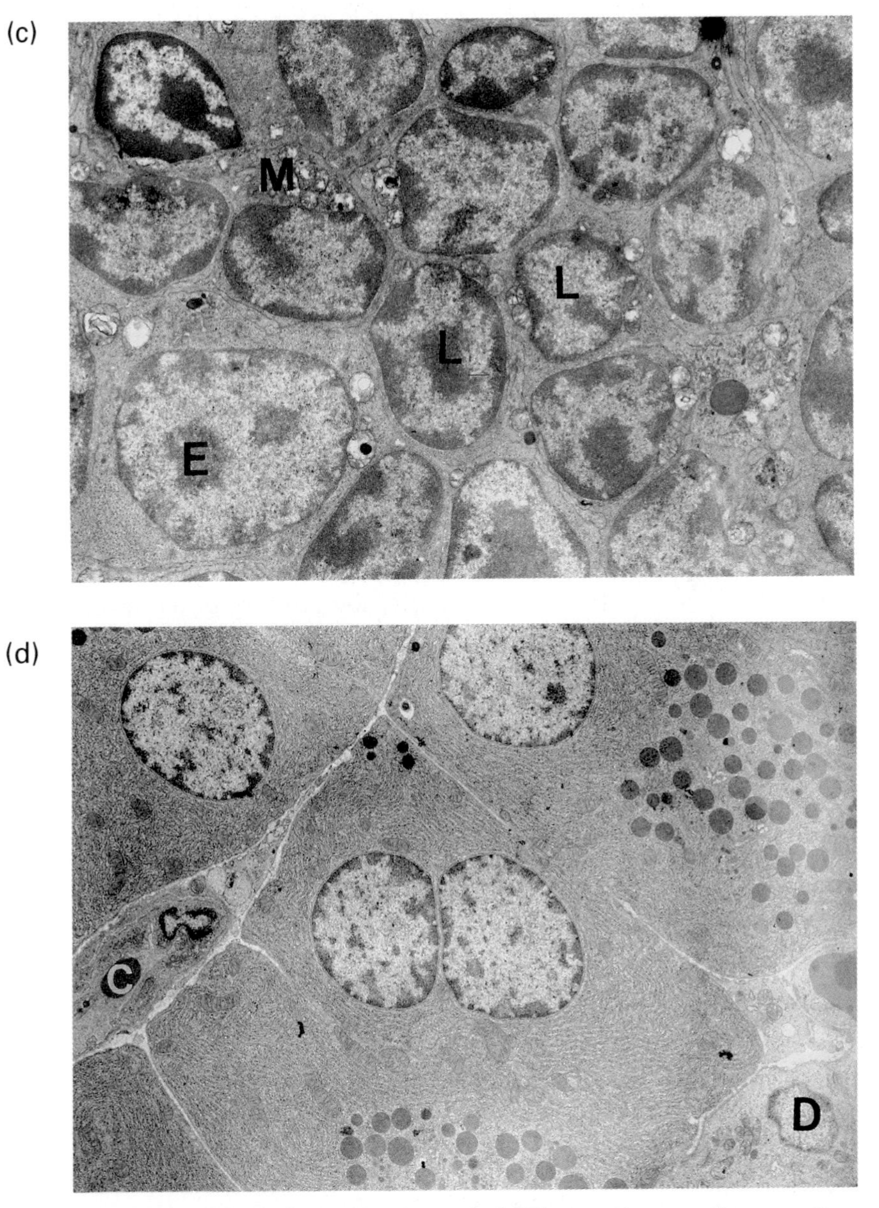

Figure 1. Electron micrographs showing the marked differences between the proportions of different cell organelles and between different cell types in a given tissue. (a) Liver showing hepatocytes (H), a Kupffer cell (K), a sinusoid endothelial cell (E), and a lipid storing (Ito) cell (F); BC, bile canaliculus; Gly, glycogen; Lip, lipid. (b) The kidney cortex showing the epithelia of the proximal (P) and distal (D) tubules and capillaries (C) (courtesy of Dr M. Dobrota). (c) The thymic cortex showing small lymphocytes (L), an epithelial cell body (E), and macrophages (M). (d) The exocrine pancreas showing acinar cells, a capillary (C), and (lower right) a duct lining cell (D).

uniform in structure. In the thymus, for example, the vast majority of cells are small lymphocytes, containing few cytoplasmic organelles. Two minor cell types present in the thymus, the epithelial cells and particularly the macrophages (*Figure 1c*) contain large numbers of cell organelles and will contribute disproportionately to the total. In the thymus, the problem is clear after the most casual glance at an electron micrograph. This is not always the case. For example, hepatocytes make up 78% of the liver by weight and Kupffer cells only 2%. However, hepatocytes contain only 56% of liver lysosomes while Kupffer cells contribute 26%. In some cases these problems may be overcome by prior separation of the different cell types, but this in itself may lead to problems. For example, in the case of the liver, isolated hepatocytes closely resemble hepatocytes in the tissue but isolated Kupffer cells differ markedly due to phagocytosis of fragments of dying cells during fractionation, while the flattened endothelial cells roll up giving an appearance quite different from that seen *in situ* (compare *Figures 1a*, *2b*, and *2c*). The properties of cell organelles may also be changed by disease, administration of xenobiotics, or physiological state. There can thus be no universal methods for cell fractionation and to interpret the results of cell fractionation studies, it is essential that one is aware of the cytology of the tissue one is about to fractionate.

2. Composition of a tissue homogenate

The first requirement when developing methods for cell fractionation is to familiarize oneself with the ultrastructure of the tissue or cell type which is to be separated. The second requirement is a method of dispersion which minimizes damage to cell organelles. These methods have been dealt with in detail in Chapter 1 of this book. It is, however, worthwhile to think again briefly about the fate of the different cell structures after homogenization. For this purpose we may think of the cell as consisting of four types of structures. First there are the relatively compact cell organelles, nuclei, mitochondria, chloroplasts, microbodies, lysosomes, and secretion granules and, in the case of plant cells, chloroplasts and the vacuole. In general, such formed structures survive homogenization intact providing that they are not too large. However, there is generally extensive loss of the outer membrane of chloroplasts and there may be some loss of the outer mitochondrial membranes. Very large organelles such as the vacuole of plant cells and the very large lysosomes formed in some pathological processes do not survive homogenization and fragment with loss of their content unless special care is taken to avoid this problem.

The second class of structures within the cell are the membrane systems of the plasma membrane, the endoplasmic reticulum, the Golgi apparatus, the endosome compartment and, in some cell types, the annulate lamellae. With the exceptions of the plasma membrane these structures consist of flattened

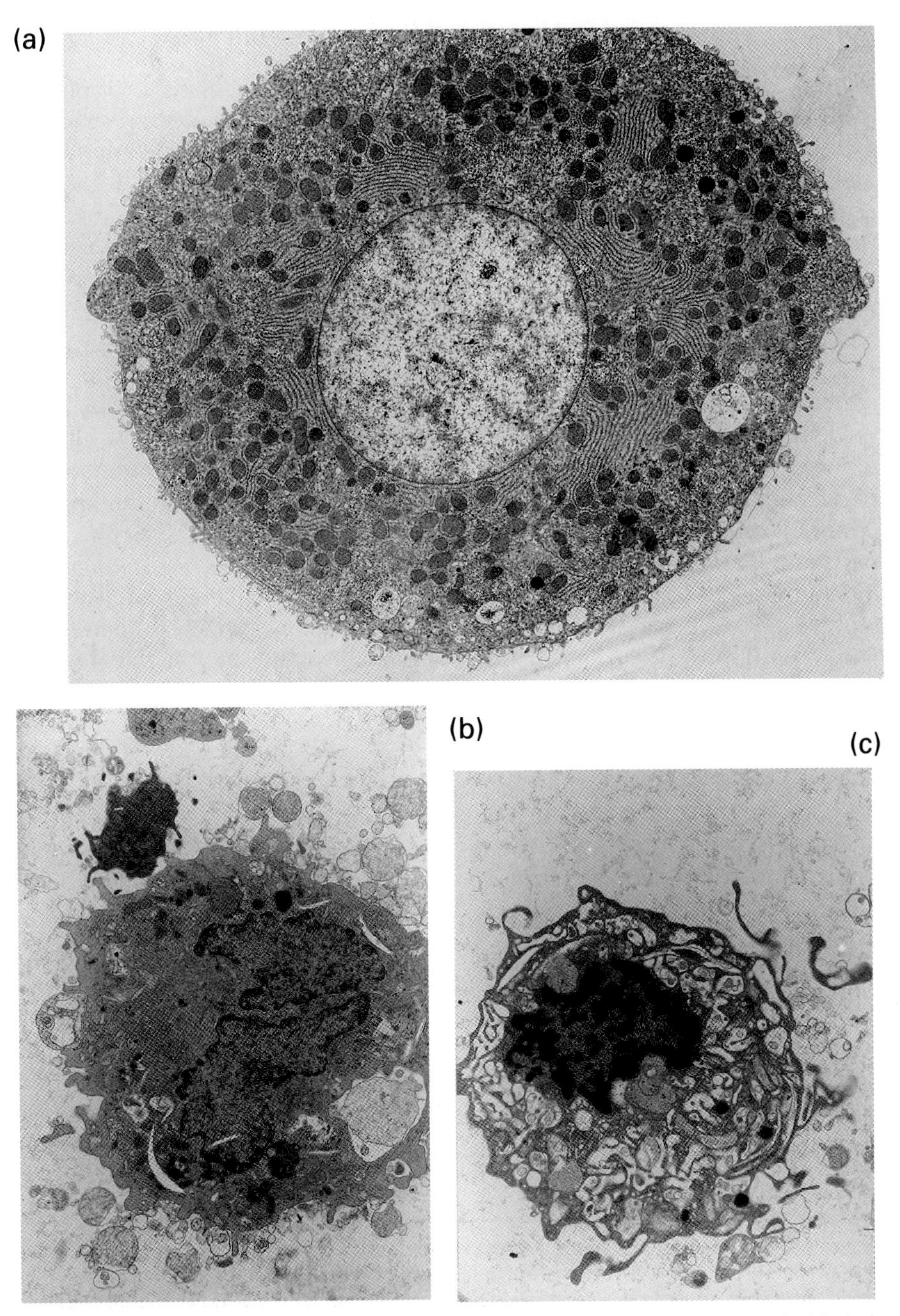

Figure 2. Cells separated from a liver dispersate. The cells should be compared with their appearance in the intact tissue (*Figure 1a*). (a) An hepatocyte. The appearance of these resembles that in the intact tissue. (b) A Kupffer cell. This is engrossed due to phagocytosis of fragments of dying cells during dispersal of the tissue. (c) A sinusoid endothelial cell. This has little resemblance to the cell *in vitro*. The thin tube which lines the sinusoid has wrapped in layers around the cell body.

cisternae which normally fragment during homogenization into small vesicles in such a way that there is minimal mixing of the cisternal contents with the cytosol (*Figure 3a*). The two exceptions to this rule are the Golgi apparatus and the plasma membrane. Under very mild homogenization conditions the central stack of the Golgi apparatus is at least partially preserved. The fate of the plasma membrane depends on the cell type and homogenization conditions. Under mild conditions, the plasma membrane of free living cells may remain more or less intact as a 'cell ghost', but normally these membranes and the membranes of cells not joined by junctional complexes fragment to vesicles which may be either 'right side out' or 'inside out' (*Figure 3b*), although the presence of high concentrations of acidic sugar side chains on the extracellular side of the plasma membrane favours the former configuration. Where, however, the sheets of cells are linked by strong junctional complexes, the plasma membranes of the face so stabilized (in the case of epithelial cells the apical face) tend to remain as large sheets while the other parts of the cell membrane fragment to vesicles (*Figure 3c*).

The third and fourth classes of intracellular components may be dismissed more briefly. Cytoskeletal elements play an exceedingly important part in the living cell but rarely attract the attention of those engaged in cell fractionation. Cells rich in cytoskeletal elements, such as many cell lines grown *in vitro*, may, however, be difficult to homogenize because of tangling of cell organelles with cytoskeletal elements. Finally there are the proteins and small particles of the cytosol. The size difference between cytosolic proteins and cell organelles means that the former do not interfere in the separation of the latter, but the small cytosolic particles such as free polysomes, ribosomes and their subunits, glycogen, ferritin granules, and crystalloid structures may contaminate preparations of small membrane-bounded fragments, confusing the interpretation of the results.

Thus, to summarize, it is possible to divide the intracellular structures with which this book is concerned into two classes. First there are the large cell organelles which survive homogenization intact and which can generally be identified by electron microscopy. Of a similar size to these structures are large sheets of plasma membrane stabilized by junctional complexes and the central stack of the Golgi apparatus, assuming that this survives homogenization. In addition the homogenate will undoubtedly contain unbroken or par-

Figure 3. Fragmentation of membrane systems during homogenization. (a) Fragmentation of the endoplasmic reticulum. The sheets and tubules present *in vivo* fragment into small vesicles with almost perfect preservation of contents. It should be noted a similar fragmentation occurs *in vivo* during mitosis. (b) Fragmentation of the plasma membrane of hepatocytes. The microvilli of the sinusoidal surface fragment into vesicles which are mainly 'right way out' due to the negatively charged glycoproteins on the cell surface. The junctional complexes keep the bile canalicular face and a part of the lateral face intact. (c) Electron micrograph of a fragment of the bile canalicular and lateral membrane. Note the intact desmosomes linking the membranes.

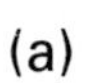

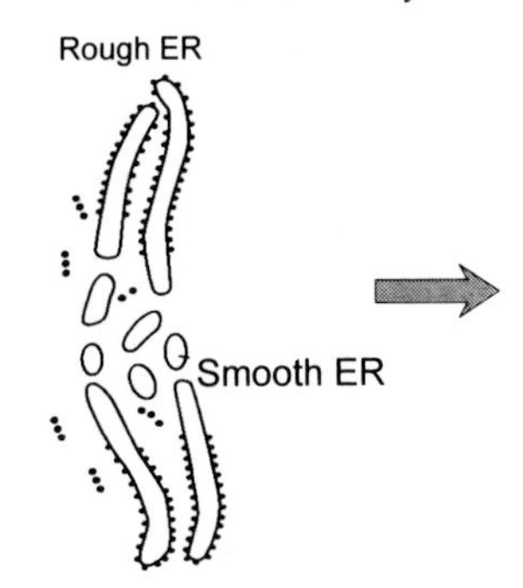

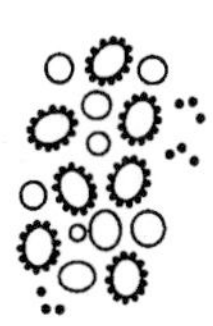

Endoplasmic Reticulum in Cell   Microsomes in Homogenate

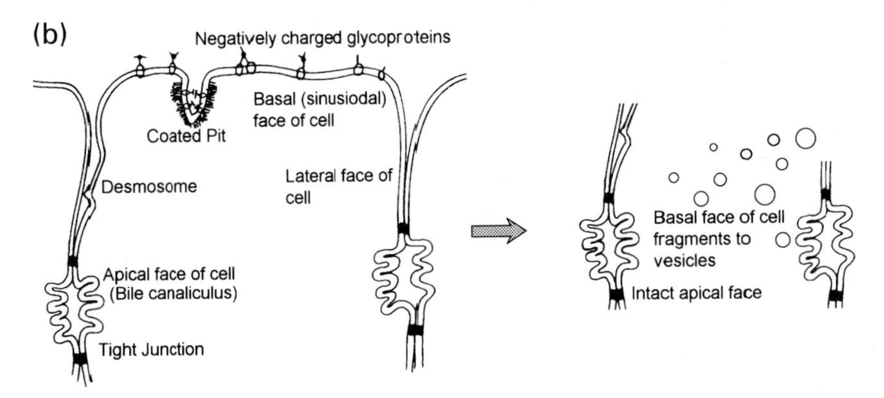

Plasma membrane of hepatocyte   Plasma membrane fragments after homogenisation

(c)

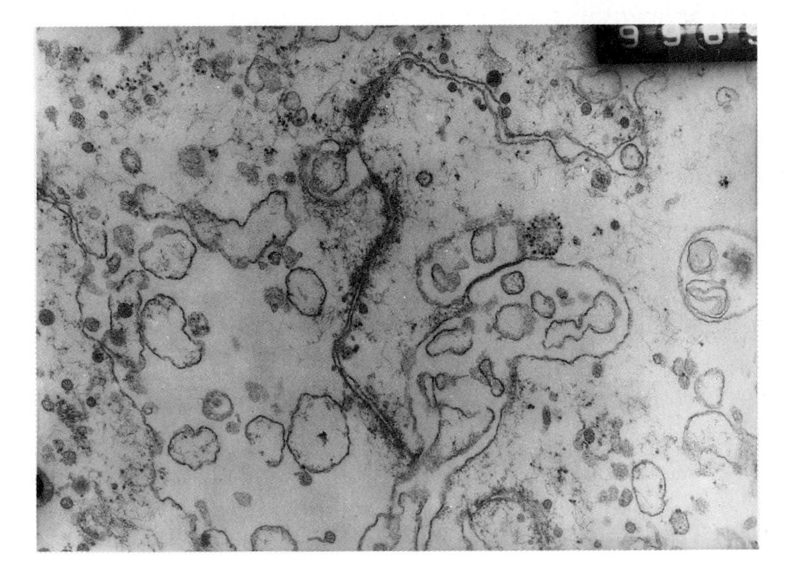

tially broken cells and may well contain aggregated material, especially if there has been any significant nuclear lysis. The second class of membrane-bounded structures in the homogenate are the small vesicles which result from the fragmentation of intracellular and plasma membranes. These are normally referred to as microsomes. In addition to the membrane systems discussed above, the microsomes will include, to a greater or lesser extent, fragments of the larger cell organelles. Because microsomes form by fragmentation, they may be quite variable in size and there is little, if any, correlation between the size of a microsome and the structure from which it derives. Also, because they are simply bubbles of membranes, microsomes deriving from sources other than the rough endoplasmic reticulum cannot be identified by electron microscopy unless antibody labelling techniques are used.

3. Properties of cell organelles

The most casual glance at an electron micrograph will show that the different cell organelles differ in size. Differences in size result in differences in rates of sedimentation in centrifugal fields and the practical procedures to exploit these differences will be described briefly in the present chapter. However, it is equally clear, first that there is considerable heterogeneity in the sizes of, for example, individual mitochondria, and secondly that there is no significant difference in the sizes of microsomes derived from different membrane systems. In practice, we always find overlap in the sizes of the different classes of cell organelle. Fortunately there are other properties of cell organelles which may be exploited. The most fundamental of these is the presence, on the outside of each cell organelle, of characteristic proteins. Consideration of the biochemical properties of cell organelles shows that such proteins must exist and that, in principle, it should be possible to recognize them by use of specific antisera or affinity ligands. As will be discussed later, the problem is that such antisera and ligands are often not currently available. Surface charge would appear a second possible property on which cell structures could be separated, but again this has found very limited use in practice. Currently the most useful property of cell organelles for the purposes of separation is the organelle density, but, because cell organelles are surrounded by semi-permeable membranes, their density is not a constant, but depends on interactions between the organelle and the surrounding media.

3.1 Factors affecting organelle density and size

The density of any structure is defined as mass divided by volume. Hence it would, at first sight, appear that the density of a cell organelle would be a simple constant. However, the methods which we employ to separate a cell organelle mean that neither mass nor volume are necessarily constant. The normal and easiest method for separating cell organelles according to their density is to form a concentration gradient of some suitable material, known as the density gradient solute, overlay some homogenate or subcellular fraction, and centrifuge until the various cell organelles and fragments have reached a liquid of their own density. A particle suspended in a liquid of its own density has no weight and therefore neither floats or sinks whatever the centrifugal field.

As mentioned earlier, the membranes of subcellular organelles are semipermeable. The structure of the membrane enclosing the organelle allows water and a small number of other molecules to pass freely, while restricting the flow of other materials. The interior of the organelle contains a mixture of non-diffusible materials, such as proteins, and of molecules which can pass through the membrane. The mass of the organelle will be the mass of its membrane, including bound structures such as ribosomes, which we can assume to be invariant, the mass of non-diffusible solutes in the interior of the organelle, and the mass of water and of diffusible solutes. As organelles in animal cells are in osmotic equilibrium with the cytosol the density of the interior of organelles of mammalian cells will be approximately equal to the density of 0.15 M KCl, plus a small contribution from the proteins. Hence if the organelle is suspended in a solution of a high molecular weight medium such as Percoll which cannot penetrate the organelle and which exerts no significant osmotic pressure, then the densities generally observed are between 1.03 g/ml and 1.10 g/ml. If, however, the medium exerts a significant osmotic pressure above that within the organelle, then the concentration of the natural solutes in the liquid within the organelle rises as increasing concentrations of the density gradient solute draw water out of the organelle. This increases the density of the material within the interior of the organelle while decreasing its size, both factors increase the density of the particle. Hence the density of the organelle increases with the concentration of the density gradient solute. However, eventually a point will be reached where the density of the gradient equals the density of the organelle (*Figure 4*), and, following centrifugation, it is at this point in the density gradient that the particle will band (i.e. at its own buoyant density in the medium). In media with these properties, such as Nycodenz, the densities of cytoplasmic organelles will vary between approximately 1.13 g/ml and 1.23 g/ml (see Section 4.2.1). The final case is where the membrane is freely permeable to the solute. This is the case with certain organelles in sucrose gradients and with all organelles in glycerol gradients. In such a situation, the density of the liquid within the organelle equals the density of the surrounding medium. This will result in the banding densities being significantly greater than those found with non-permeable solutes. Peroxisomes are permeable to many small solute molecules and consequently have very similar densities in all gradients of true solutes.

The situation is however made considerably more complex in that some

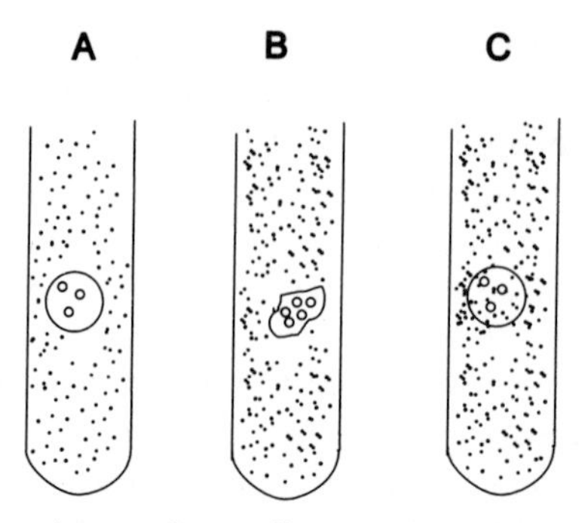

Figure 4. Diagram showing changes in a cell organelle with increases in the concentration of the suspending fluid. (A) The organelle in an isotonic solution. The small circles are solutes within the organelle which cannot pass through the membrane. (B) An increase in the concentration of an impermeant solute has resulted in an increased osmotic pressure causing the organelle to shrink, increasing the concentration of its own solutes. (C) An increase in concentration of a permeant solute means that the organelle now contains both the exogenous and its endogenous solutes again resulting in an increase in density.

organelles, such as mitochondria, have more than one aqueous compartment and that in all organelles, much of the water is not freely diffusable but bound within hydration spheres of macromolecules. Nevertheless it is clear that the isopycnic banding densities of cell organelles will depend on the properties of the solute used to form the density gradient, even though the precise relationship between solute and particle density may be difficult to predict. We have seen that the density of the organelle depends on the degree to which the solute can penetrate the membrane. Different membranes are penetrated to different extents by different solutes. Thus not only does the absolute density of an organelle vary with the nature of the solute but also the relative density. Examples both in this chapter and elsewhere in this book will show how such differences can be exploited to achieve separations which at first sight seem impossible. Hence isopycnic centrifugation, the separation of structures on the basis of their density, provides a powerful complement to separations on the basis of size.

4. Centrifugal methods for the separation of organelles

Separation of cell organelles by centrifugation depends on differences in size and in density. A particle suspended in a liquid less dense than itself will sediment. The rate of sedimentation will depend on many factors including the shape of the particle and the difference in density between the particle and the medium. However, in normal circumstances the speed of sedimentation will be influenced principally by particle size. The downward force acting on a particle will be proportional to the volume of the particle, which is dependent on the cube of the radius, the retarding force on the particle will be the viscous drag which is proportional to the square of the radius and to the particle's velocity. It is thus simple to demonstrate that the steady state velocity rises with particle size. Hence methods for separating particles according to size depend on differences in sedimentation rate. The basis of separation of particles by density is equally simple. No net force acts on a particle suspended in a liquid of its own density and hence there will be no tendency either to float or sink. The density of a liquid depends both on the nature of the solvent and on the amount of dissolved material, thus the density of water is 1 g/ml while that of a 30% (w/w) solution of sucrose is 1.129 g/ml. Hence if particles are layered over a column of liquid in which the concentration of the density gradient solute (in the example given, of sucrose) increases progressively and the column is then centrifuged, the particles will sediment until they reach a liquid of their own density and will then stop. This is referred to as isopycnic or buoyant density banding.

Overall the velocity of sedimentation of a particle (v) in a centrifugal field is described by the following equation:

$$v = \frac{d^2(\rho_p - \rho_l)}{18\,\mu} g$$

d = diameter of particle
ρ_p = density of particle
ρ_l = density of liquid
g = gravitational field (RCF)
μ = viscosity of liquid

4.1 Separation by size

Two distinct and complementary methods are used for the separation of particles on the basis of their size, simple differential pelleting and rate-zonal sedimentation.

4.1.1 Differential pelleting

In this, the simpler of the two procedures, the suspension of particles is placed in a centrifuge tube and centrifuged for sufficient time to sediment the largest group of particles. The supernatant is then poured off and recentrifuged to collect the next fraction and so on. Applied exactly as described, this would result in very impure preparations. For example, we can consider a mixture with three components, which sediment respectively at 1 cm/min, 0.5 cm/min, and 0.1 cm/min. If we centrifuge for long enough to sediment all the first class of particle, then half the second class and one-tenth of the third class will also have reached the pellet. These levels of contamination are probably too great to be tolerated. If, however, the supernatant is decanted; the pellet resuspended in new medium and recentrifuged under the same conditions, then the new pellet will contain one-half of a half of the original number of the second class of particle and one-tenth of a tenth of the third. Hence 'washing' the pellet has resulted in a very effective separation between particles differing by an order of magnitude in their sedimentation rate but very poor separation of particles differing by a factor of two. Hence differential pelleting, as this technique is termed, is highly effective at separating particles into broad size classes but not suitable for separating particles which are similar in size.

i. Separation of large particles and microsomes by differential pelleting

In Section 2 it was pointed out that the components of an homogenate of a tissue such as rat liver may be divided into three broad size classes, relatively large and intact cell organelles, the microsomes, and the soluble portion of the cell, the cytosol. As these differ significantly in size they will also differ in their rate of sedimentation and so can be separated by centrifugation. A total of four fractions is usually prepared (see *Protocol 1*). The first of these fractions (low speed pellet) contains the unbroken and partially broken cells together with the nuclei, large sheets of plasma membrane, and a proportion of the other larger cell organelles. This low speed pellet or N fraction is often referred to as the nuclear fraction, which is misleading, for nuclei generally form less than half the material in the fraction and the pelleted material is a poor source for the isolation of both nuclei and plasma membrane sheets. This low speed pellet should be 'washed' by resuspension in homogenization medium and recentrifuged under the same conditions to minimize loss of the smaller cell organelles, and the supernatant combined with the original. The combined supernatants are then recentrifuged at high speed to give the large particulate fraction or M fraction. This contains the majority of intact cell organelles. It is often termed the mitochondrial fraction, a possibly defensible term if one works only with liver, but quite inappropriate in other cases. This fraction will inevitably be contaminated by the larger microsomal fragments, for these overlap in size with the smaller cell organelles, but this contamination may be minimized by washing. Finally microsomes are sedimented in an ultracentrifuge. If the tissue has been homogenized in sucrose and contains significant amounts of rough endoplasmic reticulum there may be considerable absorption of basic cytosolic proteins on to the microsomal membranes. These proteins can be removed by 'washing' in a medium containing 0.15 M KCl.

This simple fractionation scheme is useful for several purposes. In particular it may provide a simple method for separating enzyme isoforms associated with different cell organelles (e.g. the cytochromes of the mitochondria and of the smooth endoplasmic reticulum), and provides a quick and convenient initial scheme for studies designed to identify the subcellular location of a particular cell component. However the method employed is crude, bearing the same relation to the techniques described later in this section as does solvent extraction to chromatography.

4.1.2 Rate-zonal sedimentation

The complementary approach to differential pelleting is not to centrifuge a uniform suspension of particles but to layer the particles over a denser liquid. Following centrifugation, particles would then be expected to form a series of bands depending on their size. In practice this simple approach would not work because a band of particles sedimenting through a uniform liquid is unstable but if, instead of a uniform column of liquid we layer the particles over a density gradient, then we do get an orderly separation. There are, however, a number of points to note. First, if the density of the liquid at any point in the gradient is similar to that of the particles, then these will stop, in other words there will be a separation according to density rather than size. Secondly, because bands of particles are finite in width there is a limit to the range of particle sizes which can be separated, too large a range and all the largest size particles are accumulating in a pellet before the small ones have left the starting zone. Thirdly, the amount of particles which can be loaded on the gradient is finite, too many and the sedimenting bands become unstable. Thus this second technique, known as rate-zonal sedimentation, is ideally suited to be a second stage in fractionation, following differential pelleting. It is normal to use swinging-bucket rotors for this type of fractionation to maximize the length of the gradient and so the size range of particles which can be fractionated. Rate-zonal fractionation has been used to separate nuclei, mitochondria, and plasma membrane sheets from the low speed pellet of a rat liver homogenate.

4.2 Separation by density

As mentioned earlier, separation of particles on the basis of their density is normally carried out by layering the particles over a density gradient and centrifuging to equilibrium. In principle the same separation is obtained if the particles are suspended in dense liquid and layered under the gradient or if they are distributed through the gradient, for Archimedes' principle says that if a particle is suspended in (and hence displaces) a fluid with a density greater than the particle, then that particle will float upwards. These approaches may be useful in certain specialized circumstances but in general they are to be avoided because the risk of damage to cell organelles by hyperosmotic media, although this is avoided or minimized by using an iodinated density gradient medium or Percoll (see Section 4.2.1). Damage to organelles during separation is a much greater problem in isopycnic banding than in rate-zonal centrifugation for three reasons. First, in isopycnic banding, particles are inevitably exposed to much higher concentrations of the density gradient medium than in rate-zonal sedimentation. Secondly, centrifugation times are much longer, for the rate of particle sedimentation is proportional to the difference between the density of the particle and the density of the liquid and approach to equilibrium is inevitably slow. Thirdly, in an attempt to combat the long times of centrifugation, it is common to employ the fastest possible rotors. With long gradient columns this may result in hydrostatic pressures sufficient to damage cell organelles. Hence it is generally accepted that 'vertical tube' rotors are best suited to separations by isopycnic banding, the short sedimentation distance leading to relatively rapid banding times and low hydrostatic pressures. However, separation by isopycnic banding also has advantages. Because the organelles reach an equilibrium with the gradient the technique, unlike rate sedimentation, is not sensitive to small changes in time of centrifugation or rotor design. Furthermore, while the size, and hence the sedimentation rate, of subcellular organelles varies little with the composition of the medium, the banding densities may differ markedly (see next section) so giving extra flexibility to the separation.

While continuous density gradients are required for rate-zonal separations and are strongly advised for use when developing methods based on isopycnic banding, their preparation is tedious and the separated material is normally spread over several fractions and may require concentration before further study is possible. Hence once the density range of a particular organelle in a particular medium has been established step (discontinuous) gradients are often better for preparative separations. Choice of conditions for such gradients is discussed in Section 9 of this chapter.

4.2.1 Media employed in density gradient separation

In Section 3 we described how the density of cell organelles depends on the properties of the surrounding medium as well as on the intrinsic composition of the organelle. The properties of the various chemicals used to form density gradients will now be considered in somewhat more detail. The properties of media suitable for separation of cell organelles may be stated fairly briefly:

(a) They must be very soluble in water.

(b) They must not damage the cell structures (a requirement which rules out salts such as caesium chloride).

(c) They should not be precipitated by magnesium, calcium, or other ions present in cell homogenates, and should not interfere with the assay of separated components.

(d) The viscosity of the solutions should be as low as possible.

During development of centrifugal methods a wide range of solutes were tested, but sucrose stood out as having almost all the desirable properties listed above with the additional important advantage of being cheap. Accordingly sucrose gradients remain the most usual way for separating cell organelles both on the basis of size and of density. However, more recently, there has been renewed interest in other media for use in separation by isopycnic banding. This has been driven by the increasing interest in the study of the mechanisms governing transport in the endocytic and exocytic pathways. Membranes of the endoplasmic reticulum, of the Golgi apparatus and associated structures, and of the endosome compartments fragment into vesicles which show no significant difference in size and hence can only be separated on the basis of density.

i. Sucrose

Sucrose is probably still the most widely used medium for density gradient centrifugation and, indeed, it is only rarely that any other solute is used for the fractionation of cell organelles by rate sedimentation (sucrose is quite unsuitable for use in the separation of living cells but that is a topic which lies outside the scope of this article). Sucrose has some disadvantages as a medium for isopycnic banding. Sucrose is capable of passing through the membranes of some organelles and, due to its low molecular weight, its solutions have a high osmotic pressure/density ratio. Accordingly organelles band in sucrose solutions at relatively high densities and hence high sucrose concentrations. This causes two problems. First, the organelles are exposed to solutions of very high osmolarity. This has comparatively few deleterious effects although it may make the structures liable to lysis if the solution is diluted rapidly and undoubtedly distorts their appearance under the electron microscope. Secondly, and more serious in practice is that concentrated sucrose solutions are extremely viscous, so that banding may be slow, thus increasing the risk of damage by autolysis. Sucrose, like other low molecular weight media, may interfere in the assay of enzymes and measurement of radioactivity by liquid scintillation. These effects depend on the concentration of sucrose in the assay media and are usually fully reversible.

ii. Glycerol

Glycerol gradients are widely used for the separation of large proteins and protein complexes by rate-zonal sedimentation but are rarely used for the isopycnic banding of cell organelles, although glycerol's distinctive properties may make it useful as a last resort. Glycerol, unlike any of the other media discussed in this section, passes freely through all biological membranes and hence the relative densities of cell organelles may be significantly different from those found with sucrose and the other, higher molecular weight materials described later. The permeability of organelles to glycerol means that their banding densities in gradients of this solute are very high. Concentrated glycerol solutions are also extremely viscous, so that centrifugation times to achieve banding in glycerol gradients are even longer than banding times in sucrose. Glycerol does not appear to damage cell structures and is, indeed, used as a 'cryopreservative' but does interfere in enzyme assays in the same way as sucrose.

iii. Iodinated media such as metrizamide and Nycodenz

Many years ago it was realized that the desirable properties for an X-ray contrast medium, high intrinsic density, solubility, and lack of toxicity, were identical to the requirements for a density gradient solute. A variety of media were examined but the majority proved difficult to use because, being salts of derivatives of tri-iodobenzoic acid, they precipitated in the presence of certain ions or on reduction of the pH. However, metrizamide, where the acidic residue was masked by esterification to a sugar, proved much more suitable and, after a while, the manufacturer produced a related compound, Nycodenz, especially for sale as a density gradient medium. The properties of metrizamide and Nycodenz are similar although not identical (1, 2). Neither penetrate biological membranes, although both exert a significant osmotic pressure, which is nevertheless much lower than that of sucrose. Metrizamide for example is iso-osmotic with mammalian fluids (290–300 mOsm) at a density of 1.18 g/ml. *Figure 5* compares the osmotic activity of a number of gradient media.

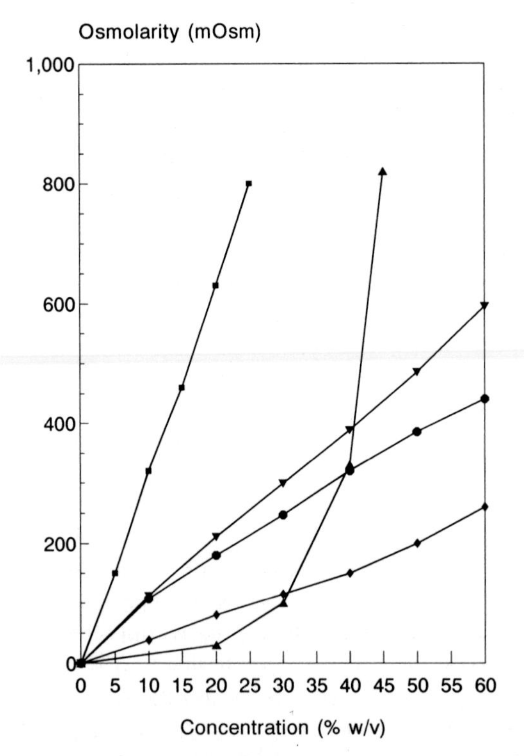

Figure 5. Osmolarity of gradient media: effect of concentration. (■) Sucrose; (▼) Nycodenz; (●) metrizamide; (♦) Iodixanol; (▲) Ficoll (courtesy of Dr J. Graham).

Table 1. Buoyant density of mammalian organelles in different gradient media[a]

Organelle	Buoyant density in gradient media (g/ml)				
	Sucrose	**Nycodenz**	**Iodixanol**	**Ficoll**	**Percoll**
Nuclei	>1.32	1.23	1.20	–	1.09
Mitochondria	1.19	1.19	1.14	1.14	1.07
Lysosomes	1.21	1.15	1.12	–	1.10
Peroxisomes	1.23	1.22	1.20	–	1.06
Plasma membranes	1.13–1.18	1.11–1.19	–	1.05	1.03

[a] Data is taken from refs 3 and 5.

Banding densities of cell organelles in gradients of these materials are, accordingly, lower than those in sucrose or glycerol and relative banding densities also differ (see *Table 1*). In addition, the high intrinsic densities of these compounds mean that lower concentrations are required to produce a solution of a given density than is the case with sucrose. Accordingly the viscosities of the metrizamide or Nycodenz gradients used in isopycnic banding are considerably lower than those of sucrose or glycerol gradients, so that centrifugation times are much shorter. As might be expected of X-ray contrast media, solutions of metrizamide or Nycodenz do not damage cell organelles. Metrizamide and Nycodenz, like other low molecular weight media, do interfere in enzyme assays (3) but, because gradients of these media are less concentrated than gradients of sucrose or glycerol the problem is less serious.

The latest addition to the range of iodinated density gradient media is Iodixanol which is essentially a dimer of Nycodenz. It therefore has approximately half of its osmotic activity (see *Figure 5*) and all organelles band at densities which are iso-osmotic (4). Osmotically active organelles tend to have marginally lower banding densities in Iodixanol compared to Nycodenz, this is particularly true of mitochondria which band at slightly hyperosmotic densities in Nycodenz (see *Table 1*), while the banding densities of peroxisomes are relatively little affected (5).

iv. Ficoll and related media

Ficoll is a high molecular weight polymer of sucrose and epichlorhydrin. It is entirely non-permeant and exerts a very low osmotic pressure. The banding densities of cell organelles in Ficoll are, accordingly, very low and relative densities may differ markedly from those observed with other media. Ficoll solutions are, however, extremely viscous so that, in spite of these low banding densities, centrifugation times are long. A further problem is that as the concentration of Ficoll rises, so an increasing amount of water is sequestered in its hydration sphere. Below 30% it exerts no significant osmotic pressure, salts or sucrose are often added to maintain tonicity. However, as Ficoll concentrations rise, so the amount of free water falls and the effective concentration of the salts or sucrose rise, causing a rapid increase in osmotic pressure which is not linearly related to Ficoll concentration (see *Figure 5*). Accordingly Ficoll gradients for the separation of cell structures which are sensitive to hypotonic solutions can be difficult to design and more specialized works than this should be consulted. Ficoll does not interfere in enzyme assays. Unlike the low molecular weight materials Ficoll cannot be removed by dialysis or ultrafiltration so that separated structures must be collected by pelleting with the consequent risk of damage. Hence if a separation involves several isopycnic steps, it is usually best that separation in Ficoll should be left to last. Although this section is entitled Ficoll, it is worth noting that gradients of dextrans and proteins have also been employed in separations. In general, these have no significant advantages and, the solutions being even more viscous than those of Ficoll, banding times can be even longer.

v. Percoll

Some time ago it was pointed out that colloidal silica could be suitable for the formation of density gradients. However silica dispersates are highly sensitive to alterations in pH so they were little used until the development of Percoll, a dispersion of extremely small silica particles 30–40 nm in diameter and coated with polyvinylpyrrolidone. Percoll particles are thus not much smaller than microsomes and it would hence seem the medium would only be suitable for fractionation of the larger cell organelles. This is not the case because centrifugation of Percoll, under suitable conditions, results in stable gradients, where downward sedimentation due to the presence of the centrifugal field is balanced by upward diffusion due to the concentration gradient. Theoretical considerations show that particles with diffusion coefficients significantly less than that of Percoll will band in such gradients. Because the viscosity of Percoll solutions does not differ significantly from that of water, banding times can be fast, but it should be noted that gradient shape is markedly affected by changes in rotor geometry as well as speed of centrifugation. Because Percoll particles are much denser than cell organelles they can, in principle, be removed by layering the suspension over a suitably dense barrier layer and high speed centrifugation but, in some cases, problems have arisen due to adhesion of Percoll particles to membranes. Because Percoll gradients form during centrifugation (20–100 000 *g* for 20–30 min), either of a Percoll solution of uniform density or of a step gradient, it is therefore not necessary to use a gradient maker. The latter approach may however be preferable if large particles are present in the sample, for these may reach the bottom of the tube and form a pellet before the gradient has had time to form. Iodixonal also forms self-generating gradients (4), but as it is a true solute it needs higher *g* forces (at least 180 000 *g* for 3 h or 350 000 *g* for 1 h).

4.3 Density perturbation

An alternative approach to the problem of separating organelles or membrane systems which overlap in size and density is to try to perturb specifically the density of one of the components. Unfortunately, like many attractive ideas, this is less easy in practice than in theory, and even when methods can be developed, results may be difficult to interpret. The major problem is that however density modification is performed, it is possible that the properties of the organelle concerned will be affected. Hence, in principle, it is better to seek to modify the properties of contaminating organelles rather than the organelle under study, but in practice this is rarely feasible.

The approaches concerned generally derive from histochemical studies. Histochemistry exploits the fact that if an enzymatic reaction results in the generation of an insoluble product, deposition of that product will occur very close to the site of enzymatic activity. Deposition of such a product will, inevitably, result in alteration of the composition, and hence density, of a structure and should therefore provide a very specific mechanism for separation of an organelle. Practical results have, however, not always lived up to expectations. Several factors are involved. First, the reagents used in the histochemical reaction may interact directly with the structures under study, an example being the interaction of lead ions used in the histochemical detection of phosphatases, with ribosomes. Secondly, vesicularization of membrane systems may result in a mixture of 'inside out' and 'right side out' vesicles, one of which may be able to hydrolyse the substrate, the other not. Thirdly, formation of the histochemical deposit may result in non-specific aggregation. These are only three of the many possibilities for interactions.

In the case of organelles in the endocytic pathway, density modifications may be made by treatment of the living cell or animal prior to homogenization. The organelle can be modified directly, as in the classic studies on lysosomes where administration of the detergent Triton WR-1339 (6) gave a marked reduction in the density of liver lysosomes. Colloidal gold complexed to antibodies against endocytic receptors can elevate the density of plasma membrane vesicles or endosomes bearing the receptor (7). Alternatively an enzyme, usually horse-radish peroxidase (8), coupled to an endocytosed ligand is used to load parts of the endocytic system, and the dense enzyme product is generated *in vitro* by incubation with diaminobenzidine and H_2O_2, after homogenization and particle fractionation. For more information on the use of this method see Chapter 7, Section 2.6.2. The results obtained have been invaluable to cell biology, but it is possible that the pre-treatments may lead either to a fundamental distortion of the endocytic system or to the selection of a subpopulation of vesicles.

In general our advice would be to use published methods for subcellular fractionation which exploit density modification only if they are readily reproducible, but to discard those methods which prove unreliable and, above all, to consider carefully the extensive time which is often involved in the development of such methods.

5. Non-centrifugal procedures

Although most procedures for the separation of subcellular organelles have involved the use of centrifuges, other methods have been investigated. Such methods include size exclusion chromatography (9), partition between liquid phases (10), and highly individual techniques aimed at the isolation of specialized structures, such as the stripping of cilia by shearing after encapsulation in an agarose gel. Separation in non-aqueous media has been used to isolate nuclei (11) without the loss of soluble proteins which seems inevitable with conventional techniques. Of more general application are methods which make use of surface proteins or surface charge, namely immunoisolation or electrophoresis.

5.1 Immunoisolation

If an antibody to a specific protein on the surface of an organelle is available, the organelle can, with care, be attached to a solid support such as cellulose or magnetic beads and thereby separated (12, 13). This approach has been more widely used for the separation of living cells than for subcellular separations but, where suitable antibodies are available, very pure preparations of specific regions of the plasma membrane or of small membrane-bounded organelles such as endosomes can be achieved. Generation of suitable antibodies is obviously the greatest limitation, but as organelle-specific proteins are identified and sequenced, peptides representing their cytoplasmic domains can be synthesized and used for antibody preparation. The labour involved may be justified in those instances where metabolic properties are under investigation, since these isolation methods generate functional immobilized organelles which can easily be retrieved from reaction mixtures during experiments. However, amounts of material separated are usually small.

An alternative to preparing an antibody to a membrane surface protein is to incorporate a foreign surface expressed antigen into the membrane. This has been executed using virally-infected cells. The proteins which make up the surface of a budding virus travel to the plasma membrane by the same route as the cell's own proteins. As the viral proteins are transmembrane, viral antigenic determinants will always be exposed on vesicles deriving from various stages of the exocytic pathway and can be recognized by antiviral antibodies. Identification of different vesicle populations may be obtained by examining the distribution of vesicles containing viral proteins at a series of times after infection. Furthermore in certain epithelial cells some viruses bud from the apical face and others from the basolateral face, so providing markers which distinguish the two exocytic pathways operating in such cells. The G protein of vesicular stomatitis virus (VSV) migrates to the basolateral face in polarized cells while the haemagglutinin of influenza virus is transported to the apical domain. Using antibodies to the viral protein cytoplasmic domain, endocytic vesicles have also been isolated after implanting VSV G protein into the plasma membrane at 4 °C followed by incubation at higher temperatures permissive to endocytosis. See refs 14 and 15 for more information.

5.2 Separation by electrophoresis

Cell organelles are charged at neutral pH due to the presence of acidic and basic groups on their surface and will hence migrate in an electric field. The rate of movement is proportional to the charge and inversely proportional to the viscous drag, hence the rate of movement is strongly influenced by particle size. The major problem is stabilizing the migrating zone of particles. This is normally done by introducing the particles into a flowing 'curtain' of buffer and the technique is accordingly termed free flow or continuous flow electrophoresis. The technique may be used for separation of particles of all sizes from single proteins to living cells. In the field of subcellular fractionation, most applications have involved separation of components of the endocytic and exocytic pathways but this probably reflects the current interests of cell biologists rather than any inherent suitability of the apparatus. Separation techniques based on free flow electrophoresis appear less reliable than those based on density gradient centrifugation, a major problem being prevention of the development of convection cells due to uneven heating of the liquid by the electric current. Moreover, the necessary dedicated equipment is expensive. Free flow electrophoresis must, therefore, be regarded as a specialized technique rather than a routine method for separating cell organelles. Moreover most of the major organelles appear to have rather similar electrophoretic mobilities and it is often necessary to resort to modification of the surface charge enzymically before a satisfactory separation can be achieved (16).

6. Identification of separated material

Subcellular fractions may be identified either by examination under the electron microscope or by biochemical techniques. Microscopy of subcellular fractions is considered elsewhere in this book (Chapter 10). That chapter should be consulted by anyone planning to use these techniques; without proper precautions electron microscopy of subcellular fractions can lead to utterly misleading results.

6.1 Marker enzymes

In general, enzymes are largely confined to a single subcellular compartment. Hence measurement of characteristic enzymes may be used to determine the composition of a cell fraction. Thus, for example, if a cell fraction shows succinate dehydrogenase activity then either intact mitochondria or fragments of the mitochondrial inner membrane are present. The qualification here is important. Succinate dehydrogenase is commonly referred to as a marker for mitochondria and this, indeed, is the case in the absence of mitochondrial damage. However, during homogenization or cell fractionation, it is possible that mitochondria will be broken. Succinate dehydrogenase is an intrinsic protein of the mitochondrial inner membrane, which is markedly different in composition from the outer membrane or, of course, from the soluble proteins of the matrix or intermembrane space. Hence the presence of succinate dehydrogenase activity, on its own, does not necessarily indicate the presence of intact mitochondria, strictly that requires either morphological evidence, or assay of markers for the outer mitochondrial membrane, or of proteins of the intermembrane space. In addition the distinction between one membrane compartment and another may be blurred by the presence of membrane vesicles shuttling between them (see Section 7.4). Hence care is needed when interpreting the results of the assay of marker enzymes but, nevertheless, this is generally the easiest method for detecting the distribution of the major cell organelles between subcellular fractions.

It is commonly assumed that the distribution of enzyme activities is identical between tissues. This is not the case. For example, the endoplasmic reticulum of hepatocytes contains a specific glucose-6-phosphatase, used in export of glucose, which provides a useful marker for these membranes. This specific enzyme is also found in the kidney but is confined to a small segment of the proximal tubule. In other cells of the liver and kidney and in cells of other tissues, the little glucose-6-phosphate breakdown detectable is catalysed not by the specific enzyme but by the non-specific phosphatases of plasma membrane, lysosomes, and the cytosol. So with these cells, glucose-6-phosphatase activity would certainly not indicate the distribution of endoplasmic reticulum fragments among subcellular fractions. Strictly, therefore, one should validate markers for each cell type of each tissue, but in practice one can often rely on enzymes whose fundamental role means that they are present in most cell types (some of these are summarized in *Table 2*).

Marker enzymes for the major organelles, mitochondria, lysosomes, and peroxisomes are perhaps the best characterized. Succinate dehydrogenase is the most common marker for mitochondria but, as mentioned earlier, the presence of succinate dehydrogenase activity in a subcellular fraction does not necessarily show that intact mitochondria are present. Monoamine oxidase is frequently used in the liver as a marker for mitochondrial outer membranes, but isoforms are also present in the inner membrane. In practice, it is probably best to use the electron microscope to determine whether the presence of succinate dehydrogenase activity indicates intact or damaged mitochondria. Fractions possessing monoamine oxidase activity but no succinate dehydrogenase activity probably contain fragments of the mitochondrial

Table 2. Marker enzymes for major cell organelles[a]

Organelle	**Marker enzyme**
Mitochondria (inner membrane)	Succinate dehydrogenase
Mitochondria (outer membrane)	Monoamine oxidase
Lysosomes	β-D-galactosidase
Endoplasmic reticulum	β-D-glucosidase
Plasma membrane	5′-Nucleotidase[b]
Golgi apparatus (*trans*-elements)	Galactosyl or sialyl transferase
Golgi apparatus (*cis*- and *mid*-elements)	No satisfactory markers
Endocytic pathway	Introduced markers

[a] The enzymes in the table are widely distributed between tissues and the different cell types within a tissue, but the activities may be rather low so that tissue-specific enzymes may be easier to measure (e.g. in liver glucose-6-phosphatase is the best marker for fragments of the endoplasmic reticulum).
[b] In epithelial tissues 5′-nucleotidase activity is generally more pronounced in the apical domain.

outer membrane. There is no perfect marker for lysosomes, for the enzymatic composition will differ between the different cell types which make up a tissue. Acid β-D-galactosidase is a useful marker as the lysosomes of all cell types must digest glycoproteins. Acid phosphatase is probably the most commonly used marker, but is probably not the most suitable as it is not uniformly spread between lysosomes, probably due to membrane association. Catalase is a marker for both large peroxisomes, as found in the liver, and for microperoxisomes.

There is no universal marker for endoplasmic reticulum. β-D-Glucosidase is probably the first-choice candidate as all cell types must process cell surface glycoproteins and this enzyme is required for that purpose. Drug metabolism enzymes are not suitable as the sole indicators of endoplasmic reticulum for they are largely confined to the smooth surfaced areas but assessment of the distribution of these along with membrane-bound RNA may together provide the best indication of the distribution of endoplasmic reticulum fragments. The choice of drug metabolizing enzyme will vary with the tissue, depending on which isoforms of cytochrome P450 are present.

There can be no single marker for the Golgi apparatus and endosome compartment for both are heterogeneous. Galactosyl transferase is the most commonly used marker for fragments of the Golgi apparatus; sialyl transferase has also been employed. These enzymes add terminal sugars to glycoproteins and hence are normally concentrated in the *trans*-elements. Galactosyl transferase activity is also found in the plasma membrane of some cell types. There would appear, at present, no other Golgi enzymes whose assay is sufficiently simple for them to be used as markers. For further information on markers which have been used for *cis*- and medial-Golgi, see Chapter 7, Section 3.1. The enzymatic composition of the endosome compartments is not well characterized and fragments are normally identified by introduced markers as described in the next section. The plasma membrane of epithelial cells is split into domains, which may differ markedly in their enzyme composition. Na^+/K^+-activated ATPase is found only in the plasma membrane, but is probably confined to the basolateral domain of epithelial cells.

5′-Nucleotidase is frequently used as a plasma membrane marker, but it is, strictly speaking, an apical membrane protein. This enzyme also appears in the basolateral plasma membrane domain of those cells in which newly synthesized apical membrane proteins are first translocated to the basolateral domain before relocation. In this respect cells vary widely. In hepatocytes, *all* newly synthesized plasma membrane proteins go first to the basolateral domain from which apical domain proteins are sorted by transcytosis. In MDCK cells, all apical domain proteins are sorted in the *trans*-Golgi network and move directly to the apical surface. In Caco-2 cells, some apical domain proteins behave like those of the hepatocyte while others behave like those of MDCK cells. For more information see refs 17 and 18.

6.2 Introduced markers for endocytic and exocytic pathways

The organelles involved in endocytosis and exocytosis of cultured cells can readily be labelled by exogenous material. In the case of the endocytic pathway labels may be introduced by binding to specific cell surface receptors. LDL and transferrin receptors are present on all cell types but many others may be usable. The introduced material may be radiolabelled for ease of measurement, gold labelled for identification under the electron microscope, or iron labelled to ease fractionation of the tissue. The exocytic pathway may also be labelled by addition of suitable radiolabelled precursors. Amino acid precursors will eventually label all parts of the exocytic pathway, while the sugar fucose will only label the *trans*-Golgi compartment and later components of the pathway. Identification of the different membrane systems can be eased by examining the distribution of radioactivity at different times after addition of the label. Inhibitors may be used to interfere with passage of material between compartments, for example lowering of the temperature prevents entry of proteins into main stack of the Golgi apparatus, but manipulations like this are likely to alter the properties of the membrane systems concerned and it may be difficult to extrapolate from the results obtained with such abnormal systems to the situation in the normal cell.

6.3 Characteristic non-enzymatic proteins

As mentioned previously a good marker for a particular cell organelle or fragments of a membrane system should both be located specifically in the one cell compartment and should be easy to assay. It is much easier to measure an enzymatic activity than to measure the concentration of a protein by immunological or electrophoretic methods although immunoblotting may be used to check the distribution of two marker proteins across a gradient. In practice therefore, almost all of the markers used in cell fractionation studies have been enzymes. However it has proved very difficult to find enzyme markers for Golgi subcompartments or for the various endosome compartments (see Section 6.2). These have traditionally been identified by introduced markers. Very recently a family of GTP proteins, known as *rab* proteins, have been discovered which do appear to localize to particular subcompartments of the exocytic and endocytic pathways and it is possible that, when these are better characterized that they will prove suitable for use as markers. For more information see ref. 19.

A second, minor, use for non-enzymatic markers is in determining the orientation of vesicles formed by the fragmentation of membrane systems. In general intracellular membranes fragment so that the orientation is preserved and the 'outside' of the vesicle is the former cytoplasmic face. However, although plasma membrane vesicles are normally orientated with the extracellular face outwards this is not always the case. As most enzyme substrates are small molecules, they may be able to cross membranes to some extent, thus enzyme assay may not be a definitive means of determining sidedness, although the interaction of glucose-6-phosphate with glucose-6-phosphatase is probably an acceptable method for the endoplasmic reticulum.

Antibodies, however, will not pass the membrane. Hence if 'sidedness' of plasma membrane vesicles is a problem, it may be useful to combine assay of a marker enzyme in the presence of a permeabilizing agent with probes with and without detergent for surface antigens, thus obtaining information both on the total amount of membrane-derived vesicles and on the proportion which display cell surface antigens. Where it is known that plasma membrane vesicles form 'right side out', a similar approach employing antisera to cell surface receptors may be employed to help identify vesicles involved in the endocytic pathway since these will have surface receptors hidden inside the vesicle.

7. Assessment of the purity of fractions

Before discussing the assessment of the purity of a fraction there are several points which need to be made. First there are many cases where overall purity is unimportant, and the aim of the cell fractionation procedure may simply be to obtain the fraction in sufficient concentration for biochemical studies. An obvious example is the study of drug metabolizing enzymes of the smooth endoplasmic reticulum. In this and similar cases, the quest for high purity is likely to be counterproductive, because the time required for further centrifugation steps will increase the risk of damage to delicate proteins such as the cytochromes P450. How to achieve the right balance between speed of separation and purity will be considered in more detail in Section 9 of this chapter. However there are occasions when purity is important, as when investigating the enzymatic composition of a particular cell compartment when knowledge of the purity of the preparation and identification of potential contaminants is essential.

7.1 Purity and purification

These terms are often confused. The purity of a preparation is the percentage contributed by the main component, in a biochemical preparation this would normally be taken to mean the percentage of the total protein. The purification of a marker is the specific activity (activity per unit of protein) in the preparation divided by the specific activity in the homogenate. To interpret purification of markers for cell organelles in terms of purity one must know the percentage of the cell occupied by a cell organelle. For example mitochondria occupy about one-fifth of the volume of a liver cell, hence a marker enzyme in a completely pure preparation of mitochondria would show a purification of five-fold, in small lymphocytes, where mitochondria make up less than 5% of the cell, then a mitochondrial preparation where marker enzymes showed a five-fold purification would be less than 25% pure.

When considering the purity of a fraction both positive evidence, enrichment of a marker for the structure in question, and negative evidence, the presence of markers for potential contaminants must be considered. In addition it is essential that one knows the percentage of the cell occupied by the various cell organelles or membrane systems. This is determined by morphometric techniques. Use of these to obtain really accurate figures is complex and time-consuming. However, for the purposes of determining fraction purity only rough estimates are required. By use of these estimates, together with data on purification of marker enzymes a rough estimate of the relative contributions of the different cell structures to the composition of the fraction can be made. Poor recoveries are frequently encountered and may arise from a number of causes. The first factor which needs eliminating is possible loss of, or activation of, markers during fractionation. If proper balance sheets have been kept this problem will have been detected. A second possibility is that a 'marker' may be less specific than is imagined, for example if the proportion of 'lysosomes' in a fraction is surprisingly high and the fraction is rich in alkaline phosphatase, then a possibility is that observed 'acid phosphatase' activity does not reflect the presence of lysosomes at all but is due to weak residual activity of the alkaline phosphatase. Assay of a second lysosomal marker should settle this point. However, in many cases explanations are not so simple and three further factors are considered below.

7.2 Problems from cell heterogeneity within tissues

In some cases the heterogeneity of cells within a tissue is so great that it is meaningless to talk about purity of cell fractions. An example here might be the kidney where both cell structure and function differ markedly along the nephron. Cell fractionation of these tissues will generally result in preparations which mix structures from different cell types in ill-defined proportions. Such preparations may be invaluable in many biochemical studies but they give little information on the composition of the intracellular compartments of the different types of cells in the kidney.

While even a cursory examination of tissues like the kidney reveals the extent of cell heterogeneity, in the liver differences are more subtle. While hepatocytes from the periportal zone appear similar to those from the centrilobular zone there are significant differences in their function and these are reflected by differences in the enzymes found in their cell organelles. Furthermore while hepatocytes make up around 90% of the mass of the liver, non-hepatocytes contain approximately 40% of the lysosomes. Hence with all solid tissues subcellular fractions will, to a greater or lesser extent be heterogeneous.

7.3 Problems arising from organelle fragmentation

It is normally assumed that large cell organelles such as mitochondria or lysosomes remain intact in the homogenate. This is largely correct when the liver of a young rodent is homogenized with a Potter–Elvehjem homogenizer but cannot be assumed for other tissues. In some cases organelles are intrinsically sensitive, for example it is extremely difficult to recover chloroplasts with an intact outer membrane. Secondly, it is not possible to homogenize tissues such as skin with a Potter–Elvehjem homogenizer and other types of homogenizer tend to cause fragmentation of the mitochondrial outer membrane and rupture of lysosomes. Thirdly, enlargement of lysosomes due, for example, to accumulation of indigestible products, may lead to a marked increase in fragility. Finally the high hydrostatic pressures found at the base of swinging-bucket tubes centrifuged at high speeds may result in damage, notably loss of the mitochondrial outer membrane.

7.4 Missorting in the exocytic and endocytic pathways

With the exception of the red blood cell there will be continuous movement of proteins along the exocytic pathway from the endoplasmic reticulum, through the Golgi apparatus to lysosomes and the plasma membrane, and from the cell surface to the endosome compartment and either back to the plasma membrane or on to lysosomes. Proteins travelling along these pathways are passed from one membrane system to the next by small 'shuttle' vesicles. When these form there will, inevitably, be some missorting so that proteins which 'should' have remained in one compartment get passed to the next. Such missorted proteins will, eventually, be retrieved and returned to their 'proper' compartment but there will, inevitably, be some proteins in the 'wrong' place. Hence the presence of very small amounts of markers for one compartment of the exocytic or endocytic pathway in a preparation of vesicles from a neighbouring compartment need not always indicate cross-contamination.

8. Fractionation problems

Very often a first attempt at a separation will not give the separation one would expect from the structure of the cells employed. Obviously it is not possible to describe all possible problems but some common observations and explanations are discussed below.

8.1 No separation

Virtually all the material in the homogenate sediments at 1000 *g* for 10 min or is gathered together in a single band in a density gradient, possibly with the formation of visible aggregates. This can be associated with:

(a) Poor breakage of cells, this can be readily identified by light microscopy. See Chapter 1 for a discussion of homogenization problems.

(b) Lysis of nuclei and formation of nucleoprotein gels. This frequently causes formation of aggregates large enough to be visible to the naked eye. The cause can be confirmed by light microscopy following staining with a dye such as trypan blue. This problem is especially common if the homogenization medium is not strictly isotonic or if tubes are contaminated with detergent. Plastic tubes tend to retain detergent contamination more readily than do glass ones. The presence of Mg^{2+} ions in the homogenization medium may reduce this problem, 5 mM is a suitable concentration for a first try, although this may itself be detrimental to the recovery of functionally intact mitochondria (see Chapter 1, Section 4 and Chapter 4).

(c) Tangling of cell organelles by cytoskeletal elements. This is a problem which is frequently encountered when fractionating homogenates of cultured cells. There is no universal cure.

(d) Trapping of material on interfaces. This is a problem especially associated with step gradients. If a compact band of material forms then this may trap denser, but more slowly sedimenting particles. A possible solution is to increase the density of the sample so that it can be placed at the bottom of the gradient and the particles allowed to separate by flotation.

8.2 Aggregation following resuspension of a fraction

This problem is most frequently observed with the low speed pellet. It may be associated with any of the situations mentioned in Section 8.1 or with the presence of erythrocytes in the pellet. This is frequently encountered in a highly vascular tissue such as liver, unless the tissue has been perfused prior to homogenization.

8.3 Poor recovery of markers

One should always calculate the recovery (i.e. the sum of the activities in the fractions divided by the activity in the starting fraction) of marker enzymes. Large differences from the ideal 100% may be due to:

(a) Damage to cell structures by the gradient medium. As markers are chosen as being robust this is not much observed in practice.
(b) Inhibition of the enzyme by the gradient medium. Low molecular weight media may inhibit enzymes to some degree and some may also interfere in measurement of radioactivity by scintillation counting.
(c) Interference of the gradient material in the assay.

The data in *Table 3* clearly show that the gradient media may have a significant but unpredictable inhibition of enzyme activity when they are included at 15–30% concentration in the assay medium. This may arise through a direct effect on the enzyme or some interference in the assay procedure. The data from *Table 4* on the other hand indicate that if the media do

Table 3. Effect of gradient media on enzyme activity[a]: % inhibition

Enzyme	Gradient medium			
	Nycodenz	Metrizamide	Sucrose	
			15%	30%
5′-Nucleotidase	20	0	58	65
Adenylate cyclase	36	27	13	nd
NADPH cyt *c* reductase	50	47	53	nd
Glucose-6-phosphatase	16	4	nd	50
Succinate dehydrogenase	0	6	0	62
Acid phosphatase	6	0	nd	28
Catalase	0	0	0	nd
ATPase	14	7	0	41

[a] Enzyme assayed in the presence of 15% (w/v) Nycodenz or metrizamide and 15% and 30% (w/v) sucrose. Data taken from refs 20 and 21. nd: not determined.

Table 4. Effect of gradient media on enzyme activity[a]: % inhibition

Enzyme	Gradient medium		
	Iodixanol	Nycodenz	Sucrose
5′-Nucleotidase	0	0	0
NADPH cyt *c* reductase	0	0	0
Succinate dehydrogenase	0	17	19
Acid phosphatase	0	0	9
Catalase	0	0	0
ATPase	25	23	7

[a] Membrane incubated for 3 h at 4°C in the presence of 30% (w/v) gradient medium and enzyme assayed at approx. 5% (w/v) gradient medium. Data taken from ref. 4.

have a direct inhibitory effect on the enzyme, this can be largely reversed by suitable dilution of the gradient solute by the assay medium.

8.4 Damage to cell structures

This may be detected by electron microscopy or by functional analysis. It is normally associated with problems during homogenization which have been considered in Chapter 1 but may also be associated with:

(a) Damage during resuspension of pelleted fractions. It is important to use gentle means for this purpose, e.g. a few gentle strokes of the pestle of a loose-fitting Dounce homogenizer.

(b) Damage during collection of fractions from a gradient. The normal method of concentration fractions from a density gradient is by dilution and centrifugation.

When low molecular weight gradient materials such as sucrose are employed, dilution will result in a rapid change in osmotic pressure which may damage the organelles. Although with some of the modern iso-osmotic media (Nycodenz, Iodixanol, or Percoll) this may not be a problem.

(a) Damage due to trace contaminants of the gradient material. Normal laboratory sucrose, for example, contains very significant amounts of ribonuclease.

(b) High hydrostatic pressure during centrifugation. This can cause problems if isopycnic banding is carried out in swinging-bucket tubes.

Enzymes leaked from lysosomes, proteases in particular, may attack surface proteins on other structures unless suitable inhibitors are added. Standard inhibitor cocktails are recommended in Chapter 1.

9. A systematic approach to cell fractionation

Development of a method for separating cell organelles from a previously uncharacterized tissue falls into four phases. If, as is generally the case, the method is based on density gradient centrifugation these will be:

(a) Preliminary studies. These include development of a method for homogenization which gives good dispersion while minimizing damage to organelles and validation of markers for the different cell structures.

(b) Investigation of the relative sedimentation rates and banding densities of the different components of the homogenate and development of an S–ρ diagram.

(c) Method development.

(d) Simplification of the method for routine separations.

Obviously these must be modified if immunoaffinity labelling or free flow electrophoresis is to be employed but, in practice, these techniques are generally used as a final 'clean-up' after a centrifugal fractionation.

9.1 Preliminary studies

The development of methods for tissue dispersion has been considered in Chapter 1 of this book while selection and validation of markers has been discussed earlier in this chapter. Neither problem is trivial. Cultured cells, in particular, can be very difficult to homogenize and validation of markers is a vexed question. Ideally this validation should be carried out by examining the distribution of enzymes through the cell by cytochemical techniques, but the poor resolution of the light microscope and the risk of loss of enzyme activity in preparing samples for electron microscopy may make this difficult. Hence frequently, all one can do, is to show that a group of enzymes believed to be associated with a single cell compartment show identical distribution under a range of fractionation conditions. This is not a phase in development to be skipped over lightly.

9.2 Determination of the properties of components of the homogenate

With the exception of polysomes, ribosomes and their subunits, cell organelles are heterogeneous both in size and in banding density. Conventionally the size and density of the different types of cell organelles are represented on a two-dimensional chart called an S–ρ diagram (*Figure 6*) in which size or sedimentation rate is plotted on one axis while banding density is plotted on a second. The first step in any cell fractionation study is to establish this diagram. The approach which we would recommend is shown in *Protocol 2*. While sucrose is the obvious gradient material to use in the rate separation step, the choice of initial solute for the isopycnic separations is much less obvious. If the aim of the study is to analyse the distribution of material over a range of cell organelles, for example when following transport of material through the endocytic pathway, then sucrose or an iodinated material such as Nycodenz or Iodixanol is probably the best starting point, sucrose is much cheaper, but banding times are longer. If, on the other hand, the aim of the study is to develop a method for separating an organelle which does not vesiculate during homogenization, then Percoll gradients may be preferred on grounds of ease of use.

The experiments outlined in *Protocol 2* are sufficient to draw up an S–ρ diagram and although the process is time-consuming, the information it provides is very beneficial to the design of subsequent routine isolation procedures. While it is possible to calculate 'sedimentation coefficients' for membrane-bound cell organelles this is of little use in practice and it is much better to use an arbitrary scale for the 'size' axis. When the diagram is prepared, the different cell organelles are represented by ellipses and it will rapidly become clear whether complete separation will be a possibility. Where there is partial overlap, as between lysosomes and mitochondria in the example, then the reason for the heterogeneity should be considered. Heterogeneity may be due to random variation, for example in the size of the fragments formed by fragmentation of the large tubular mitochondria of living cells, or due to heterogeneity; for example, between the lysosomes of hepatocytes and of Kupffer cells. Obviously, in the former case, preparations which excluded organelles in the overlap zone would be fully representative, in the latter case they might not.

A very different situation exists when there is a major overlap of organelle populations, as with lysosomes and peroxisomes in the example given in *Figure 6*. In such cases the isopycnic banding should be repeated using a different gradient solute. As discussed earlier, the relative banding densities of cell organelles alter with changes in the medium. In the case of the liver homogenate, which has been used as an example, the banding densities of lysosomes and peroxisomes are similar in sucrose gradients but very dissimilar in gradients of metrizamide or Nycodenz. Comparison of the S–ρ diagrams in *Figure 7* makes this point clearly.

As already mentioned, the microsomes formed by fragmentation of the membrane systems within cells are very heterogeneous in size and rate sedimentation does not normally give usable separations. In some cases fragments of a given membrane system are sufficiently distinct in density to permit separation in a single step; fragments of the rough endoplasmic reticulum

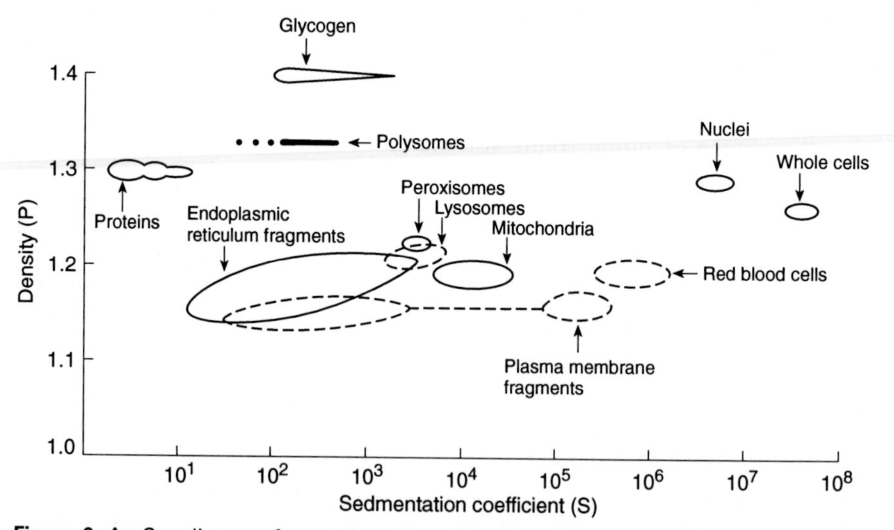

Figure 6. An S–ρ diagram for rat liver. The diagram shows the distribution of different components present in the liver homogenate. The densities of materials such as nucleoproteins are estimates (from ref. 20).

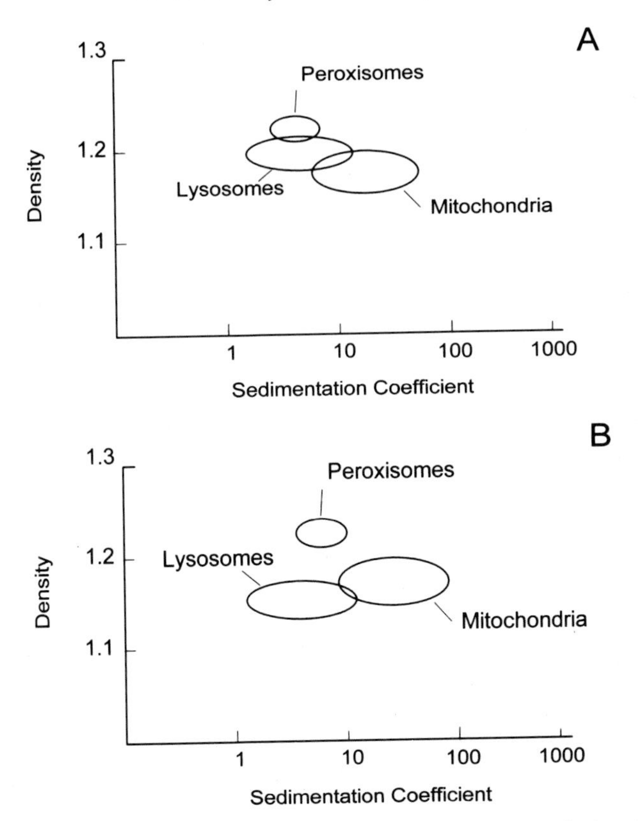

Figure 7. An S–ρ diagram for the major components of the large particulate fraction from rat liver in gradients of (A) sucrose and (B) metrizamide. Note the marked difference in the banding density of lysosomes.

are, for example, far denser than other cell membranes. Generally, however, there is sufficient overlap in the density of the fragments of the different membrane systems to prevent complete separation in a single step. In such circumstances it may be necessary to band sequentially using two different solutes. The results of such studies may be displayed on a two-dimensional diagram similar to an S–ρ diagram (technically a ρ–ρ diagram); again actually plotting such a diagram will help visualize which classes of microsome are potentially separable.

9.3 Method development

The preliminary studies described in the previous section may suffice if the aim of the study is solely to determine the distribution of some component between cell organelles, but will not be suitable if the aim is to purify some subcellular structure for further examination. However the preliminary studies

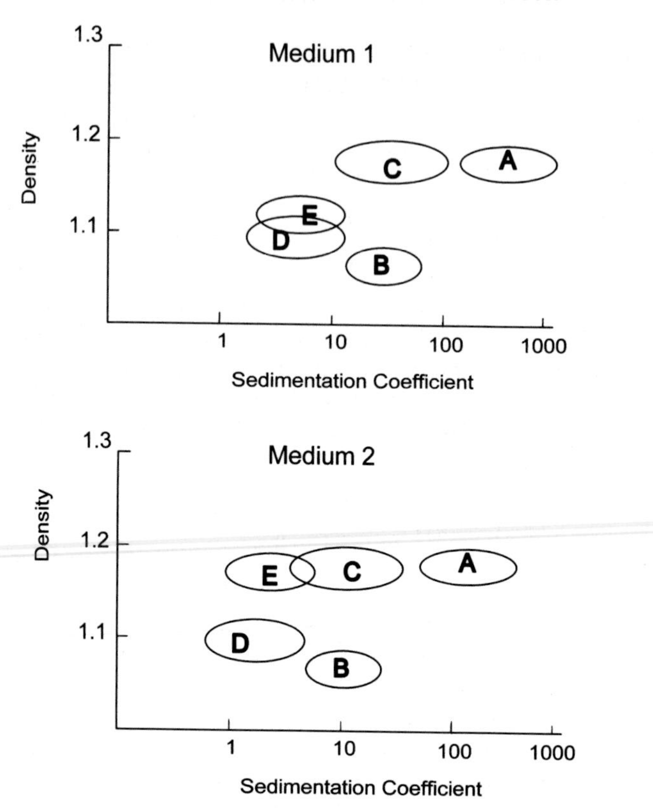

Figure 8. S–ρ diagrams of five imaginary components in two different gradient media. See text for discussion.

do provide the data needed for rational method development. The first step is to use the S–ρ and other diagrams to decide on the basis of the separation. Model data is shown in *Figure 8*. If the structure in question has a unique size (component A) or density (component B) then the basis of separation is obvious. If there is a unique combination of size and density as with component C then sequential rate and isopycnic steps are required. Other organelles will require sequential isopycnic banding steps (component D) or even rate separation followed by two isopycnic banding steps (component E). Other considerations when designing separations are the amount of sample which requires separation and the fragility of the organelles. With a large sample, one will normally first fractionate the homogenate by differential pelleting, as described in Section 4.1.1, but if only a very small amount of sample is present or if organelles are fragile, making pelleting undesirable then one can proceed directly to rate-zonal separation.

As stated, the first step in purification of a cell organelle is normally a preliminary fractionation by differential pelleting. Usually the conditions given in *Protocol 1* will prove satisfactory, for reasons discussed earlier it is rarely worth 'fine tuning' separations by differential pelleting. When separations require both rate-zonal and isopycnic banding it is normal to carry out the rate separation first, the fractions containing the separated material can normally then be layered directly over the gradient for the isopycnic separation without any need for pelleting if gradients of sucrose or iodinated media are used for the isopycnic banding step. On the other hand if gradients of Percoll or Ficoll are used for the isopycnic banding the sample applied to the gradient should be suspended in the isotonic solution used to prepare the density gradient. In fact with Percoll or Ficoll it may be worthwhile to consider carrying out the isopycnic banding first. Similarly, if separating by isopycnic banding then it should be remembered that while the presence of small amounts of a high molecular weight solute such as Ficoll or Percoll will not affect the banding density of membrane-bounded particles in a gradient of a lower molecular weight solute, the converse is not true. Hence it is generally preferable to start with the higher molecular weight solute.

Design of separations is thus normally just a matter of inspection of S–ρ and, if needed, ρ–ρ diagrams, followed by selecting the sequence for rate and isopycnic banding steps which avoids pelleting of particles between steps or at least minimizes the number of such pelletings. There are, however, a number of special situations where 'tricks' may be appropriate. The first problem is when the isopycnic banding density of some component of the homogenate is so low that it falls within the range of the gradient being used for the rate separation. Myelin fragments in brain homogenates are an especial problem. It may be necessary to remove such material by flotation prior to further fractionation. The second situation is where the presence of large amounts of soluble material results in overloading of the gradient and the fragility of some component of the homogenate makes collection of particulates by pelleting undesirable. There is no simple solution to this problem but approaches such as passage of the sample through columns of a large pore gel filtration material such as Sephacryl may prove useful.

9.4 Simplification of the separation

The methods described above depend on the use of continuous density gradients. Preparation of these can be time-consuming and requires special apparatus and, in rate separations, limits on the amount of sample which may be loaded on to a gradient keep yields low. When simplifying a separation, the first thing to consider is whether the greatest possible purity is required or whether some contaminants are harmless. For example, it is unlikely that peroxisomes will interfere in studies on the import of proteins into mitochondria, but lysosomes are likely to leak hydrolases and so are most undesirable. The next stage is to replace continuous gradients by step gradients in all separations which involve isopycnic banding. The biggest problem with step gradients is aggregation as small, denser, particles try to pass through an interface on which large amounts of material have already collected. Following the steps laid out in *Protocol 3* will normally avoid the problem.

While the continuous gradients used in isopycnic banding may be replaced by step gradients it is generally not possible to simplify rate-zonal separations. Although the resolution of differential pelleting may be improved by increasing the number of washing steps the repeated pelleting and resuspension increase the risk of damage to organelles. Theoretically attempting rate separations using step gradients should be utterly disastrous with all transfer occurring by 'sedimentation in droplets'; in practice the density gradients which form around interfaces stabilize sedimentation to some degree and particle movement is sufficiently orderly for use in student practicals. However, we would not recommend the technique for use in research. Basically if particles must be separated on the basis of their sedimentation rates and these differ by much less than an order of magnitude then there is no alternative to properly performed rate-zonal centrifugation.

Protocol 1. Separation of the four basic subcellular fractions

Equipment and reagents

- Centrifuge and rotors
- Homogenization medium
- 1.5 M KCl

Method

1. Homogenize tissue in a suitable medium and adjust concentration so that 1 ml of homogenate contains less than 0.2 g of tissue.
2. Centrifuge homogenate for 10 min at 1000 *g* in a swinging-bucket rotor.
3. Decant supernatant and retain. Resuspend pellet in fresh homogenization medium and recentrifuge for the same speed and time. Decant supernatant and combine with supernatant from first spin. Resuspend pellet (low speed pellet or N fraction) and label.
4. Centrifuge the combined supernatants for 15 min at 10 000 *g* in a fixed-angle rotor. Decant the supernatant. Do not worry if some of the loosely packed material on the surface of the pellet comes away, it will be removed in the next step in any case. Resuspend the pellet in fresh homogenization medium and recentrifuge under the same conditions.
5. Decant supernatant and combine with supernatant from the previous spin. Resuspend pellet (large particulate or M fraction) and label.
6. Spin combined supernatants for 60 min at 120 000 *g* in a fixed-angle rotor. Decant the supernatant and label cytosol. Resuspend the pellet

(microsomes). If contamination by cytosolic proteins is not likely to present a problem, resuspend in a suitable volume of homogenization medium. If contamination is likely to be a problem and sucrose has been used as a homogenization medium, resuspend in homogenization medium supplemented with 0.15 M KCl and recentrifuge under the same conditions as the first spin. Resuspend the pellet in homogenization buffer and label microsomes.

(Note: for best preservation of the activity of the cytochromes P450 resuspend in a buffer containing 10% glycerol and store at −80 °C.)

Protocol 2. Determination of the sedimentation rate and density of particles

Equipment and reagents

- Centrifuge and rotors
- Homogenization medium
- 0.3 M, 1 M and 2 M sucrose
- Percoll, Iodixanol, Nycodenz, or metrizamide

A. *Particle size*

1. Homogenize tissue in a suitable medium so that the concentration is no more than 0.2 g tissue/ml homogenate.
2. Pipette into a centrifuge tube for a swinging-bucket rotor approx. 1 ml of 2 M sucrose.
3. Overlay the sucrose with a continuous sucrose gradient ranging from 0.3 M to 1 M.
4. Apply the sample and centrifuge for approx. 15 min at 5000 *g*.
5. Recover the gradient and assay for the distribution of marker enzymes.
6. If necessary adjust the speed or time of centrifugation and repeat.

B. *Particle density*

1. The approach here depends on whether the density of a particular organelle is likely to depend on its size. This will be the case with, for example, kidney lysosomes where the large 'droplets' at the base of the cells of the proximal tubule are significantly denser than other lysosomes in the tissue.
2. If it can be assumed that the organelles are homogeneous in density then the homogenate should be layered over a density gradient which spans the density range of organelles (i.e. 1.03–1.1 for Percoll, 1.1–1.23 for Iodixanol, Nycodenz, metrizamide, or sucrose) in the tubes of a vertical tube rotor and centrifuged for 60 min at 240 000 *g*.

Protocol 2. *Continued*

The gradient can then be fractionated and the distribution of marker enzymes determined.

3. If it cannot be assumed that the distribution of marker enzymes is uniform with respect to density then the sample should first be fractionated by rate-zonal centrifugation as described above. It will probably be necessary to run several samples to obtain sufficient material. Fractions should then be pooled. If the same gradient solute is being used for the rate and isopycnic steps then the samples may be layered over gradients which extend from the density of the sample to the highest density at which cell organelles band in that medium. If a different gradient solute is employed then the organelles should be collected by dilution and pelleting, then the pellet resuspended and layered over gradients as described in part B, step 2.

Protocol 3. Design of a step gradient for routine separation of a well-characterized organelle

1. By inspection of the results of the continuous gradients identify the concentration or density of the gradient solute at the light and dense side of the particle band.
2. Adjust the sample by addition of a concentrated solution of the gradient material until the density corresponds to the light end of the sample band.
3. Prepare a solution of the density gradient solute with a concentration equal to the dense end of the sample band.
4. Prepare a solution (termed the cushion) of the density gradient solute with a density greater than any organelle in the sample.
5. Pipette some of very dense solution prepared in step 4 into a centrifuge tube, overlay with the solution prepared in step 3, followed by the sample solution prepared in step 2, and overlay this with a light solution termed the overlay—typically the homogenization medium.
6. Centrifuge and collect the material gathered at the interface between solutions 2 and 3 with an L-shaped pipette. Problems may arise from material gathered at other interfaces sticking to the wall of the tube. This may be reduced if the interfaces (other than the critical one between solutions 2 and 3, are slightly stirred prior to centrifugation to make 'mini-gradients'.

This system will deal with most problems. If, however, much material collects on the interface between layers 2 and 3 it may trap more slowly sedimenting particles. If this is the case, then the density of the sample should be adjusted to that of the dense end of the density range of the particles of interest and should be overlaid with a solution corresponding to the light end of the range. Cushion and overlay should be used as in the example given above.

References

1. Rickwood, D. (1983). In *Iodinated density gradient media: a practical approach* (ed. D. Rickwood), p. 1. Oxford University Press, Oxford, UK.
2. Ford, T., Rickwood, D., and Graham, J. (1983). *Anal. Biochem.*, **128**, 232.
3. Rickwood, D., Ford, T., and Graham, J. (1982). *Anal. Biochem.*, **123**, 23.
4. Ford, T., Graham, J., and Rickwood, D. (1994). *Anal. Biochem.*, **220**, 360.
5. Graham, J., Ford, T., and Rickwood, D. (1994). *Anal. Biochem.*, **220**, 367.
6. Leighton, F., Poole, B., Beaufay, H., Baudhuin, P., Coffey, J. W., Fowler, S., *et al.* (1968). *J. Cell Biol.*, **37**, 482.
7. Beaumelle, B. D., Gibson, A., and Hopkins, C. R. (1990). *J. Cell Biol.*, **111**, 1811.
8. Courtoy, P. J., Quintart, J., and Baudhuin, P. (1984). *J. Cell Biol.*, **98**, 870.
9. Nagasawa, M., Koide, H., Ohsawa, K., and Hoshi, T. (1992). *Anal. Biochem.*, **201**, 301.
10. Walter, H., Johansson, G., and Brooks, D. E. (1991). *Anal. Biochem.*, **197**, 1.
11. Jakob, R. (1986). *Prep. Biochem.*, **22**, 1.
12. Richardson, P. J. and Luzio, J. P. (1986). *Appl. Biochem. Biotech.*, **13**, 133.
13. Salamero, J., Sztul, E. S., and Howell, K. E. (1990). *Proc. Natl. Acad. Sci. USA*, **87**, 7717.
14. Gruenberg, J. E. and Howell, K. E. (1986). *EMBO J.*, **5**, 3091.
15. Wandinger-Ness, A., Bennett, M. K., and Antony, C. (1990). *J. Cell Biol.*, **111**, 987.
16. Graham, J. M. (1993). In *Methods in molecular biology* (ed. J. M. Graham and J. A. Higgins), Vol. 19, p. 41. Humana Press, Totowa, NJ, USA.
17. Bartles, J. R. and Hubbard, A. L. (1988). *Trends Biol. Sci.*, **13**, 181.
18. Matter, K. and Mellman, I. (1994). *Curr. Opin. Cell Biol.*, **6**, 545.
19. Novick, P. and Brennwald, P. (1993). *Cell*, **75**, 597.
20. Dobrota, M. and Hinton, R. (1992). In *Preparative centrifugation: a practical approach* (ed. D. Rickwood), p. 77. Oxford University Press, Oxford, UK.
21. Hartman, G. C., Black, N., Sinclair, R., and Hinton, R. H. (1974). In *Subcellular studies; methodological developments in biochemistry* (ed. E. Reid), Vol. 4, p. 93. Longman Group, London.

3

Isolation and characterization of nuclei and nuclear subfractions

DAVID RICKWOOD, ANTHEA MESSENT, and DIPAK PATEL

1. Introduction

The nucleus is the centre of a wide range of diverse activities that are essential for the survival of the cell. The development of methods for the isolation and purification of functional nuclei, nuclear subfractions, and individual components have been central to understanding how the nucleus functions in the cell. However, given the space limitations of a single chapter, it is only possible to focus on a selection of the methods used for isolating and characterizing the various components of the nucleus and analyzing their functions. The reader will be referred to other, relevant practical manuals for details of other types of technique as appropriate.

The nuclei of vertebrate cells tend to be fairly similar in terms of their composition, structural characteristics, and microscopic appearance. In contrast, at the level of the lower eukaryotes such as the ciliates there is a broad diversity of nuclear structure and organization. In addition, some cell types are very difficult to homogenize satisfactorily without damaging the nuclei. These two variables have combined to generate a wide range of methods for the isolation of nuclei from the different types of cells and organisms.

2. Methods for preparing purified nuclei

2.1 Types of cells and tissue samples

The vast majority of methods used to isolate nuclei relate to soft animal tissues. However, there is a general requirement to be able to isolate nuclei from a whole range of cell types from single celled lower eukaryotes to mammals. The actual method used in terms of homogenization and centrifugation conditions depends on the nature of the sample material and the following sections describe the range of methods that have been used. Probably the main aspect is that wherever possible the sample of cells must be fresh, as storage, particularly freezing can release significant amounts of intracellular hydrolytic enzymes which subsequently interfere with the isolation of good quality nuclei.

For non-aqueous separations some authors (1, 2) have advocated the use of freeze-dried samples. However, work carried out in the authors' laboratory has compared the use of fresh, and freeze-dried material from tissue that had either been quick-frozen in liquid nitrogen or frozen in the presence of cryoprotectants (e.g. dimethyl sulfoxide (DMSO) or glycerol). These studies indicated that, when using formamide or ethylene glycol-based non-aqueous isolation methods (see Section 2.4.2), the best results are obtained using fresh material in terms of the yield and purity of nuclei.

2.2 Homogenization media

Chapter 1 has described in detail the methods used to homogenize different cell types. Aqueous homogenization media used for animal cells have varied from dilute citric acid (3) to 2.4 M sucrose (4). The former was found to be useful for the isolation of nuclear RNA while the latter has the advantage that it minimizes the leaching of nuclear proteins into the homogenizing medium. However, the homogenizing medium most frequently used is isotonic (0.25 M) sucrose buffered to about pH 7.4 with 10 mM Tris–HCl and containing either 3 mM Ca^{2+} or 5 mM Mg^{2+}; both ions stabilize nuclei but the former can inhibit some enzyme activities (5). Instead of divalent cations, polyamines such as spermine and spermidine are also added to some homogenization media to stabilize the nuclei particularly when isolating nuclei from plants and lower eukaryotes (6, 7). In some cases, sucrose has been supplemented or even replaced by Ficoll, dextran, or gum arabic (8, 9).

In some procedures it has become usual to add enzyme inhibitors in the homogenization medium. For example, the addition of 1 mM phenylmethylsulfonyl fluoride (PMSF) is useful in reducing the amount of proteolysis during preparation of nuclei (10). Nuclease inhibitors (11) can also be added if required.

All of the media described previously relate to aqueous isolation procedures. However, nuclei can be isolated using non-aqueous procedures such as by homogenization of lyophilized cells or tissues in dried glycerol (12) or other organic solvents (1, 2, 13). However, the technical difficulties of these methodologies have tended to limit exploitation of non-aqueous separation methods to a group of enthusiasts except where the use of non-aqueous media is obligatory. However, new, less hazardous procedures have been devised in the authors' laboratory. In this case the wet tissue is homogenized in 19 volumes of either ethylene glycol or formamide, containing 1 mM $MgCl_2$; unlike the solvents used for earlier methods, neither of these solvents dissolves membrane lipids. The procedures described in this chapter will mainly relate to aqueous isolation and fractionation methods and the non-aqueous methods devised in the authors' laboratory. The reader is referred to the references given previously for details of other non-aqueous isolation procedures for nuclei.

2.3 Homogenization methods

Homogenization is a crucial step in the isolation of nuclei. Detailed descriptions of the various homogenization procedures are given in Chapter 1. The method chosen must be efficient, ideally capable of lysing more than 90% of the cells, yet at the same time disrupting neither the structure, organization, nor contents of the nuclei. Similar homogenizers can be used for aqueous and non-aqueous nuclei. The vast majority of methods described for isolating nuclei relate to animal cells, either as tissues or cultured cells. For soft animal tissues homogenization in a Potter–Elvehjem homogenizer using a loose (0.1 mm clearance) Teflon pestle works well with essentially all cells lysed after six to eight strokes at 1000 r.p.m.; tight pestles should be avoided as nuclei may be fragmented by the high shear forces. The use of blenders for animal tissues is best avoided if possible as the blades can damage the nuclei. Tissue culture cells can be sheared in a Dounce homogenizer after they have been swollen in hypotonic medium (14). In the case of other cell types different types of homogenization media and centrifugal techniques will be required. For preparing nuclei from fungal cells such as yeast quite different approaches are required reflecting the altered nature of the nucleus and the nature of the cell. The preferred method for fungi is to remove the cell wall by enzymic digestion to give protoplasts and then to disrupt the protoplasts by homogenization (15); isolation procedures based on breaking open fungal cells by grinding tend to give very low yields of nuclei. The nuclei of ciliates such as *Tetrahymena*, *Pseudomonas*, and *Stylonychia* are complex. Homogenization is achieved by blending in media based on either gum arabic–octanol or detergents such as Nonidet P-40 (8). A variety of compounds are used as stabilizers in homogenization media for these organisms. An added problem is that nuclei from ciliates can fragment when frozen. *Table 1* summarizes the composition of a selection of homogenization media.

2.4 Centrifugation conditions

2.4.1 Aqueous nuclear preparations

Nuclei are the largest and densest organelles of cells and the ease with which nuclei can be purified by centrifugation has caused other possible fractionation methods to be neglected. As described in Chapter 2, simple differential pelleting (600 *g* for 10 min) can be used to give crude nuclear fractions. Some methods such as homogenization in citric acid can inactivate most enzyme activities and thus are only useful for preparing RNA or other nuclear components. However, when using media such as isotonic sucrose solutions, the nuclei pelleted at 600 *g* are contaminated with large mitochondria and sheets of membranes, and should not be considered to be pure enough for most pur-

Table 1. Homogenization media and methods for lower eukaryotic, fungal, and plant cells

Organism	Homogenization medium	Lysis method	Nuclear stabilizers	Reference
Amoeba proteus	10 mM Mes, 0.03% 2-ethyl-1-hexanol, 0.2 mM dithiothreitol, 1 mM KCl, 0.4 mM $Mg(OAc)_2$, 50 mM $CaCl_2$ pH 5.9	Dounce homogenizer	Ca^{2+}, Mg^{2+}	16
Paramecium aurelia	0.17 M sucrose, 4.8 mM $CaCl_2$, 0.22% sodium deoxycholate (DOC), 0.33% Nonidet P-40, 150–175 μg/ml spermidine–HCl pH 7.1	'Tri-R-Stir' with Teflon pestle	Spermidine, Ca^{2+}	6
Tetrahymena pyriformis	10 mM Tris–HCl pH 7.4, 2 mM $CaCl_2$, 0.5 mM $MgCl_2$, 0.5 M sucrose, 0.2% (w/v) Triton X-100		Ca^{2+}, Mg^{2+}	17
Neurospora crassa	0.35 M mannitol	Mortar and pestle with acid washed sand		18
Physarum polycephalum	0.01 M Tris–HCl pH 7.0, 0.25 M sucrose, 0.01 M $MgCl_2$, 1% (w/v) Triton X-100	Waring blender	Mg^{2+}	19
Physarum polycephalum	10 mM Tris–HCl pH 7.0, 0.25 M sucrose, 10 mM $CaCl_2$ or $MgCl_2$, 0.1–0.4% (w/v) Triton X-100	Waring blender	Ca^{2+} or Mg^{2+}	20
Yeast (whole cells)	1 M sorbitol, 20% glycerol, 5% polyvinylpyrrolidone (M_r 40 000)	French press		21
Yeast protoplasts	8% polyvinylpyrrolidone (M_r 40 000), 1 mM $MgCl_2$, 0.02 M KH_2PO_4 pH 6.5, 0.02% (w/v) Triton X-100	Potter–Elvehjem homogenizer	Mg^{2+}	15
Plants	2.3% acacia, 0.4 M sucrose, 2 mM $CaCl_2$, 4 mM octanol-1, 20 mM Tris–acetate pH 7.6	Waring blender	Octanol-1	22

poses. There are three different ways of purifying nuclei, either on the basis of their density by sedimentation through dense (2.0–2.4 M) sucrose (5, 14) or banding them in a gradient (23), or by dissolving the membrane contaminants by washing the nuclei with isotonic sucrose (0.25 M sucrose, 6 mM $MgCl_2$, 10 mM Tris–HCl pH 7.4) solution containing 0.5% of the non-ionic detergent Triton X-100 (24); a combination of these methods can also be used to purify nuclei. The use of Triton X-100 has two significant disadvantages in that the detergent can activate endogenous nuclease activity several-fold and nuclei tend to aggregate and become more fragile as a result of the loss of their nuclear membranes. The method found by the authors to be applicable for most types of soft animal tissues is based on the method of Widnell and Tata (5) and is given in *Protocol 1*.

Protocol 1. Purification of animal cell nuclei from soft tissues

Equipment and reagents

- Muslin
- Motor-driven Potter–Elvehjem homogenizer with loose-fitting Teflon pestle
- 0.25 M sucrose, 5 mM $MgCl_2$, 10 mM Tris–HCl pH 7.4
- High-speed centrifuge with fixed-angle or swinging-bucket rotor capable of generating 80 000 *g*
- 2.2 M sucrose, 1 mM $MgCl_2$, 10 mM Tris–HCl pH 7.4

Method

1. Dissect out the required tissue, weigh it, and chop finely with scissors in cold homogenizing medium (0.25 M sucrose, 5 mM $MgCl_2$, 10 mM Tris–HCl pH 7.4).
2. Pour off the homogenizing medium and then suspend the chopped tissue in 9 vol. of fresh homogenizing medium. Homogenize the suspension of chopped tissue using a Potter–Elvehjem homogenizer with a loose-fitting (0.1 mm clearance) Teflon pestle driven at 1000 r.p.m. Use eight or nine up-and-down strokes to obtain complete cell breakage. If the glass homogenizer vessel is held in the hand use a glove as this both helps to reduce warming of the homogenizer and reduces risk of injury if the homogenizer breaks.
3. Filter the homogenate through four layers of muslin taking care not to squeeze the muslin to obtain the last few drops of liquid. Centrifuge the homogenate at 600 *g* for 10 min at 5 °C.
4. Carefully pour off the supernatant and resuspend the pellet in half the original volume of homogenizing medium. Centrifuge the resuspended nuclei as described in step 3 and again discard the supernatant leaving a pellet of crude nuclei.
5. Add 9 vol. of 2.2 M sucrose, 1 mM $MgCl_2$, 10 mM Tris–HCl pH 7.4 to the pellet of crude nuclei. Resuspend and homogenize the pellet in the 2.2 M sucrose solution in a Potter–Elvehjem homogenizer using five or six strokes of a ball pestle driven at 1000 r.p.m. This step is critical in removing membrane contamination from the nuclei.
6. Centrifuge the suspension of nuclei at 60–80 000 *g* for 80 min at 5 °C in a swinging-bucket rotor.
7. After centrifugation, the nuclei should form a cream coloured pellet at the bottom of the tube, any membrane contaminants tend to give the pellet a pinkish tinge.
8. Remove any skin that may have formed at the top of the dense sucrose using a stainless steel spatula and empty the tubes by rapid inversion. Keeping the tubes inverted, wipe out any remaining sucrose solution with a tissue wrapped around a glass rod or spatula.

Protocol 1. *Continued*

9. Resuspend the nuclei in homogenizing medium or other medium of your choice. When stained with 1% methylene blue the nuclei should appear clean without any membrane adhering to the outside of the nucleus. The nuclei can be frozen down in homogenizing medium at −20 °C.
10. The purified nuclei of step 9 still retain their nuclear membrane. If necessary this can be removed by washing the nuclei twice in 10–20 ml of homogenizing medium containing 0.5% Triton X-100 followed by two washes with homogenizing medium to remove the Triton X-100. The nuclei can be resuspended in homogenizing medium and stored at −20 °C.

The procedure described in *Protocol 1* was originally developed for liver, and for this tissue and similar tissues such as kidney it works very well giving nuclei of very high purity. However, centrifugation through dense sucrose does cause a significant decrease in yield as the nuclei with membrane material attached do not pellet. This method also works quite well with tissue culture cells and muscle cells provided that efficient cell lysis has been achieved. In this context, when tissue culture cells have been lysed by hypotonic lysis it is important to readjust the medium to isotonicity as soon as possible as otherwise it is quite difficult to obtain clean nuclei. A particular problem in purifying nuclei from liver from *Xenopus laevis* is the presence of melanin granules which are very dense and pellet with the nuclei through dense sucrose. The nuclei can be purified by separation in a solution of 58% metrizamide by centrifugation at 10000 *g* for 20 min (25); the nuclei with a density of about 1.22 g/ml float to the top and the melanin granules pellet. Iodixanol (OptiPrep™) (23) gradients are also able to give a similar type of purification.

A particular problem with plants is the contamination of nuclei pelleted by centrifugation at 600 *g* for 10 min with starch grains, chloroplasts, and cell wall fragments arising from disruption of the cell using a blender or Polytron-type of homogenizer. As in the case of animal cell nuclei, plant nuclei can be purified by washing them with Triton X-100 or by pelleting them through dense sucrose. In the authors' laboratory some success has been achieved by banding nuclei in discontinuous Iodixanol gradients by centrifugation at 10000 *g* for 20 min.

2.4.2 Non-aqueous nuclear preparations

As mentioned previously, nuclei can be isolated using non-aqueous procedures. The original methods used hazardous solvents such as benzene and carbon tetrachloride most of which are not compatible with plastic centrifuge tubes. The procedures developed in the authors' laboratory use ethylene glycol or formamide containing 1 mM $MgCl_2$. The full procedure is described in *Protocol 2*.

Protocol 2. Isolation of liver nuclei under non-aqueous conditions

Equipment and reagents

- Muslin
- Motor-driven Potter–Elvehjem homogenizer with loose-fitting Teflon pestle
- Homogenizing medium of ethylene glycol containing 1 mM $MgCl_2$
- High-speed centrifuge with fixed-angle or swinging-bucket rotor capable of generating 80 000 *g*.
- 10% and 40% Nycodenz solutions in homogenizing medium

Method

1. Dissect out the liver, weigh it, chop finely with scissors, and add 19 vol. of homogenizing medium.
2. Homogenize the suspension of chopped tissue using six to nine strokes of a Potter–Elvehjem homogenizer with a loose-fitting Teflon pestle. If the glass homogenizer vessel is held in the hand use a glove as this both helps to reduce warming of the homogenizer and reduces risk of injury if the homogenizer breaks.
3. Filter the homogenate through four layers of muslin and centrifuge the homogenate at 1000 *g* for 10 min at 5 °C.
4. Discard the supernatant and resuspend the pellet of crude nuclei in 10% Nycodenz in ethylene glycol, 1 mM $MgCl_2$.
5. Load the nuclei from approx. 1 g of liver on to a 12 ml continuous gradient of 10%–40% Nycodenz in ethylene glycol containing 1 mM $MgCl_2$ and centrifuge at 80 000 *g* for 90 min at 5 °C.
6. The nuclei will be found as a light scattering band in the centre of the gradient at a concentration of about 22% Nycodenz. Remove the band of nuclei using a Pasteur pipette, dilute it with 2 vol. of ethylene glycol, and centrifuge at 1000 *g* for 10 min at 5 °C.

An example of mouse liver nuclei isolated by this method is shown in *Figure 1*. Experiments have shown that nuclei with good morphology can also be obtained if formamide is used instead of ethylene glycol.

2.5 Assays of nuclear purity

Nuclei are surrounded by a double membrane, the outer one of which is continuous with the endoplasmic reticulum; the actual amount of endoplasmic reticulum varies depending on the type of cell. Hence, unless the nuclear membranes are removed by detergent treatment, there are always contaminants present. The purity of nuclei can be assessed on the basis of their lack of activity of marker enzymes for other subcellular fractions. Nuclei isolated in aqueous media lose a number of enzymes (e.g. DNA polymerase) through

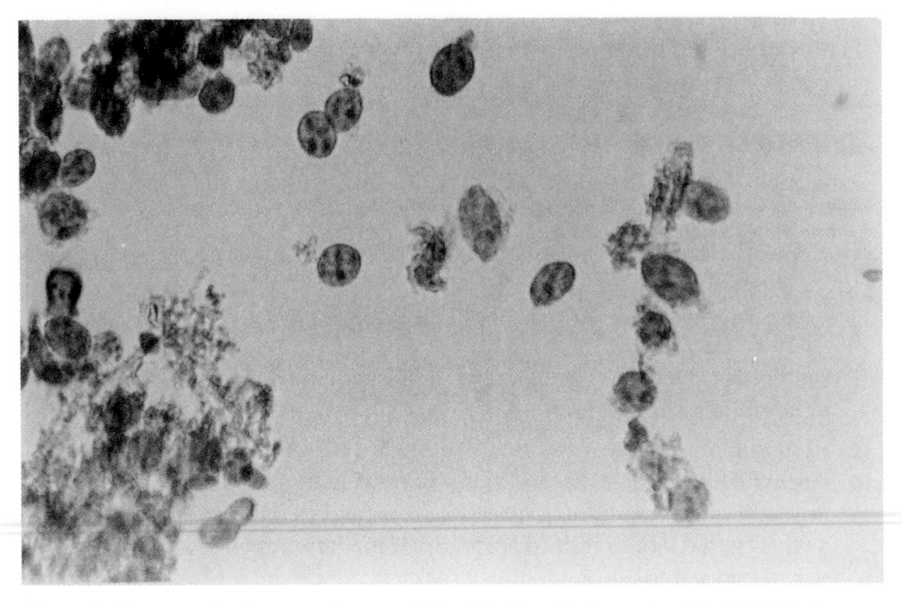

Figure 1. An example of mouse liver nuclei isolated by ethylene glycol-based non-aqueous isolation methods.

leaching. leaching can be minimized by homogenizing tissues in concentrated (2.2 M) sucrose solutions (5,14) and abolished by using non-aqueous isolation methods (1,2,12,13). Light microscopy of stained nuclei can indicate the presence of non-nuclear material as can phase-contrast microscopy of unstained nuclei, but electron microscopy is required in order to obtain a definitive assessment of the purity of nuclear preparations. Details of the procedures involved in electron microscopy are described in detail in Chapter 10.

3. Methods for purifying metaphase chromosomes

The metaphase stage of cell division is important in that it is the one time in the cell cycle when the chromatin appears in the cell as compact chromosomes which are visible in the light microscope. Metaphase chromosomes can be isolated from cells which have been arrested in metaphase by the addition of a mitotic blocking agent such as colcemid. The conditions used must ensure that the structure and composition of chromosomes is retained; the range of procedures used have been reviewed (26,27). Early methods used either acid pH or Zn^{2+} to stabilize the chromosomes but the former tends to extract basic proteins from the chromosome and the latter makes cells resistant to homogenization procedures. The procedure devised by Wray (28) which uses Ca^{2+} and hexylene glycol combined with Pipes–NaOH pH 6.8 buffer to lyse the cells and stabilize the chromosomes is described in *Protocol 3*.

Protocol 3. Isolation of metaphase chromosomes

Equipment and reagents

- Low and high speed centrifuges
- 22 Gauge hypodermic needle
- 1 M hexylene glycol, 0.5 mM $CaCl_2$, 0.1 mM Pipes–NaOH pH 6.8
- Colcemid
- 10% and 50% sucrose, or 28% and 60% metrizamide, in hexylene glycol buffer

Method

1. Arrest the cultured cells in metaphase by the addition of 0.06 μg/ml of colcemid to the culture medium and incubate for 3 h.
2. Selectively detach the metaphase cells by gentle shaking and cool the detached cells for 20 min at 4 °C prior to pelleting them by centrifugation at 300 *g* for 5 min.
3. Wash the cells by suspending them in 1 M hexylene glycol, 0.5 mM $CaCl_2$, 0.1 mM Pipes–NaOH pH 6.8 and pellet them by centrifugation as in step 2.
4. Resuspend the cells in the same buffer (cell concentration <2 mg/ml) and incubate them for 10 min at 37 °C before lysing the cells by passing the cell suspension several times through a 22 gauge hypodermic needle or nitrogen cavitation at 250 psi (17 bar). Monitor cell lysis using phase-contrast microscopy.
5. Pellet the chromosomes by centrifugation at 2000 *g* for 10 min.
6. To purify the chromosomes further either:
   (a) Centrifuge them through a 10–50% sucrose gradient in the hexylene glycol lysis buffer at 1500 *g* for 30 min (28).
   (b) Flotation in a 28–60% metrizamide gradient in lysis buffer by centrifugation at 16 300 *g* for 10 min (29).

The metaphase chromosomes can be stained by one of the banding methods described in the literature (30).

While the method described in *Protocol 3* should work with most cultured animal cells, modifications may be required for other types of cell (27). In addition, for flow karyotype analysis it is desirable to be able to prepare larger amounts of chromosomes than can be conveniently prepared using the procedure of Wray (28). A procedure suitable for the large-scale preparation of chromosomes has been described which uses polyamines to stabilize the chromosomes and digitonin to lyse the cells (31).

4. Isolation of nuclear subfractions

4.1 Preparation of nucleoli

Nucleoli can be readily viewed as intranuclear bodies that stain intensely when nuclei are stained with a basic stain such as 0.1% methylene blue and they are present as discrete bodies within the nucleus. There are a number of reviews in the literature on methods for the isolation of nucleoli (e.g. 32, 33). However, over the years the use of sonication to release nucleoli from the nucleus has become most widely used. The method used is described in *Protocol 4*. This method is based on the procedures developed for the isolation of nucleoli from rat liver (34). Other types of cells and tissues may require some modification of this procedure.

Protocol 4. Purification of nucleoli

Equipment and reagents

- Refrigerated low speed centrifuge with swinging-bucket rotor
- 0.34 M sucrose, 0.05 mM $MgCl_2$, 10 mM Tris–HCl pH 7.5
- Sonicator
- 0.88 M sucrose, 0.05 mM $MgCl_2$, 10 mM Tris–HCl pH 7.5

Method

1. Suspend nuclei, purified by pelleting through 2.2 M sucrose (see *Protocol 1*), in 0.34 M sucrose, 0.05 mM $MgCl_2$, 10 mM Tris–HCl pH 7.5.
2. Sonicate the nuclear suspension in ice for 10–15 sec followed by a cooling period of 30–40 sec in ice. Monitor the extent of nuclear breakage by light microscopy using 0.1% methylene blue as a stain. Typically 40-–60 sec total sonication time is sufficient to disrupt all of the nuclei.
3. Underlayer the sonicated nuclear suspension with 0.5 vol. of 0.88 M sucrose, 0.05 mM $MgCl_2$, 10 mM Tris–HCl pH 7.5 and centrifuge at 2000 *g* for 20 min at 4°C.
4. The nucleoli form a tight pellet with the vast majority of the nuclear material remaining in the upper 0.34 M sucrose layer. If the purity is less than required resuspend the nucleoli in 0.25 M sucrose and recentrifuge them through 0.88 M sucrose. The yield of nucleoli is typically 50–70%.

The procedure given in *Protocol 4* will work with nuclei of the vast majority of higher eukaryotic cells. One exception to this are the nuclei of *Xenopus laevis* liver which contain large numbers of melanin granules. In such cases an alternative to centrifugation through sucrose is to band the nucleoli in a gradient of metrizamide (35); Nycodenz or Iodixanol gradients should also give satisfactory purification.

4.2 Preparation of nuclear membranes

The nuclear membrane that surrounds the nucleus has unique permeability properties but is not responsible for physically containing the nuclear chromatin. The properties of the nuclear membrane have been extensively reviewed (e.g. 36, 37). Over the years a number of different methods have been devised for the isolation of nuclear membranes.

A summary of the different methods used for preparing nuclear membranes is given in *Table 2*. It is striking that nearly all methods developed for isolating nuclear membranes have used liver nuclei. While such nuclei probably provide a suitable model for most mammalian cell nuclear membrane, it is possible that other types of nuclei, particularly of lower eukaryotic cells may behave differently with some methods giving both poorer quality and lower yields of nuclear membrane. In such situations the best approach is to try out

Table 2. Summary of methods used for the isolation of nuclear membranes

Cell type	Ionic strength	Comments	Reference
Rat liver	High	Takes about 3 h, reproducible yields of 50–60% with low levels of DNA. Uses high concentration of Mg^{2+} to disrupt chromatin.	38
Rat liver	High	Rapid method giving high yields of nucleus-sized hollow spheres with intact membranes, lamina, and pores. Little or no contamination. Useful for studying the relationships between nuclear matrix and envelope (see *Protocol 5* for details).	39
Various liver	Low	Rapid ($1^1/_2$ h) method giving reasonable yields with well preserved pore complexes and fibrous lamina. Nuclear envelopes take the form of broken sheets (size of fragments range from single pore complex to whole torn 'nuclear ghost'). Chromatin disrupted by DNase treatment. Low levels of chromatin contamination (see *Protocol 6* for detail).	40
Various liver	Low	Rapid method giving high yields with well preserved pore complexes and fibrous lamina. Nuclear envelopes take the form of broken sheets (size of fragments range from single pore complex to whole torn 'nuclear ghost'). Chromatin disrupted by DNase treatment. Method preserves enzyme activity.	41
Various liver	Low	Heparin disruption of chromatin. Resulting envelopes take the form of resealed vesicles useful for permeability studies.	42

each type of method to find out which is most suitable for the particular type of cell that is being used.

As can be seen from *Table 2*, different types of study require different preparation methods; some methods are suitable for studies of the envelope composition while others are suitable for enzyme or permeability studies. Basically the isolation methods can be divided into two types that use either high or low ionic strength. One of the earliest methods for the isolation of nuclear membranes was that described by Monneron *et al.* (38). This method uses high (0.3–0.7 M) magnesium ion concentrations to disrupt undigested chromatin, however, the composition of the membranes, as reflected by their density in a sucrose gradient, varies depending on the concentration of magnesium ions used. However, this method does give high yields of membrane.

Another method which uses high ionic strength to disrupt the DNA in the nucleus to isolate nuclear membranes is that described in *Protocol 5* (39). It also gives high yields (approximately 45%) and is simple and fairly rapid to carry out.

Protocol 5. High ionic strength isolation of nuclear membranes (39)

Equipment and reagents

- Refrigerated low speed centrifuge
- 0.25 M sucrose, 5 mM $MgCl_2$, 50 mM Tris–HCl pH 7.5
- 100 mM phenylmethylsulfonyl fluoride (PMSF) in isopropanol
- DNase I
- RNase
- 10 mM Tris–HCl pH 7.4, 0.2 mM $MgCl_2$
- 2.0 M NaCl, 0.2 mM $MgCl_2$, 10 mM Tris–HCl pH 7.4 containing 1 mM PMSF
- 2-mercaptoethanol
- 20% (v/v) glycerol, 1 mM EDTA, 10 mM Tris–HCl pH 7.5

Method

1. Prepare the nuclei by centrifugation through dense (2.2 M) sucrose as described in *Protocol 1* but omitting any detergent treatment. In addition, all solutions used for preparing nuclei contain 1 mM phenylmethylsulfonyl fluoride (PMSF); to do this prepare a 100 mM PMSF in isopropanol, add 1/100 vol. just before use, and mix immediately.
2. Suspend the nuclei in 0.25 M sucrose, 5 mM $MgCl_2$, 50 mM Tris–HCl pH 7.5, at a concentration of 5–8 mg/ml of DNA and add a 1/100 vol. of 100 mM PMSF.
3. To the nuclear suspension add DNase I and RNase both to final concentrations of 250 μg/ml and incubate for 60 min at 4°C. Pellet the nuclei by centrifugation at 1000 *g* for 10 min.
4. Resuspend the digested nuclei at a concentration of 5–8 mg/ml of DNA in 10 mM Tris–HCl pH 7.4, 0.2 mM $MgCl_2$, and add 1/100 vol. of 100 mM PMSF, followed by dropwise addition of 4 vol. of 2.0 M NaCl,

0.2 mM $MgCl_2$, 10 mM Tris–HCl pH 7.4 containing 1 mM PMSF. Continue stirring at 4°C and add 2-mercaptoethanol to a final concentration of 1% (v/v).

5. Incubate the suspension for a further 15 min at 4°C and pellet the nuclear envelopes by centrifugation at 1600 *g* for 30 min at 4°C.

6. Repeat steps 4 and 5 but omitting the addition of 2-mercaptoethanol.

7. Resuspend the nuclear envelopes in 20% (v/v) glycerol, 1 mM EDTA, 10 mM Tris–HCl pH 7.5 and store at −20°C.

For step 3 it has been found that the incubation conditions can be modified to 20 μg/ml of DNase and RNase incubated at 37°C for 20 min (37). The morphology of the nuclear envelopes prepared using the methods described in *Protocol 5* is excellent with most of the envelopes in the form of hollow spheres with both nuclear membranes, nuclear pores, and laminar largely intact. The other attraction of this method is that the procedure is similar to that used for making nuclear matrix preparations (Section 4.3) and so it is useful for joint studies of these two nuclear components.

Alternatively, nuclear envelopes can be isolated using methods that avoid the use of high ionic strength (40–42). These methods give nuclear envelopes that appear to be morphologically intact and retain high levels of enzyme activity. *Protocol 6* is an example of the procedure that can be used.

Protocol 6. Isolation of nuclear envelopes at low ionic strength (40)

Equipment and reagents

- High speed centrifuge with fixed-angle rotor capable of generating 38 000 *g*
- 0.1 mM $MgCl_2$
- DNase I
- 20% (v/v) glycerol, 1 mM EDTA, 10 mM Tris–HCl pH 7.5
- 0.3 M sucrose, 0.1 mM $MgCl_2$, 5 mM 2-mercaptoethanol, 10 mM triethanolamine–HCl pH 8.5
- 0.3 M sucrose, 0.1 mM $MgCl_2$, 5 mM 2-mercaptoethanol, 10 mM triethanolamine–HCl pH 7.4

Method

1. Prepare purified nuclei by pelleting through dense sucrose as described in *Protocol 1* but omitting treatment with Triton X-100.

2. Suspend the nuclei at a DNA concentration of 3–4 mg/ml in 0.1 mM $MgCl_2$ and add DNase I to a final concentration of 5 μg/ml. Add 4 vol. of 0.3 M sucrose, 0.1 mM $MgCl_2$, 5 mM 2-mercaptoethanol, 10 mM triethanolamine–HCl pH 8.5.

3. Incubate the suspension for 15 min at 22°C and then add an equal volume of ice-cold distilled water. Centrifuge the diluted suspension at 38 000 *g* for 15 min at 4°C in a fixed-angle rotor.

Protocol 6. *Continued*

4. Resuspend the pellet in 5 vol. of 0.3 M sucrose, 0.1 mM $MgCl_2$, 5 mM 2-mercaptoethanol, 10 mM triethanolamine–HCl pH 7.4, add DNase I to a final concentration of 1 μg/ml, and incubate the suspension for 15 min at 22°C.
5. Terminate the incubation by adding an equal volume of ice-cold distilled water as before and centrifuge the diluted suspension at 38 000 *g* for 15 min at 4°C in a fixed-angle rotor.
6. Resuspend the pellet in 20% (v/v) glycerol, 1 mM EDTA, 10 mM Tris–HCl pH 7.5 and store at −20°C.

Instead of using DNase to disrupt the chromatin an alternative is to use heparin which dissociates the basic histones from the DNA (42).

4.3 Isolation of nuclear matrix

The original work on isolation and identification of the nuclear matrix (also known as the nuclear scaffold) was surrounded by controversy as to whether the isolated nuclear matrix was simply an artefact of the preparation method. It is now accepted that a protein scaffold does exist within the nucleus and subsequent work has indicated that it probably plays an important role in transcription (43).

Since the original protocol of Berezney and Coffey (44) a number of different methods have been devised for the isolation of nuclear matrices (43,45,46). *Protocol 7* was devised with the aim of avoiding the apparent aggregation of fibrils that occurs with some other methods.

Protocol 7. Isolation of nuclear matrices (43)

Equipment and reagents

- Refrigerated low-speed centrifuge
- STEM buffer: 0.25 M sucrose, 50 mM Tris–HCl pH 7.4, 5 mM $MgCl_2$, 1 mM EGTA, 1 mM PMSF
- Sodium tetrathionate
- *N*-ethylmaleimide
- DNase I (RNase-free grade)
- LS buffer: 0.2 mM $MgCl_2$, 1 mM EGTA, 1 mM PMSF, 10 mM Tris–HCl pH 7.4
- 70% (w/v) sucrose in LS buffer
- 60%, 50%, and 40% (w/v) sucrose in LS buffer containing 2 M NaCl

Method

1. Purify nuclei by centrifugation through dense sucrose as described in *Protocol 1* except that all solutions should contain 1 mM PMSF (see *Protocol 5*) and the step of washing the nuclei with Triton X-100 should be omitted. This procedure assumes that the nuclei were prepared from 50 g of liver.

2. Suspend the nuclei in 20 ml (DNA concentration of 2–3 mg/ml) of 0.25 M sucrose, 50 mM Tris–HCl pH 7.4, 5 mM $MgCl_2$, 1 mM EGTA, 1 mM PMSF (STEM buffer) and add sodium tetrathionate to a final concentration of 0.2 mM. Incubate the nuclei for 20 min at 4 °C.
3. Add *N*-ethylmaleimide to a final concentration of 5 mM and then pellet the nuclei by centrifugation at 1000 *g* for 10 min at 4 °C.
4. Wash the nuclei three times, by resuspension in 20 ml of STEM buffer and pelleting by centrifugation at 1000 *g* for 10 min as before.
5. Suspend the nuclei in 20 ml of STEM buffer containing 20 μg/ml of DNase I (RNase-free grade) and incubate for 20 min at 37 °C. Cool the nuclear suspension in ice and centrifuge at 1000 *g* for 10 min.
6. Suspend the pellet in 20 ml of 0.2 mM $MgCl_2$, 1 mM EGTA, 1 mM PMSF, 10 mM Tris–HCl pH 7.4 (LS buffer) cooled to 4 °C, and incubate for 10 min at 4 °C. Pellet by centrifugation at 1000 *g* for 10 min as before.
7. Resuspend the pellet in 1 ml 70% (w/v) sucrose in LS buffer. Overlayer the suspension with three 2 ml aliquots of 60%, 50%, and 40% (w/v) sucrose in LS buffer containing 2 M NaCl and centrifuge at 800 *g* for 1 h.
8. The nuclear matrices are found at the interface between the 40% and 50% sucrose steps. Wash the matrices by resuspension in 10 ml of STEM buffer and pelleting by centrifugation at 1000 *g* for 10 min.

4.4 Preparation of nucleoids

Nucleoids contain most of the elements found within matrix preparations, however, due to the method of preparation (cells are lysed directly) they contain all of the cellular DNA, which is undamaged and protected by a filamentous protein network derived from the cytoskeleton. They are obtained when nuclei are treated with high salt releasing the DNA as a nucleoid structure. The techniques have been devised by Cook and co-workers (47, 48) primarily for HeLa cells. *Protocol 8* gives the procedure that can be used for the isolation of nucleoids from tissue culture cells.

Protocol 8. Isolation of nucleoids from HeLa cells

Equipment and reagents

- Refrigerated high-speed centrifuge
- 15% (w/v) sucrose in 1.95 M NaCl, 10 mM Tris–HCl pH 8.0, 1 mM EDTA
- 30% (w/v) sucrose in 1.95 M NaCl, 10 mM Tris–HCl pH 8.0, 1 mM EDTA
- Siliconized, wide-bore Pasteur pipette
- Phosphate-buffered saline: 0.15 M NaCl, 3 mM KCl, 9 mM Na_2HPO_4, 2 mM KH_2PO_4 pH 7.2
- Lysis buffer: 2.6 M NaCl, 2.7 mM Tris–HCl pH 8.0, 133 mM EDTA, and 0.67% (v/v) Triton X-100

Protocol 8. *Continued*

Method

1. Harvest approx. 2×10^8 HeLa cells growing in suspension and resuspend in 4.5 ml of phosphate-buffered saline (0.15 M NaCl, 3 mM KCl, 9 mM Na_2HPO_4, 2 mM KH_2PO_4 pH 7.2).
2. To lyse the cells add 3 vol. of lysis buffer (2.6 M NaCl, 2.7 mM Tris–HCl pH 8.0, 133 mM EDTA, and 0.67% (v/v) Triton X-100).
3. Overlayer the suspension on to three sucrose gradients prepared by underlayering 25 ml of 15% (w/v) sucrose in 1.95 M NaCl, 10 mM Tris–HCl pH 8.0, 1 mM EDTA with 5 ml 30% (w/v) sucrose in the same buffer. Allow 10 min on ice for cells to lyse and centrifuge at 7500 *g* for 25 min at 4°C.
4. Harvest the fluffy white aggregate of nucleoids from the top of the 30% sucrose layer using a siliconized wide-bore Pasteur pipette. Between 0.5–1×10^8 nucleoids are isolated from a single step gradient.

To prevent artefactual cross-linking of protein, dithiothreitol (1–10 mM) may be added to the lysis and gradient buffers. Solutions may also be supplemented with 1 mM PMSF or 5 mM vanadyl ribonucleoside complex to inhibit protease and ribonuclease activities respectively. Modifications of this protocol consist of varying the salt concentration used in the lysis and gradient solutions. HeLa nucleoids may be prepared at any salt concentration in the range 0.4–2 M NaCl (48) using the centrifugation conditions described. However, in comparison to nucleoids prepared in 0.4 M NaCl, those prepared in 2 M NaCl have 'robust cages' which protect the fragile DNA within from mechanical damage even during vigorous pipetting.

5. Isolation of nucleoprotein complexes

5.1 Isolation of polynucleosomes of chromatin

Chromatin is a functional term for the complex of DNA and protein that was originally defined as the insoluble residue remaining after extraction of whole cells or nuclei with physiological saline. Extraction was aimed at removing soluble nuclear proteins and ribosomal material of the nucleolus. Much of the early work was carried out on calf thymus, a tissue in which the nucleus occupies almost all of the cell, and most experimental protocols were essentially empirical and involved extracting whole tissue with 0.14 M NaCl containing EDTA. However, subsequently, methods have been modified to ensure that the isolated chromatin reflects the composition and structure of the complex *in vivo*. It is best to prepare chromatin from purified nuclei using one of the appropriate methods described in Section 2 of this chapter; the addition of protease inhibitors such as PMSF to the solutions used for the preparation of nuclei helps to ensure the integrity of the chromosomal proteins.

Current methods utilize the release of chromatin from nuclei by limited nuclease digestion followed by extraction using a salt solution. Enzymes that break both DNA strands (e.g. micrococcal nucleases or restriction nucleases) are preferred since they predominantly cut the linker DNA between nucleosomes. The medium used for extracting digested chromatin varies, the usual composition includes 0.1 M NaCl or KCl, millimolar amounts of EDTA or EGTA to inhibit nucleases, a protease inhibitor such as PMSF, and polyamines such as spermine and spermidine (49, 50). In some cases, the presence of sucrose (0.15–0.35 M) in the extraction medium appears to be beneficial. The inclusion of polyamines helps to compensate for the loss of divalent cations. An example of the method used for preparing chromatin is described in *Protocol 9*.

Protocol 9. Preparation of chromatin

Equipment and reagents

- Low speed centrifuge
- Digestion buffer: 0.2 M sucrose, 80 mM NaCl, 1 mM $CaCl_2$, 5 mM Tris–HCl pH 7.5, 0.1 mM PMSF
- Micrococcal nuclease
- 80 mM NaCl, 5 mM Tris–HCl pH 7.5, 5 mM EDTA, 0.1 mM PMSF

Method

1. Wash sucrose purified nuclei (*Protocol 1*) by suspending them in digestion buffer (0.2 M sucrose, 80 mM NaCl, 1 mM $CaCl_2$, 5 mM Tris–HCl pH 7.5, 0.1 mM PMSF) and pelleting them by centrifugation at 1000 *g* for 10 min.
2. Resuspend the nuclei at a concentration of 10 mg/ml of DNA in digestion buffer and add micrococcal nuclease to a final concentration of 500 U/ml.
3. Incubate the nuclei in ice with occasional shaking for 20 min.
4. Terminate the digestion by the addition of an equal volume of ice-cold 80 mM NaCl, 5 mM Tris–HCl pH 7.5, 5 mM EDTA, 0.1 mM PMSF.
5. Incubate the diluted nuclear suspension a further 5 min in ice and pellet the nuclear debris by centrifugation at 4000 *g* for 5 min.
6. Chromatin, released as polynucleosomes, is in the supernatant and can be carefully removed using a wide-bore pipette. The concentration of the chromatin can be determined from the absorption at 260 nm.

The actual digestion conditions for step 2 can be variable and it is often useful to examine a range of conditions in terms of the temperature and length of incubation to optimize the release of chromatin from the nuclei. Instead of using micrococcal nuclease it is possible to digest nuclei with a restriction nuclease.

The isolated chromatin can be separated on the basis of size by rate-zonal centrifugation of sucrose gradients (51–54) or gel electrophoresis (52,54,55).

5.2 Ribonucleoproteins

The RNA is present in nuclei complexed with protein. The RNA–protein complexes that form along the RNA chain, termed ribonucleoprotein (RNP) particles, serve to condense and package each RNA transcript. Packaging of RNA by nuclear proteins is thought to guide the primary RNA transcripts through the subsequent RNA processing and transport events. Virtually all types of nuclear RNA are packaged in this way at some stage in their existence and a number of methods have been developed for isolating and studying them. As preribosomal particles comprise the most abundant class of nuclear RNP particles, a procedure that can be used for their isolation from nucleoli is described (*Protocol 10*).

Protocol 10. Isolation of preribosomal ribonucleoprotein particles (56)

Equipment and reagents

- High speed and ultracentrifuges capable of generating 25 000 *g* and up to 185 000 *g*
- 10 mM Tris–HCl pH 7.4, 10 mM EDTA, 10 mM dithiothreitol
- Sodium deoxycholate
- Brij-58
- 10%, 15%, 30%, and 55% (w/v) sucrose in 10 mM NaCl, 10 mM Tris–HCl pH 7.4, 10 mM EDTA

Method

1. Resuspend isolated nucleoli (see *Protocol 4*) from 20 g of tissue in 0.5 ml of 10 mM Tris–HCl pH 7.4, 10 mM EDTA, 10 mM dithiothreitol and agitate gently for 15 min at 0 °C.
2. Centrifuge the suspension at 25 000 *g* for 10 min at 0 °C.
3. Resuspend the pellet in 0.5 ml of 10 mM Tris–HCl pH 7.4, 10 mM EDTA, 10 mM dithiothreitol, and agitate gently for 10 min at 0 °C.
4. Add sodium deoxycholate and Brij-58, both to a final concentration of 0.5% (w/v) and centrifuge the suspension as in step 2. The RNP particles remain in the supernatant.
5. To purify the RNP particles further either:
   (a) Centrifuge them through a 12 ml gradient of 10–30% (w/v) sucrose in 10 mM NaCl, 10 mM Tris–HCl pH 7.4, 10 mM EDTA at 185 000 *g* for 160 min (56).
   (b) Centrifuge them through a 30 ml gradient of 15–55% (w/v) sucrose in 10 mM NaCl, 10 mM Tris–HCl pH 7.4, 10 mM EDTA at 85 000 *g* for 16 h (56).

This method is less effective if the nuclei from which the nucleoli were isolated had been prepared in the presence of Ca^{2+}; Mg^{2+} must be used in buffers at the early stages. Furthermore, the sonication step in the isolation of nucleoli should be performed in the presence of polyvinyl sulfate (PVS) (5–10 μg/ml) tominimize RNase activity.

The ribonucleoprotein complexes can be isolated from nuclei in the form of either heterogeneous ribonucleoprotein (hnRNP) or small nuclear ribonucleoprotein (snRNP) particles; the range of procedures used have been reviewed (57, 58). In addition, a number of practical manuals covering all aspects on RNPs have been published, including procedures for the isolation of ribosomes from bacteria, plants, and animals (59–61).

6. Isolation of nuclear macromolecules

6.1 Isolation of nuclear proteins

The DNA of all organisms is closely associated with a wide variety of different DNA-binding proteins. These proteins act as structural proteins to organize the DNA in the cell nucleus, and act as enzymes to mediate such important functions as RNA synthesis. Traditionally nuclear proteins have been classified as histones and non-histone proteins on the basis of their acid solubility.

6.1.1 Histones

Histones are distinguished from other nuclear proteins by their very basic nature. As a result of the basic nature of these proteins they can be preferentially extracted with dilute acid. Early experiments involved the isolation of histones from whole calf thymus tissue (62), however, purer preparations in most cases can be obtained from salt extracted nuclei as described in *Protocol 11*.

Protocol 11. Isolation of total histone proteins

Equipment and reagents

- Motor-driven (500 r.p.m.) Potter–Elvehjem homogenizer with loose-fitting Teflon pestle
- Centrifuge capable of generating 15 000 *g*
- Ethanol/1.25 M HCl (4:1, v/v)
- 0.14 M NaCl, 0.05 M Tris–HCl pH 5, 5 mM EDTA, 0.1 mM PMSF
- Acetone

Method

1. To prepare chromatin, suspend isolated nuclei (see *Protocol 1*; include treatment with 1% Triton X-100) in 30 vol. of ice-cold 0.14 M NaCl, 0.05 M Tris–HCl pH 5, 5 mM EDTA, 0.1 mM PMSF, and homogenize in a Potter–Elvehjem homogenizer (use five or six up-and-down strokes). Stir the suspension for 20 min at 0°C and then centrifuge at 15 000 *g* for 15 min. Repeat this procedure twice.

2. Extract the histones from the chromatin pellet with 20 vol. of ethanol/1.25 M HCl (4:1, v/v) at 4°C. Centrifuge at 15 000 *g* for 15 min at 4°C.
3. Add acetone to the supernatant to precipitate the histone proteins. Pellet the precipitate by centrifugation as before. Wash the precipitate three times in acetone to remove traces of acid.

While the above method is used for the isolation of total histones, methods for subfractionation of histone proteins have been reviewed elsewhere (63,64).

6.1.2 Chromatin non-histone proteins

The non-histone proteins are an extremely heterogeneous group of proteins which include a variety of classes of proteins all of which are closely associated with the DNA. In fact some distinct groups of proteins can be identified in terms of their properties. Included in these groups are the high mobility group (HMG) proteins and the loosely bound chromatin proteins. The isolation of non-histone proteins is given in *Protocol 12*.

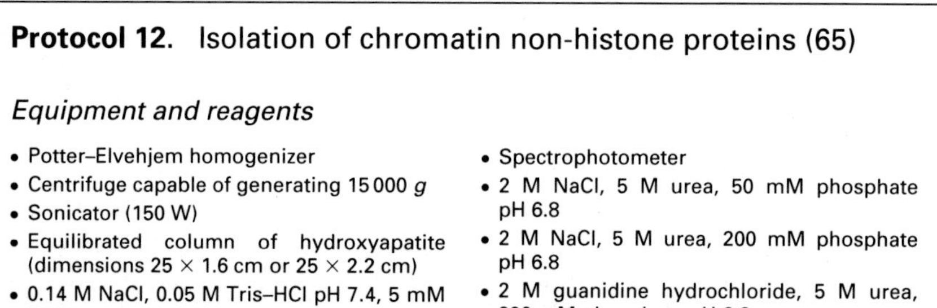

Protocol 12. Isolation of chromatin non-histone proteins (65)

Equipment and reagents

- Potter–Elvehjem homogenizer
- Centrifuge capable of generating 15 000 *g*
- Sonicator (150 W)
- Equilibrated column of hydroxyapatite (dimensions 25 × 1.6 cm or 25 × 2.2 cm)
- 0.14 M NaCl, 0.05 M Tris–HCl pH 7.4, 5 mM EDTA, 0.1 mM PMSF
- 2 M NaCl, 5 M urea, 1 mM sodium phosphate, 2 mM Tris, 0.1 mM PMSF pH 6.8
- Spectrophotometer
- 2 M NaCl, 5 M urea, 50 mM phosphate pH 6.8
- 2 M NaCl, 5 M urea, 200 mM phosphate pH 6.8
- 2 M guanidine hydrochloride, 5 M urea, 200 mM phosphate pH 6.8
- 2 M guanidine hydrochloride (or 2 M NaCl), 5 M urea, 500 mM phosphate pH 6.8

Method

1. To prepare chromatin, suspend isolated nuclei (see *Protocol 1*; include treatment with 1% Triton X-100) in 30 vol. of ice-cold 0.14 M NaCl, 0.05 M Tris–HCl pH 7.4, 5 mM EDTA, 0.1 mM PMSF, and homogenize in a Potter–Elvehjem homogenizer (use five or six up-and-down strokes). Stir the suspension for 20 min at 0°C and then centrifuge at 15 000 *g* for 15 min. Repeat this procedure twice.
2. Homogenize the chromatin pellet in sufficient 2 M NaCl, 5 M urea, 1 mM sodium phosphate pH 6.8, 2 mM Tris, 0.1 mM PMSF pH 6.8, at 4°C, to give about 0.5 mg DNA/ml.
3. Centrifuge at 15 000 *g* for 15 min at 4°C.
4. Homogenize the pellet as described in step 2 and centrifuge as described in step 3. Combine the two extracts and sonicate for two

periods of 15 sec each. Centrifuge at 15000 *g* for 15 min to remove traces of residual material.

5. Apply the supernatant (OD_{260} 5–10; 50–100 ml) on to an equilibrated column of hydroxyapatite (65) and allow to enter the column at a flow rate of 5–10 ml/h. Solutions containing up to 15 mg of DNA are normally applied to a 25 × 1.6 cm column, and those containing between 15 and 50 mg are normally applied to a column of dimensions 25 × 2.2 cm. With concentrated chromatin solutions it is sometimes necessary to stir the top of the column gently with a Pasteur pipette to maintain the flow rate.
6. Once the sample has entered the column, add several millilitres of the salt–urea solution to the top of the column.
7. Allow the salt–urea solution to run through the column until, as seen by monitoring fractions at 280 nm, the unretained histone fraction (HAP1) has eluted.
8. Elute the non-histone proteins from the column by elution with 2 M NaCl, 5 M urea, 50 mM phosphate pH 6.8 (fraction HAP2), then 2 M NaCl, 5 M urea, 200 mM phosphate pH 6.8 (fraction HAP3), then 2 M guanidine hydrochloride, 5 M urea, 200 mM phosphate pH 6.8 (fraction HAP4). For the final fraction (HAP5) which contains predominantly DNA, elute with 2 M guanidine hydrochloride (or 2 M NaCl), 5 M urea, 500 mM phosphate pH 6.8.

The non-histone proteins can then be characterized by two-dimensional gel electrophoresis. The specific fractionation and isolation of HMG proteins has been reviewed by Goodwin *et al.* (66).

6.2 Isolation of nuclear RNA

One of the most important aspects in the isolation of RNA is to prevent any degradation of the RNA during the isolation procedure. Since RNA breakdown is a universal feature of all cells during the processing and metabolism of RNA, all cells possess a wide range of nucleases with both specific and non-specific activities. Both nuclei and lysosomes are particularly rich in nucleases but nuclease activity is not restricted to these organelles. Some tissues such as pancreatic tissue have particularly high nuclease activity. Given the wide distribution of nucleases and the potential damage that they can inflict on RNA during isolation, it is most important to ensure that nuclease activity is totally inhibited, a problem made all the more difficult by the robust nature of many nucleases.

A few golden rules should be followed if the isolation procedure is to be successful; all too often the first attempts to isolate RNA leave the novice with only degraded fragments of RNA. First, it is important to ensure that all

Table 3. Nuclease inhibitors

Agent	Active concentration	Method of ribonuclease inactivation	Comments
Aurintricarboxylic acid (ATA)	10 μM	Complexes to a wide range of nucleases	
Bentonite	3 mg/ml	Inactivation by adsorbing to nucleases	Purified suspension of clay particles
Diethylpyrocarbonate (DEPC)	0.1% (v/v)	Alkylates proteins disrupting protein structure	Toxic, can modify bases of RNA
EDTA	1–10 mM	Chelates divalent cations needed for activity of some ribonucleases	Not all nucleases need divalent ions for activity
Guanidinium hydrochloride	8 M	Denatures proteins	Toxic
Guanidinium thiocyanate	4 M	Denatures proteins	Toxic, strong agent for ribonuclease inactivation
Heparin	0.5 mg/ml	Binds to basic ribonucleases	Can strip proteins from DNA
8-Hydroxyquinoline[a]	0.1% (w/v)	Inactivates ribonucleases	Toxic
Macaloid	0.015% (w/v)	Adsorbs to ribonucleases	
Phenol/chloroform	50% (v/v)	Extracts and denatures ribonucleases	Toxic
Placental RNase inhibitor	1 U/μl	Protein inhibitor of RNases	RNasin™ (Promega), Inhibit-Ace™ (5 prime-3 prime, Inc), RNAguard™ (Pharmacia)
Polyvinyl sulfate (PVS)	1–10 μg/ml	Complexes to basic nucleases	
Proteinase K	100–200 μg/ml	Hydrolysis of proteins	Pre-digest to remove any nuclease contamination
Sodium dodecyl sulfate (SDS)	0.1–1% (w/v)	Disrupts protein structures	Hazardous in powder form
Ribonucleoside vanadyl complex	10 mM	Binds to active site of ribonuclease	Reversible inhibition

[a] 8-Hydroxyquinoline/8-hydroxy-1-azanaphthalene.

of the reagents used are as free as possible of nucleases. All solutions should be made from either fresh material or material whose history is known using autoclaved double distilled water and, where possible, autoclaving the final solution. Solutions used in the initial extraction (during which nucleases are released) are often supplemented with nuclease inhibitors such as 0.1% (w/v) 8-hydroxyquinoline (*Table 3*). A pH range of pH 6.8 to pH 7.2 for solutions is normally a necessity to ensure that no RNA degradation occurs as RNA is degraded at alkaline pHs as low as pH 9. Glassware should be oven baked at 200°C for 4 to 12 hours whereas plastic is usually autoclaved. Skin is a well known source of nucleases and so it is essential to wear disposable plastic or latex gloves during all manipulations.

RNA is, for the most part, complexed with proteins some of which have nuclease activity. Hence it is important that any isolation method ensures rapid inactivation of cellular proteins. This can be achieved by using nuclease inhibitors such as hydroxyquinoline but usually better results can be obtained by denaturing the proteins in the sample using strong detergents, chaotropic solutions, or other protein denaturants (*Table 3*).

All nuclear RNA isolation methods include three steps, the inactivation of nucleases, the dissociation of RNA from proteins, and the separation of the RNA from other macromolecules especially DNA. Usually it is best if the first two steps are immediate and concurrent so that degradation of RNA is minimal. Total RNA may then be fractionated to give the particular type of RNA required, for example, hnRNA, mRNA, or rRNA. Irrespective of the type of RNA required, a good general rule is to use as few steps as possible for purification procedures.

6.2.1 Phenol extraction methods

This procedure was one of the first published methods to prepare nucleic acids free of contaminating proteins (67). The procedure is based on the ability of organic molecules to denature and precipitate proteins in solution while not affecting the solubility of the nucleic acids. A number of procedures have been developed based around phenol containing various additives to improve the inhibition of nuclease activity (e.g. 8-hydroxyquinoline) or to improve the deproteinization efficiency of phenol (e.g. chloroform).

Phenol/chloroform extraction is carried out by addition of an equal volume of buffer-saturated phenol/chloroform/isoamyl alcohol (25:24:1) to the sample, vigorous mixing to form an emulsion, and separation of the denser phenol from the aqueous layer by centrifugation. High purity phenol is available commercially but it is also possible to redistil the phenol before use although the procedure can be hazardous. Phenol/chloroform/isoamyl alcohol (25:24:1) is prepared by equilibrating the pH of the acidic phenol/chloroform/isoamyl alcohol solution with an autoclaved buffer of the required pH (e.g. 0.1 M Tris buffer pH 7.0). An equal volume of buffer is mixed into the phenol/chloroform/isoamyl alcohol by vigorous shaking. The mixture is allowed to settle and the upper layer (aqueous buffer) then removed. Equilibration with buffer is repeated until the required pH is reached.

The basis of the method (*Protocol 13*) is to suspend the cells in a solution containing a detergent which will lyse the cells and dissociate nucleoprotein complexes; various lysis buffers have been reviewed (68).

This basic phenol extraction procedure has been modified in a number of ways by the inclusion of additives such as 8-hydroxyquinoline or *m*-cresol to phenol/chloroform/isoamyl alcohol to enhance RNA extraction (69). Other approaches have been to modify the temperature of phenol extraction to 60 °C ('hot phenol extraction') increasing the efficiency of the process, particularly for nuclear and bacterial RNA. An extension of this has been to perform the series of phenol extractions at different temperatures and to obtain different fractions of RNA (70), but this method is hardly ever used now.

Protocol 13. Phenol extraction method

Equipment and reagents

- Polytron or Potter–Elvehjem homogenizer
- Orbital incubator
- Microcentrifuge capable of generating 14 000 *g*
- High speed centrifuge capable of generating 60 000 *g*
- Lysis buffer suitable for the cells being used
- Ethanol
- Buffer-saturated phenol/chloroform/isoamyl alcohol (25:24:1)
- 70% ethanol
- Isopropanol
- 3 M sodium acetate pH 5.2
- Sterile, double distilled water treated with 0.1% (w/v) diethylpyrocarbonate (DEPC)
- Chloroform/isoamyl alcohol (24:1)

Method

1. Homogenize ground plant tissue in lysis buffer, using two high speed treatments with a Polytron homogenizer, each for 5 sec. For tough animal tissue such as pancreas homogenize for 30–60 sec at high speed, however, for soft animal tissue use a Potter–Elvehjem homogenizer.
2. Leave homogenate for 15 min at room temperature to allow hydrolysis of proteins.
3. To remove protein add an equal volume of buffer-saturated phenol/chloroform/isoamyl alcohol (25:24:1) and mix well either by hand or in orbital incubator for 5–10 min.
4. Centrifuge the emulsion at 10 000 *g* for 10 min; the phenol/chloroform separates out as a dense bottom layer with the proteins either dissolved in the phenol/chloroform layer or precipitated at the interface, leaving the RNA in the top aqueous layer.
5. Recover the top aqueous phase and repeat steps 3 and 4; usually three extractions are sufficient to yield a clear interface.
6. Precipitate the RNA in the aqueous phase by adding sodium acetate pH 5.2 to a final concentration of 0.3 M followed by the addition of either 2 vol. of 70% ethanol or 1 vol. of isopropanol. Leave at −20 °C for at least 1 h for RNA to precipitate.
7. Pellet the RNA by centrifugation at 10 000 *g* for 20 min at 4 °C; if only submicrogram amounts of RNA are being precipitated it is better to

use a higher centrifugal force (e.g. 60 000 *g* for 1 h) in order to ensure a high recovery of RNA.

8. After centrifugation, remove the aqueous fraction by aspiration and wash the pellet with ice-cold 70% (v/v) ethanol to remove excess salt. Redissolve the RNA pellet in approx. 0.5 ml sterile, double distilled water (treated with DEPC) per 5 g starting material and transfer to a sterile 0.5 ml microcentrifuge tube.
9. Mix the redissolved RNA with an equal volume of phenol/chloroform/isoamyl alcohol to remove any residual protein. Centrifuge at 10 000–14 000 *g* for 3–5 min at room temperature, using a microcentrifuge, to separate the phenol and aqueous layers.
10. Recover the aqueous layer, add an equal volume of chloroform/isoamyl alcohol (24:1), and mix well. Centrifuge as described in step 9; this step removes traces of phenol.
11. Precipitate the RNA with 2 vol. of ethanol and 1/10 vol. of 3 M sodium acetate pH 5.2. Leave at −20 °C for at least 2 h to precipitate the RNA and then centrifuge at 10 000–14 000 *g* for 10–20 min at 4 °C.
12. Redissolve the RNA in either DEPC treated water, 10% sucrose in formaldehyde, or 10% sucrose in 8 M urea.

A major drawback with phenol extraction is that, depending on the nature of the starting material, the RNA can be contaminated with significant amounts of both DNA and polysaccharides as both can be present in the aqueous layer after phenol extraction. Both contaminants are precipitated along with the RNA by ethanol or isopropanol and, depending on the subsequent use of the RNA, they may affect further analyses. DNA and polysaccharides can be removed from the RNA by enzymic degradation either by including a digestion step before the last phenol/chloroform/isoamyl alcohol extraction of the protocol or after extraction. DNA can be removed by addition of DNase I to a final concentration of 2 μg/ml and incubated for 1 h at 37 °C.

It is important that any enzymes used are chromatographically purified before use to remove RNases. After enzymatic digestion of contaminating material, the RNA must be deproteinized again to remove these enzymes. This is accomplished by re-extraction with phenol/chloroform/isoamyl alcohol and reprecipitation with ethanol or isopropanol before it is used for further preparative analytical procedures. An alternative method to purify the phenol extracted RNA is to pellet it through a cushion of 5.7 M CsCl, 0.01 M EDTA pH 7.5 in an ultracentrifuge. This procedure will remove both DNA and polysaccharides as they are not dense enough to pellet through 5.7 M CsCl. Centrifugation methods for the isolation of RNA are described in Section 6.2.3.

6.2.2 Acid guanidinium thiocyanate method

The phenol/chloroform/isoamyl alcohol method has been largely superseded by the acid guanidinium thiocyanate method (71). This method uses the chaotropic agent guanidinium thiocyanate in which the cation and the anion are both strongly chaotropic. This allows a very fast inactivation of cellular ribonucleases (RNases) upon cell disruption. A fast inactivation is often essential making this the method of choice. The tissue types used with this method are highly diverse, bacterial, plant, and animal tissue have all successfully had RNA extracted from them.

Protocol 14. Acid guanidinium thiocyanate method (71)

Equipment and reagents

- Hand-held Polytron or Potter–Elvehjem homogenizer
- Orbital incubator
- Microcentrifuge capable of generating 14 000 *g*
- High speed centrifuge capable of generating 60 000 *g*
- GTC solution: 4 M guanidinium thiocyanate, 25 mM sodium citrate pH 7.0, 0.5% sarcosyl, 0.1 M 2-mercaptoethanol
- Buffer-saturated phenol/chloroform/isoamyl alcohol (25:24:1)
- Ethanol
- 70% ethanol
- Isopropanol
- 3 M sodium acetate pH 5.2
- Sterile, double distilled water treated with 0.1% (w/v) diethylpyrocarbonate (DEPC)
- Chloroform/isoamyl alcohol

Method

1. Homogenize cells in 4 vol. of GTC solution using a hand-held Polytron or Potter–Elvehjem homogenizer.
2. Add an equal volume of buffer-saturated phenol/chloroform/isoamyl alcohol (25:24:1) and shake to form an emulsion.
3. Centrifuge the emulsion at 10 000 *g* for 20 min at room temperature; the phenol/chloroform separates out as a dense bottom layer with the proteins either dissolved in the phenol/chloroform layer or precipitated at the interface with most of the DNA, leaving the RNA in the top aqueous layer.
4. Recover the top aqueous phase and repeat the phenol extraction as described in steps 2 and 3; usually three extractions are sufficient to yield a clear interface.
5. Follow *Protocol 13*, steps 6–11. Finally, redissolve the RNA in an appropriate solution.

6.2.3 Isolation of RNA using ultracentrifugation

Due to the hazards inherent in working with phenol a number of other methods for the isolation of RNA have been investigated. Efforts have also been made to devise a simple method that will avoid the problem of DNA contamination. Both of these objectives can be achieved by using centrifugation for isolating RNA. RNA species of defined size such as transfer RNA (tRNA) and small nuclear RNA (snRNA) can be isolated from total RNA preparations by rate-zonal density gradient centrifugation on some gradients or by preparative gel electrophoresis.

In solutions containing high concentrations of caesium salts RNA has a much higher density than DNA, protein, or polysaccharide. Under the appropriate centrifugation conditions, RNA will pellet while the contaminants remain in the supernatant. One of the original methods involved lysis of cells with sodium dodecyl sulfate followed by the addition of CsCl solution and ultracentrifugation (72), however, caesium dodecyl sulfate formed during this procedure is insoluble and must be removed before ultracentrifugation.

The most popular preparative centrifugation procedure for preparing total cellular RNA is described below (*Protocol 15*).

Protocol 15. Preparative centrifugation for the preparation of total cellular RNA

Equipment and reagents

- Homogenizer
- Ultracentrifuge with a swinging-bucket rotor capable of generating 250 000 *g*
- 4 M guanidinium thiocyanate, 30 mM sodium acetate, 0.14 M 2-mercaptoethanol
- 5.7 M CsCl
- 70% (v/v) ethanol, 0.3 M sodium acetate
- Sterile double distilled water treated with 0.1% (w/v) diethylpyrocarbonate (DEPC)

Method

1. Homogenize the tissue in a solution of 4 M guanidinium thiocyanate, 30 mM sodium acetate, 0.14 M 2-mercaptoethanol (70).
2. Layer the cell homogenate over 0.5 vol. of 5.7 M CsCl and centrifuge in a swinging-bucket rotor at 250 000 *g* for 3 h at 22 °C; the DNA, polysaccharides, and proteins will remain above the density barrier while the RNA will pellet giving a yield of about 90%.
3. Wash the RNA pellet twice with 70% (v/v) ethanol, 0.3 M sodium acetate and then dissolve in sterile double distilled water treated with 0.1% (w/v) DEPC.

6.3 Isolation of DNA

The plethora of methods for the isolation of DNA are due mainly to the susceptibility of the nucleic acids to degradation by, and ubiquity of, nucleases. Fortunately, unlike ribonucleases, deoxyribonucleases are labile to heat and so can be eliminated from buffers by autoclaving, denatured by phenol and detergents like SDS, and inhibited effectively by including a chelating agent such as EDTA in the buffers. The precautions outlined for the isolation of RNA in Section 6.2 are also very much applicable to the isolation of DNA.

The methods for isolating DNA are so varied in their detail that a comprehensive discussion of them would be of limited use, especially as many of the methods have been devised for particular types of cell or tissue. *Protocol 16* gives a method which can be used to isolate DNA from mammalian cells.

Protocol 16. Isolation of genomic DNA from cultured mammalian cells (73)

Equipment and reagents

- Rubber policeman
- Low-speed and high-speed centrifuges with swinging-bucket rotors
- Erlenmeyer flask
- Water-baths at 37 °C and 50 °C
- Orbital table
- Tris-buffered saline (TBS): 25 mM Tris base, 0.2 M NaCl, 2.5 mM KCl, 15 μg/ml phenol red pH 7.4, autoclaved
- TE: 10 mM Tris, 1 mM EDTA
- Lysis solution: 10 mM Tris–HCl pH 8.0, 0.5% SDS, 0.1 M EDTA, 20 μg/ml pancreatic RNase
- Proteinase K (20 mg/ml)
- Buffer-saturated phenol pH 8.0
- 10 M ammonium acetate (filtered)
- Ethanol
- 70% ethanol

Method

1. Gently wash monolayers of cells twice with 5 ml ice-cold TBS buffer. Decant TBS leaving 0.5 ml in the culture dish and harvest the cells by scraping a rubber policeman over the monolayers. Transfer the cells to a tube on ice. Wash the dish with 1 ml ice-cold TBS to recover any cells remaining and add this wash to the cells on ice. For suspension cultures, pellet the cells and wash with TBS.
2. Pellet the cells by centrifugation at 1500 *g* for 10 min at 4 °C and resuspend in 1 ml ice-cold TBS. Pellet cells as before and then resuspend the cells in TE buffer to a final concentration of 5×10^7 cells/ml.
3. Transfer the cells to an Erlenmeyer flask (50 cm^3/1 ml of cells) and add 10 ml lysis solution/ml of cell suspension. Incubate for 1 h at 37 °C. Add 1/200 vol. of proteinase K and mix into the cell suspension using a glass rod. Incubate for 3 h at 50 °C with occasional mixing.
4. Cool the solution to room temperature and phenol extract three times with an equal volume of phenol. Mix the phases by gentle inversion to form an emulsion and then centrifuge at 8000 *g* for 15 min at room temperature using a swinging-bucket rotor. Recover the aqueous top phase with a wide-bore Pasteur pipette, avoiding the proteinaceous interface.
5. Precipitate the DNA with 0.2 vol. of 10 M ammonium acetate and 2.5 vol. of ethanol and swirl to mix. The DNA precipitate will be a visible strand/lump which can be removed to a fresh tube using a thin glass rod or spatula.

6. Wash the DNA precipitate with 70% ethanol and allow to air dry at room temperature until the ethanol evaporates leaving a moist DNA lump; do not allow to dry completely. Redissolve the DNA in 1 ml TE/ 5×10^6 cells by agitation on an orbital table at 20 r.p.m. for 12–24 h.

The advantage of this method is that it can also be used for the isolation of plant genomic DNA. However, the reader should be aware that genomic DNA extractions are prone to shearing of the very long DNA molecules during isolation.

7. Functional assays of nuclei

7.1 Analysis of DNA-binding proteins

The interaction of nucleic acids and proteins is essential to many molecular processes including gene transcription and DNA stability. The interaction between DNA and proteins can be located to a specific area of the DNA sequence using gel retardation and DNA footprinting.

7.1.1 Gel retardation assays

A solution containing protein–nucleic acid complexes and unbound molecules can be separated by electrophoresis on the basis of size. Free radiolabelled DNA molecules will migrate rapidly through the gel and constitute the 'free' band, while the protein–nucleic acid complexes migrate more slowly through the gel. Gel retardation of DNA–protein complexes is described in *Protocol 17*.

As complex formation is usually in flux with association and dissociation continually occurring and that a mixture of protein will bind to almost any DNA, it is important to define conditions under which only specific interactions will occur. These conditions are usually defined by salt concentrations. In addition, nucleic acid-binding proteins are usually only partially active (e.g. 5–75% of the binding protein is active) and so an excess of protein is included in the assay to compensate for the shortfall. DNA, usually from plasmid is used for binding, while non-specific binding is inhibited by the addition of poly(dI–dC).

Protocol 17. Gel retardation of DNA–protein complexes

Equipment and reagents

- Polyacrylamide gel plates (150 × 150 × 2 mm)
- Electrophoresis equipment
- Whatman 3MM filter paper
- Cling-film
- Heated gel drier
- Autoradiography equipment
- 2 × binding buffer: 40 mM Hepes–NaOH pH 7.6, 80 mM NaCl, 1 mM DTT, 8% Ficoll, 10 mM $MgCl_2$, 0.2 mM EDTA
- 5′[^{32}P] end-labelled DNA
- Nuclear protein (0.5–10 mg/ml)
- Poly(dI–dC) (0.3 mg/ml) in 50 mM NaCl, 10 mM Tris–HCl pH 7.6, 1 mM EDTA

Protocol 17. *Continued*

- [^{32}P]pUC19/pBS plasmid (1 mg/ml)
- Acrylamide stock solution (deionized 29% acrylamide, 1% bisacrylamide)
- 5 × TBE: 0.45 Tris, 0.45 M boric acid, 10 mM EDTA pH 8.3
- Sterile double distilled water
- 10% ammonium persulfate (APS) (freshly prepared)
- *N,N,N′,N′*-tetramethylethylenediamine (TEMED)
- Ethanol/ether (1:1, v/v)

Method

1. For the binding reaction, set up a series of reactions varying protein and poly(dI–dC) concentration as follows: 10 μl of 2 × binding buffer, 1 μl (5 fmol (20 000 c.p.m.)) [^{32}P] end-labelled DNA, 1–9 μl of nuclear protein, and 9–1 μl of poly(dI–dC). Incubate on ice for 1 h. A negative control of [^{32}P] plasmid DNA can be used to replace end-labelled DNA (1 μl).
2. Prepare a 5% acrylamide gel in 0.2 × TBE by mixing 12.5 ml acrylamide stock, 3 ml 5 × TBE, and 59.5 ml water. Degas the mixture for 10 min and add 400 μl APS and 30 μl TEMED to initiate polymerization. Pour the solution into clean and assembled gel plates and leave at room temperature to polymerize; the gel plates must be cleaned with detergent, rinsed with distilled water, and wiped with ethanol/ether (1:1, v/v).
3. Load complex on to the polyacrylamide gel and electrophorese at 150 V for 3 h at 4 °C.
4. Separate the plates and lay a sheet of Whatmann 3MM paper over the gel. Pull the paper back with the gel adhering to the paper and cover with cling-film. Dry the gel on a gel drier at 80 °C.
5. Expose dried gel to X-ray film for at least one day.

7.1.2 DNA footprinting assays

DNA footprinting focuses upon the specific bases of DNA with which a binding protein interacts. This method is used to define the area of DNA that the protein binds. In addition, it may allow the determination of which face of the helix is interacting with the various binding proteins (74). The DNA binding proteins that are usually analysed are ones which interact with promoter sequences, however, this technique has also been used to analyse restriction nucleases (75). The procedure for DNA footprinting is described in *Protocol 18*.

Protocol 18. DNA footprinting using DNase I

Equipment and reagents

- Centrifuge capable of generating 15 000 *g*
- Water-bath at 90 °C
- 10 mM dithiothreitol (DTT) (fresh)
- 5′[^{32}P] end-labelled DNA

- 10 × Hepes buffer: 200 mM Hepes–NaOH pH 7.6, 400 mM NaCl, 20 mM $CaCl_2$, 50 mM $MgCl_2$
- Nuclear protein (0.5–10 mg/ml)
- Poly(dI–dC) (0.3 mg/ml) in 50 mM NaCl, 10 mM Tris pH 7.6, 1 mM EDTA
- 50% glycerol
- Sterile double distilled water
- DNase I (20 ng/μl)
- Stop buffer (fresh): 2% SDS, 10 mM EDTA pH 8.0, 1 mg/ml tRNA
- Phenol containing 0.5% hydroxyquinoline
- Chloroform/isoamyl alcohol (24:1)
- Ethanol
- Sample buffer: 95% deionized/fresh formamide, 0.02% xylene cyanol FF, 0.025% bromophenol blue

Method

1. For the binding reaction, set-up a duplicate series of reactions varying protein and poly(dI–dC) concentration as follows: 10 μl of 10 × Hepes buffer, 10 μl of 10 mM DTT, 2 μl (5 fmol (10 000 c.p.m.)) [^{32}P] end-labelled DNA, 0–50 μl of nuclear protein, 50–0 μl of poly(dI–dC), 20 μl of 50% glycerol, and 8 μl of water. Incubate on ice for 1 h.
2. Add 1 μl of DNase I to each reaction and digest one set for 15 sec and one set for 30 sec at room temperature. Terminate reaction by adding 100 μl of stop buffer.
3. Extract samples twice with 100 μl of phenol and once with 100 μl of chloroform/isoamyl alcohol. Precipitate with 2.5 vol. ice-cold ethanol for 30 min at −20 °C, and then centrifuge at 15 000 *g* for 15 min at 4 °C.
4. Dissolve the pellet in 10 μl of sample buffer and heat denature for 2 min at 90 °C. Snap cool in ice prior to electrophoresis. Run sample on a sequencing gel against a sequence of the DNA fragment under analysis (76).

7.2 Transcription assays

For many years it has been known that nuclei can synthesize RNA when incubated *in vitro* under the correct conditions. Incubation of purified nuclei *in vitro* with nucleotide triphosphates (NTPs) and Mg^{2+} allows elongation and termination of existing RNA chains, but effectively no initiation of new RNA molecules. There are a number of *in vitro* transcriptional systems used for analysing a range of factors involved in the control of gene expression. In some cases, for example when using isolated nuclei as a transcription system, there is already a large amount of pre-existing RNA and so the RNA that is synthesized *in vitro* will only represent a tiny percentage of the total RNA. In order to characterize the newly-synthesized RNA it must be separated from the pre-existing RNA. All of the methods devised for the isolation of newly-synthesized RNA depend on the incorporation of modified nucleotides which can be used as a means of distinguishing and isolating the RNA that has been synthesized *in vitro*.

7.2.1 Incorporation of nucleotide analogues into RNA

One approach is to produce labelled transcripts *in vitro* using mercurated nucleotides and subsequently select these using sulfhydryl–Sepharose affinity chromatography (77). Alternatively, *in vitro* labelling of polynucleotides with phosphothiate analogues of nucleoside triphosphates has been used in conjunction with organomercury affinity chromatography to study RNA chain initiation, capping, and enzyme–polynucleotide interactions (78); 4-thiouridine (4-TU) and 6-thioguanosine (6-TG) triphosphates are normally used for this purpose. The newly-synthesized RNA is usually also radiolabelled by the addition of 10–40 kBq/ml of RNA precursor simultaneously with 4-TU or 6-TG as a marker for uptake and incorporation into newly-synthesized RNA, and used to monitor RNA recovery during subsequent isolation and enrichment steps.

Protocol 19 describes a procedure used for the isolation of sulfhydryl labelled RNA.

Protocol 19. Isolation of sulfhydryl labelled RNA by affinity chromatography

Equipment and reagents

- Water-bath at 65 °C
- *p*-Hydroxymercuribenzoate agarose (e.g. Affi-Gel 501, Bio-Rad)
- 0.15 M NaCl, 4 mM EDTA, 0.1% SDS, 50 mM sodium acetate pH 5.5
- TE (10 mM Tris, 1 mM EDTA) containing 0.5 M NaCl
- TE (10 mM Tris, 1 mM EDTA) containing 10 mM 2-mercaptoethanol

Method

1. Isolate *in vitro* synthesized RNA (see *Protocols 13* and *14*) and dissolve in 0.15 M NaCl, 4 mM EDTA, 0.1% SDS, 50 mM sodium acetate pH 5.5
2. Heat the sample to 65 °C for 5 min to denature secondary structure, and then cool rapidly on ice to minimize the re-formation of helical structures.
3. The denatured RNA is then batch adsorbed to *p*-hydroxymercuribenzoate agarose (e.g. Affi-Gel 501, Bio-Rad) at 25–400 μg RNA/ml.
4. Elute the non-specifically bound RNA with TE containing 0.5 M NaCl. Discard this fraction.
5. Elute the thio-substituted RNA with TE containing 10 mM 2-mercaptoethanol. The purity of the newly-synthesized RNA can be quantitated by determining the specific activity of ^{3}H-labelled RNA in c.p.m./μg RNA.

7.2.2 Incorporation of biotinylated oligonucleotides into RNA

Another technique for the isolation of specific RNA synthesized *in vitro* uses small, reversibly biotinylated RNAs for affinity chromatography (79). An anchor DNA probe with sequences complementary to the RNA to be isolated is synthesized *in vitro*. The probe is linked to molecules containing biotin at carbon-5 of a uridine residue via a linker containing a disulfide bond. The biotinylated anchor DNA is attached to the solid support via succinylavidin molecules each of which is capable of binding four biotin molecules. When the total RNA mixture is transferred into the column, the RNA of interest is hybridized to the anchor DNA via its complementary sequence. The RNA that has not hybridized may be eluted by washing the column with buffer. The specifically bound RNA can then be eluted with buffer containing dithiothreitol (DTT) which reduces the disulfide bonds linking biotin to the anchor DNA.

References

1. Allfrey, V. G., Stern, H., Mirsky, A. E., and Saetren, H. (1951). *J. Gen. Physiol.*, **35**, 529.
2. Siebert, G. (1963). *Exp. Cell Res. Suppl.*, **9**, 389.
3. Dounce, A. L. (1955). In *The nucleic acids* (ed. E. Chargaff and J. N. Davidson), Vol. 2, p. 93. Academic Press, New York.
4. Chauveau, J., Moule, Y., and Rouiller, C. (1956). *Exp. Cell Res.*, **11**, 317.
5. Widnell, C. C. and Tata, J. R. (1964). *Biochem. J.*, **92**, 313.
6. Price, C. A. (1974). In *Methods in enzymology* (ed. S. Fleischer and L. Packer), Vol. 31A, p. 501. Academic Press, London.
7. Cummings, D. J. (1972). *J. Cell Biol.*, **53**, 105.
8. Gorovsky, M. A., Yao, M.-C., Keevert, J. B., and Pleger, G. L. (1975). In *Methods in cell biology* (ed. D. M. Prescott), Vol. IX, pp. 311–27. Academic Press, London.
9. Dunham, V. L. and Bryant, J. A. (1983). In *Isolation of membranes and organelles from plant cells* (ed. J. L. Hall and A. L. Moore), pp. 237–75. Academic Press, London.
10. Evans, W. H. (1992). In *Preparative centrifugation: a practical approach* (ed. D. Rickwood), pp. 233–70. IRL Press at Oxford University Press, Oxford.
11. Rickwood, D. (1992). In *Preparative centrifugation: a practical approach* (ed. D. Rickwood), pp. 143–86. IRL Press at Oxford University Press, Oxford.
12. Gurney, Jr., T. and Foster, D. N. (1977). In *Methods in cell biology* (ed. G. Stein, J. Stein, and L. J. Kleinsmith), Vol. XVI, pp. 45–68. Academic Press, London.
13. Dounce, A. L., Tishkoff, G. H., Barnett, S. R., and Freep, R. M. (1950). *J. Gen. Physiol.*, **33**, 629.
14. Smuckler, E. A., Koplitz, M., and Smuckler, D. E. (1976). In *Subnuclear components: preparation and fractionation* (ed. G. D. Birnie), pp. 1–57. Butterworths, London.
15. Rozijn, T. H. and Tonino, G. J. M. (1964). *Biochim. Biophys. Acta*, **91**, 105.
16. Tautvydas, K. J. (1971). *Exp. Cell Res.*, **68**, 299.
17. Byfield, J. E. and Lee, Y. C. (1970). *J. Protozool.*, **17**, 445.
18. Reich, E. and Tsuda, S. (1961). *Biochim. Biophys. Acta*, **53**, 574.
19. Mittermayer, C., Braun, R., and Rusch, H. P. (1966). *Biochim. Biophys. Acta*, **114**, 536.
20. Mohberg, J. and Rusch, H. P. (1971). *Exp. Cell Res.*, **66**, 305.

21. Bhargava, M. M. and Halvorson, H. O. (1971). *J. Cell Biol.*, **49**, 423.
22. Stout, J. T. and Hurley, C. K. (1977). In *Methods in cell biology* (ed. G. Stein, J. Stein, and L. J. Kleinsmith), Vol. XVI, pp. 87–96. Academic Press, London.
23. Graham, J., Ford, T., and Rickwood, D. (1994). *Anal. Biochem.*, **220**, 367.
24. Hoffmann, P. and Chalkley, R. (1978). In *Methods in cell biology* (ed. G. Stein, J. Stein, and L. J. Kleinsmith), Vol. XVII, pp. 1–12. Academic Press, London.
25. Risley, M. S., Gambino, J., and Eckhardt, R. A. (1979). *Dev. Biol.*, **68**, 299.
26. Hanson, C. V. (1975). In *New techniques in biophysics and cell biology* (ed. R. H. Pain and B. J. Smith), p. 43. John Wiley, London.
27. Young, B. D. (1986). In *Nuclear structures: isolation and characterization* (ed. A. J. MacGillivray and G. D. Birnie), pp. 74–85. Butterworths, London.
28. Wray, W. (1976). In *Biological separations in iodinated density gradient media* (ed. D. Rickwood), pp. 57–69. IRL Press, London.
29. Rickwood, D. and Ford, T. (1983). In *Iodinated density gradient media: a practical approach* (ed. D. Rickwood), p. 69. IRL Press, London.
30. Comings, D. E. (1978). In *Methods in cell biology* (ed. G. Stein, J. Stein, and L. J. Kleinsmith), Vol. XVII, pp. 115–32. Academic Press, London.
31. Sillar, R. and Young, D. B. (1981). *J. Histochem. Cytochem.*, **29**, 74.
32. Loening, U. E. and Baker, A. M. (1976). In *Subnuclear components: preparation and fractionation* (ed. G. D. Birnie), pp. 107–27. Butterworths, London.
33. Beebee, T. J. C. (1986). In *Nuclear structures: isolation and characterization* (ed. A. J. MacGillivray and G. D. Birnie), pp. 100–17. Butterworths, London.
34. Higashinakagawa, E., Muramatsu, M., and Sugano, H. (1972). *Exp. Cell Res.*, **71**, 65.
35. Higashinakagawa, T., Sezaki, M., and Kondo, S. (1979). *Dev. Biol.*, **69**, 601.
36. Fry, D. J. (1976). In *Subnuclear components: preparation and fractionation* (ed. G. D. Birnie), pp. 59–105. Butterworths, London.
37. Agutter, P. S. (1986). In *Nuclear structures: isolation and characterization* (ed. A. J. MacGillivray and G. D. Birnie), pp. 34–46. Butterworths, London.
38. Monneron, A., Blobel, G., and Palade, G. E. (1972). *J. Cell Biol.*, **55**, 104.
39. Kaufmann, S. H., Gibson, W., and Shaper, J. H. (1983). *J. Biol. Chem.*, **258**, 2710.
40. Kay, R. R., Fraser, D., and Johnston, I. R. (1972). *Eur. J. Biochem.*, **30**, 145.
41. Harris, J. R. and Milne, J. F. (1974). *Biochem. Soc. Trans.*, **2**, 1251.
42. Bornens, M. and Courvalin, J. C. (1978). *J. Cell Biol.*, **76**, 191.
43. Comerford, S. A., Agutter, P. S., and McLennan, A. G. (1986). In *Nuclear structures: isolation and characterization* (ed. A. J. MacGillivray and G. D. Birnie), pp. 1–13. Butterworths, London.
44. Berezney, R. and Coffey, D. S. (1974). *Biochem. Biophys. Res. Commun.*, **60**, 1410.
45. Long, B. H., Huang, C.-Y., and Pogo, A. O. (1979). *Cell*, **18**, 1079.
46. Kaufmann, S. H., Coffey, D.S., and Shaper, J. H. (1981). *Exp. Cell Res.*, **132**, 105.
47. Cook, P. R. and Brazell, I. A. (1975). *J. Cell Sci.*, **19**, 261.
48. Levin, J. M., Jost, E., and Cook, P. R. (1978). *J. Cell Sci.*, **29**, 103.
49. Hewish, D. R. and Burgoyne, L. A. (1973). *Biochem. Biophys. Res. Commun.*, **52**, 504.
50. Panyim, S., Jensen, R. H., and Chalkley, R. (1968). *Biochim. Biophys. Acta*, **160**, 255.
51. Renz, M., Nehls, P., and Hozier, J. (1977). *Proc. Natl. Acad. Sci. USA*, **74**, 1879.

52. Butler, P. G. J. and Thomas, J. O. (1980). *J. Mol. Biol.*, **140**, 505.
53. Bates, D. L., Butler, P. G. J., Pearson, E. C., and Thomas, J. O. (1981). *Eur. J. Biochem.*, **119**, 469.
54. Allan, J., Rau, D. C., Harborne, N. R., and Gould, H. J. (1984). *J. Cell Biol.*, **98**, 1320.
55. McGhee, J. D., Rau, D. C., Charney, E., and Felsenfeld, G. (1980). *Cell*, **22**, 87.
56. Narayan, K. S. and Birnstiel, M. L. (1969). *Biochim. Biophys. Acta*, **190**, 470.
57. Knowler, J. T., McGregor, C. W., and Islam, Z. (1986). In *Nuclear structures: isolation and characterization* (ed. A. J. MacGillivray and G. D. Birnie), pp. 118–29. Butterworths, London.
58. MacGillivray, A. J. (1986). In *Nuclear structures: isolation and characterization* (ed. A. J. MacGilivray and G. D. Birnie), pp. 130–62. Butterworths, London.
59. Spedding, G. (ed.) (1990). *Ribosomes and protein synthesis: a practical approach.* IRL Press at Oxford University Press, Oxford.
60. Higgins, S. J. and Hames, B. D. (ed.) (1993). *RNA processing: a practical approach*, Vol. I. IRL Press at Oxford University Press, Oxford.
61. Higgins, S. J. and Hames, B. D. (ed.) (1993). *RNA processing: a practical approach*, Vol. II. IRL Press at Oxford University Press, Oxford.
62. Johns, E. W. (1964). *Biochem. J.*, **92**, 55.
63. Johns, E. W. (1976). In *Subnuclear components: preparation and fractionation* (ed. G. D. Birnie), pp. 187–208. Butterworths, London.
64. Spring, T. G. and Cole, R. D. (1977). In *Methods in cell biology* (ed. G. Stein, J. Stein, and L. J. Kleinsmith), Vol. XVI, pp. 227–40. Academic Press, London.
65. Rickwood, D. and MacGillivray, A. J. (1975). *Eur. J. Biochem.*, **51**, 593.
66. Goodwin, G. H., Walker, J. M., and Johns, E. W. (1978). In *The cell nucleus* (ed. H. Busch), Vol. VI, pp. 181–219. Academic Press, London.
67. Kirby, K. S. (1996). In *Methods in enzymology* (ed. L. Grossman and K. Moldave), Vol. XIIB, pp. 87–92. Academic Press, London.
68. Jones, P. G., Qiu, J., and Rickwood, D. (1994). *RNA: isolation and analysis.* Bios Scientific Publishers, Oxford.
69. Qureshi, S. A. and Jacobs, H. T. (1993). *Nucleic Acids Res.*, **21**, 811.
70. Wilkinson, M. (1991). In *Essential molecular biology: a practical approach* (ed. T. A. Brown), Vol. I, pp. 69–87. IRL Press at Oxford University Press, Oxford.
71. Chomczynski, P. and Saachi, N. (1987). *Anal. Biochem.*, **162**, 156.
72. Baulcome, D. C. and Buffard, D. (1983). *Planta*, **157**, 493.
73. Blin, N. and Stafford, D. W. (1976). *Nucleic Acids Res.*, **3**, 2303.
74. Yonezawa, A. and Sugiura, Y. (1994). *Biochim. Biophys. Acta*, **1219**, 369.
75. Latham, K. A., Dodson, M. L., and Lloyd, R. S. (1994). *J. Cell. Biochem.* S18c sic:148 (meeting abstract).
76. Murphy, G. and Kavanagh, T. A. (1988). *Nucleic Acids Res.*, **16**, 5198.
77. Feist, P. L. and Danna, K. J. (1981). *Biochemistry*, **20**, 4243.
78. Gilmore, R. S. (1984). In *Transcription and translation: a practical approach* (ed. B. D. Hames and S. J. Higgins), pp. 131–52. IRL Press at Oxford University Press, Oxford.
79. Soh, J. and Pestka, S. (1993). In *Methods in enzymology* (ed. R. Wu), Vol. 216, p. 186. Academic Press, London.

4

Subcellular fractionation of mitochondria

J. E. RICE and J. G. LINDSAY

1. Introduction

Mitochondria are ubiquitous organelles which are present in the cytoplasm of the vast majority of animal and plant cells. They are also one of the best characterized organelles and indeed were the first membrane-bound particles to be isolated in an intact functional state from living tissues. Together with chloroplasts, they are unique in containing the only sites of extranuclear DNA, both possessing small, separate genomes which are capable of independent transcription and translation.

In 1890, Altman carried out a series of cytological studies and observed, using the light microscope, subcellular bodies or organelles which he postulated to be the basic unit of cellular activity. Earlier in 1886, MacMunn discovered the presence of cytochromes and proposed that these were haem-containing compounds functioning in the transfer of oxygen. It took another 40 years, however, before the work of Keilin who, by studying the oxidation–reduction and spectral properties of cytochromes in aerobic cells, confirmed their involvement in respiration. It is now widely accepted that there are several differing types of cytochromes in the mitochondrial inner membrane which are major constituents of the mitochondrial respiratory chain.

The identification of the mitochondrion as a locus for energy metabolism in the cell was pioneered through the early work of Claude (1). By applying the technique of differential centrifugation to cell homogenates, he succeeded in separating large and small subcellular granules. On studying the enzymatic characteristics of the isolated granules he concluded that the large granules were mitochondria since this fraction possessed both cytochrome *c* oxidase and succinate oxidase activities. However, it is now evident that the saline medium employed in his purification process caused extensive damage to mitochondrial ultrastructure. A major breakthrough was achieved when sucrose-containing media were adopted to provide osmotic support for the isolated organelles. The ability to isolate relatively pure suspensions of intact mitochondria, permitted analysis of their involvement in complex biochemical processes for the first time. The identification of a major role for mitochondria in cellular energy metabolism was demonstrated by Kennedy and Lehninger (2) who showed that the enzymes involved in fatty acid oxidation and the citric acid cycle, were localized in the mitochondrial matrix. Moreover, these isolated organelles were found to be capable of catalysing the complete process of oxidative phosphorylation, namely the oxidation of the products of fat and carbohydrate metabolism linked to the concomitant generation of ATP.

The dimensions of isolated mitochondria are sufficiently large (1.0–2.0 μm long: 0.5–1.0 μm wide), to permit their detection under the light microscope in appropriately stained cells. However, the advent of the electron microscope, through studies performed initially by Palade (3), prompted more detailed examination of mitochondrial ultrastructure. *Figure 1* shows an electron micrograph of mitochondria in a typical mammalian liver cell. The organelle is bounded by a double membrane comprising a smooth, featureless outer membrane and a highly convoluted inner membrane separated from each other by the intermembrane space. The outer membrane is freely permeable to small molecules owing to the presence of porin, an oligomeric protein, forming relatively non-selective channels in the bilayer which permits the passage of small molecules of M_r value less than 5000. The inner membrane, which borders the matrix compartment, adopts a highly invaginated formation (cristae) which serves to increase the surface area, thereby enhancing the respiratory capacity of the organelle by permitting the incorporation of more respiratory chain assemblies and ATP synthetic machinery into the lipid bilayer. In contrast to the outer membrane, the inner membrane is impermeable to the passage of most molecules, especially protons, except through the action of specific carriers. This is particularly important in maintaining the electrochemical gradient required for oxidative phosphorylation (see Section 5.2). *Figure 2* details a schematic representation of the mitochondrion and shows some of the metabolic processes which are catalysed within this organelle.

The overall aim of this chapter is to provide a practical guide to some of the more general methods employed in the routine isolation and standard analysis of mitochondrial function. For a comprehensive review of the literature the reader is directed to a more extensive treatment of this topic (4).

2. Purification of mitochondria from various eukaryotic sources

2.1 Introduction

In principle, it is possible to obtain samples of isolated mitochondria from a variety of diverse cell types and sources by the application of differential cen-

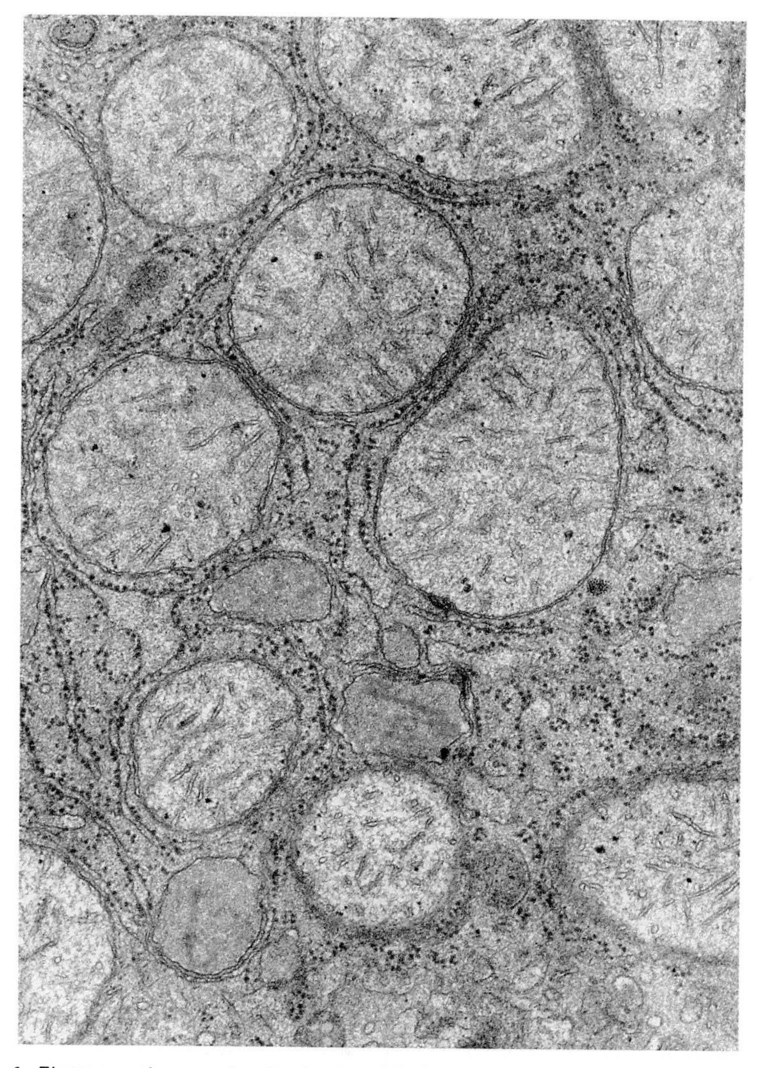

Figure 1. Electron micrograph of mitochondria in mammalian liver. Note the double membrane system and cristae formed by infoldings of the inner membrane. Tissue was fixed in Millonig's phosphate-buffered glutaraldehyde and osmium fixative, and lead stained. By courtesy of Dr Ian Montgomery, Institute of Biomedical and Life Sciences, University of Glasgow, UK.

trifugation procedures to prepared homogenates. In practice however, it is not feasible to apply a single isolation procedure uniformly: this necessitates that techniques be devised to suit individual circumstances as dictated by the nature of the tissues which are to be fractionated. Several factors which can influence the success or failure of the isolation procedure must be considered; for example, the method of tissue disruption depending on cell type, the most

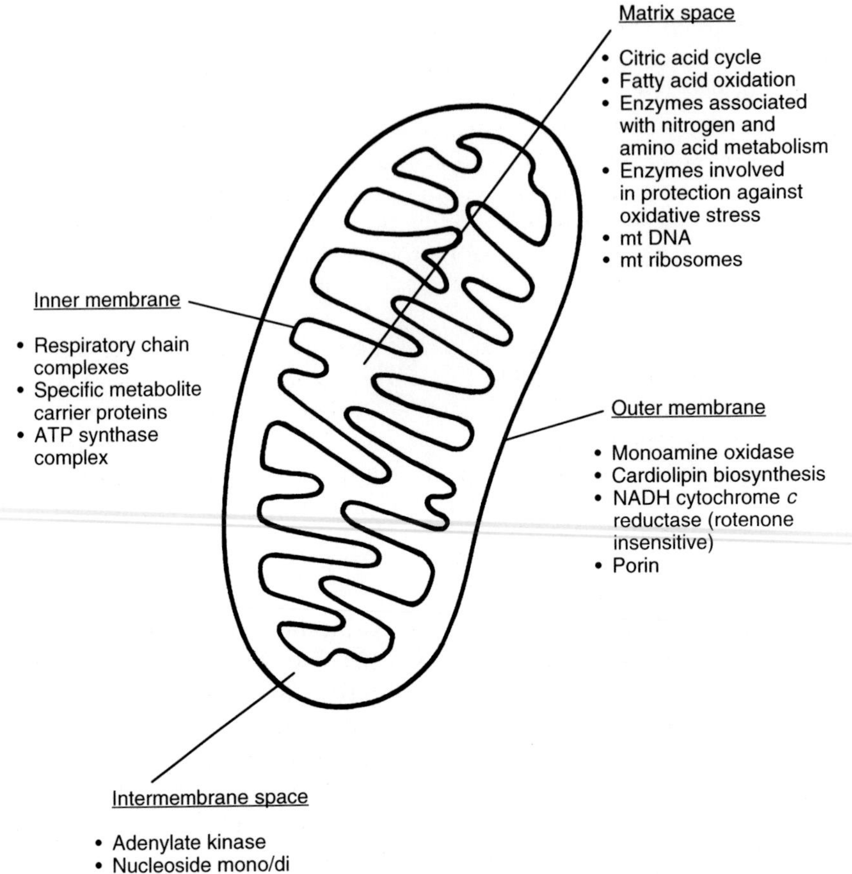

Figure 2. Schematic representation of a 'typical' mitochondrion. The figure describes some of the common enzyme activities associated with particular subregions of the mitochondrial structure. mt = mitochondrial.

appropriate buffers, and any further additions (Section 2.1.1) which may be necessary to optimize the isolation medium. The following sections are intended to provide guidelines for the purification of mitochondria from a variety of sources. However, as is true of most experimental procedures, these methods may be adapted to suit a particular purification protocol.

2.1.1 Isolation medium

The sedimentation behaviour of isolated mitochondria and their metabolic properties are affected by several basic properties of the suspending fluid during purification such as its pH, tonicity, and chemical nature. The principal osmotic support can be either of an ionic or non-ionic nature.

i. Non-ionic media

Non-ionic media include sucrose, mannitol, or sorbitol (mixtures of these can also be used). Molarities in the order of 250 mM are used for most mammalian tissues, with isotonic sucrose being the routine fractionation medium for the isolation of mitochondria from liver and other sources. The use of sucrose and other non-electrolyte solutions (mannitol), improves the purity of mitochondria obtained as it is thought to promote their separation from other structural elements of the cell cytoskeleton.

ii. Ionic media

Examples of ionic media include potassium chloride (approx. 100–150 mM) which can be used for tissues that may assume a viscous consistency on homogenization. It should be borne in mind, however, that high ionic strength of buffers may cause dissociation of peripheral proteins from the mitochondria such as cytochrome *c* or creatine kinase. Occasionally, so-called 'physiological' buffers such as potassium aspartate (approx. 160 mM) can be substituted.

iii. Buffers

The pH of isolation media used in the purification of mitochondria is normally close to physiological values, i.e. pH 7.5. The presence of suitable buffers is desirable to minimize large alterations in pH and control acid formation during purification which results from lactate production as a result of glycolysis in concentrated anaerobic homogenates. Buffers employed on a regular basis include MOPS (3-*N*-morpholinopropanesulfonic acid), Hepes (*N*-2-hydroxyethyl piperazine-*N*′-2-ethanesulfonic acid), and triethanolamine with molarities usually in the range 5–25 mM.

iv. Other additions

Binding agents are routinely added to isolation media to control the presence of contaminating ions which can act as enzymatic inhibitors or activators. For example, EDTA (1.0–0.5 mM) or EGTA (1.0–0.5 mM) are included to chelate free calcium which can be transported into the organelle in an energy-linked fashion inducing mitochondrial swelling or act as a cofactor for specific phospholipases, e.g. A_2, the activity of which damages the integrity of the lipid bilayer. EGTA is the preferred chelating agent as it is the weaker ligand and more specific for Ca^{2+} ions. Inclusion of EDTA in the medium can lead to damage to the inner mitochondrial membrane by chelating tightly bound Mg^{2+} ions which are essential to its integrity.

Bovine serum albumin (0.1–1.0 mg/ml) is also a standard ingredient in most media because of its ability to bind endogenous fatty acids, acyl CoA esters, lysophospholipids, or other compounds with detergent-like properties.

Dithiothreitol, cysteine, 2-mercaptoethanol, or other sulfhydryl compounds (0.1–2.0 mM) are often included to protect certain enzymes and integral membrane proteins with reactive sulfhydryl groups which are essential to their overall activity.

Anticoagulants such as heparin or polyethylene sulfonate (PES) can be added to prevent the development of gelatinous consistency and agglutination of organelles in non-ionic media. This particular problem may need to be addressed in the purification of mitochondria from muscle and brain tissue.

2.1.2 Tissue disruption

The method used for tissue disaggregation is dictated both by the tissue type and the nature of the intracellular connections. A few examples of methods available are given below.

i. Soft vertebrate tissue

Tissues of this type (e.g. brain, kidney) can be easily homogenized using a Potter–Elvehjem glass–Teflon homogenizer (approx. 1000 rev/min) or a Dounce hand homogenizer.

ii. Tough vertebrate tissues

Inclusion of a protease is often necessary to promote breakdown of the cellular structure (e.g. heart). Commonly used proteases include trypsin, collagenase, or Nagarse, a mixture of bacterial proteolytic enzymes; however, problems can arise with these procedures as it is often difficult to achieve the correct level of enzymic degradation. Where a tissue is either high in fat, rich in endogenous proteases, or requires pre-incubation with degradative enzymes to promote tissue disaggregation, rapid separation of the organelles from the starting homogenate is essential at this stage to minimize damage to the mitochondria.

In addition to chemical means of disrupting cellular structure, mechanical methods are sometimes employed. Mincers and blenders such as the Waring commercial blender or the Polytron tissue blender are often used in obstinate cases where large amounts of connective material are present; glass beads can be included in the homogenizing mixture for particularly difficult cases. Cell presses such as the French press can be used to break open cells with rigid cell walls by extruding them through a narrow aperture under high pressure. However, these regimes, in particular the French press, have a tendency to cause extensive disruption of intracellular structures, including mitochondria, so should be employed with caution under strictly controlled conditions. For more information on homogenization procedures see Chapter 1.

2.2 Protocols for purification of mitochondria from several eukaryotic sources

2.2.1 Mitochondrial isolation from rat liver

Rat liver is commonly used in many biochemical studies as it is readily available and can be homogenized immediately without prior treatment with degradative enzymes. Hepatocytes constitute 90% of liver volume; thus a single cell type dominates liver function. These cells are responsible for a diverse range of metabolic and biosynthetic activities and are rich in mitochondria (approx. 1000 per cell) which are estimated to represent 20% of total cellular protein.

Rat liver mitochondria are prepared by differential centrifugation according to the method of Chance and Hagihara (5), although it should be noted that several modifications of the original procedure now exist including the use of Percoll density gradients (6). This modification may be useful in studies which require rapid preparation of functionally intact and relatively pure mitochondria. Rats are normally starved overnight prior to the experiment to lower the levels of endogenous fatty acids. Wherever possible, all steps of the purification should be performed at 0–4 °C.

Protocol 1. Purification of mitochondria from rat liver

Equipment and reagents

- Potter–Elvehjem homogenizer
- Dounce homogenizer
- High-speed centrifuge and rotors
- Isolation medium: 0.225 M mannitol, 0.075 M sucrose, 0.5 mM EGTA, 2 mM MOPS pH 7.4

Method

1. Starve two rats (150–200 g) overnight and kill by cervical dislocation. Remove the livers into ice-cold isolation medium, 0.225 M mannitol, 0.075 M sucrose, 0.5 mM EGTA, 2 mM MOPS pH 7.4. Wash briefly to remove any remaining traces of blood.
2. Add a small amount of fresh isolation medium to the livers after chopping finely using scissors. Transfer contents to a glass homogenizing vessel (50 ml capacity) and homogenize the livers in isolation medium (approx. 30 ml per liver), with a tight-fitting Potter–Elvehjem homogenizer (approx. 1000 rev/min). Six to ten passes with the homogenizer may be required to ensure maximal cell disruption although this will vary with cell type.
3. Distribute the liver homogenate equally to 4 × 50 ml polypropylene centrifuge tubes or equivalent, adding more isolation medium to fill the tubes if necessary. Nuclei, cell debris, and red blood cells are sedimented by centrifugation at 800 *g* for 7 min.
4. An additional low speed spin may be included at this stage, particularly if a large amount of red blood cell contamination is apparent. If this is the case, carefully decant the supernatant fractions into fresh 4 × 50 ml polypropylene tubes and recentrifuge at 800 *g* for 7 min.
5. After centrifugation decant the supernatant fluid carefully into fresh

Protocol 1. *Continued*

4 × 50 ml polypropylene tubes and spin at 6500 *g* for 15 min to sediment a crude mitochondrial pellet.

6. Resuspend the mitochondrial pellet in a small volume of isolation medium and disaggregate by hand (Dounce) homogenization. Distribute the suspension equally between 2 × 50 ml centrifugation tubes filled to capacity and centrifuge at 6500 *g* for 7 min.
7. Repeat step 6 once more using only 1 × 50 ml centrifuge tube for the sample.
8. Resuspend the final mitochondrial pellet in 3–5 ml of isolation medium (approx. 20–30 mg/ml).

2.2.2 Mitochondrial isolation from bovine heart

The use of heart tissue in preference to liver for the purification of mitochondria offers several distinct advantages:

(a) It is possible to obtain large amounts of mitochondria (1.5–1.8 g mitochondrial protein from 300 g tissue).

(b) Mitochondria prepared from heart tend to be more tightly 'coupled' and can be stored for longer periods owing to the lower background levels of contaminating proteases and the slower release of fatty acids.

(c) Heart mitochondria are convenient sources for purifying components of the electron transport chain and ATP synthetic machinery which are three- to fourfold more abundant than in liver mitochondria.

The method described is a modification of the original protocol as described by Smith (7) where all steps should be performed at 0–4°C. As always, it is a case of quality versus quantity, this method represents a large scale purification of mitochondria of reasonable quality.

Protocol 2. Purification of mitochondria from beef heart

Equipment and reagents

- Commercial mincer
- Muslin cloth
- Waring blender
- High-speed centrifuge and rotors
- 250 mM sucrose, 10 mM Tris–HCl pH 7.8
- 2 M Tris base
- TESS: 250 mM sucrose, 1 mM succinate, 0.2 mM EDTA, 10 mM Tris–HCl pH 7.8

Method

1. Fresh heart used for this preparation should be carefully cleaned of all fat and connective tissue and cut into 3–5 cm cubes. The diced

tissue (600 g) is passed once through a commercial mincer which is pre-cooled to and maintained at 4 °C.

2. Add the minced meat to 800 ml of buffer containing 250 mM sucrose, 10 mM Tris–HCl pH 7.8.
3. While stirring the minced suspension, adjust pH of the solution to approx. 7.8 by addition of 2 M Tris base.
4. Strongly squeeze the minced tissue through a double layer of muslin cloth to recover as much of the sucrose solution as possible.
5. Resuspend the minced tissue in 800 ml of buffer containing 250 mM sucrose, 1 mM succinate, 0.2 mM EDTA, 10 mM Tris–HCl pH 7.8 (TESS).
6. Pour half of this solution into a Waring commercial blender and blend at high speed for 20 sec. Take the blended material and re-adjust the pH to 7.8 using 2 M Tris base. Return to the blender and homogenize at high speed for a further 60 sec.
7. Repeat step 6 with the other half of the minced material.
8. Pool the two portions of blended heart tissue (obtained from steps 6 and 7), and dilute with fresh TESS medium to a volume not greater than 2.4 litres.
9. Centrifuge this suspension in 250 ml centrifuge buckets at 800 *g* for 20 min.
10. Remove the supernatant fraction containing the mitochondria and centrifuge in polypropylene tubes at 26 000 *g* for 20 min.
11. Take the crude mitochondrial pellets and resuspend in 300 ml fresh TESS buffer. Centrifuge at 26 000 *g* for 20 min.
12. Resuspend the resultant mitochondrial pellet in approx. 20–30 ml of fresh TESS buffer.

2.2.3 Isolation of mitochondria from rat brain

This method involves the isolation of non-synaptosomal mitochondria from rat brain by centrifugation using discontinuous Percoll density gradients (8). Previous methods result in an overall yield of mitochondria between 3–11% of the total content (9). Application of these techniques to a single brain or subregion of brain, often results in lower recoveries being achieved. The protocol detailed below is for studies using at least 500 mg of tissue and yields a relatively pure preparation of mitochondria free from contaminating synaptosomes. A slight modification of the method also exists which is more rapid and is particularly suitable for obtaining mitochondria which are well 'coupled', i.e. actively phosphorylating, but are likely to be slightly contaminated with synaptosomes.

Protocol 3. Purification of mitochondria from rat brain

Equipment and reagents

- Dissection equipment
- Dounce homogenizer
- High-speed centrifuge and rotors
- Pasteur pipettes
- 23 gauge needle and syringe
- Isolation medium: 0.32 M sucrose, 1 mM EDTA (K^+), 10 mM Tris–HCl pH 7.4
- 40%, 23% and 15% (v/v) Percoll in isolation medium
- Fatty acid-free bovine serum albumin

Method

All steps performed at 0–4 °C.

1. Starve rats (150–200 g) overnight and kill next day by decapitation. Rapidly remove the cerebellum and underlying structures, and use the remaining material as forebrain.
2. Place dissected brain into ice-cold isolation medium containing 0.32 M sucrose, 1 mM EDTA (K^+ salt), 10 mM Tris–HCl pH 7.4.
3. Wash tissue three times with isolation buffer and mince finely using scissors in a small amount of the same buffer.
4. Hand homogenize the tissue (10%, w/v) using a Dounce homogenizer. Perform five up-and-down strokes using a loose-fitting pestle followed by a further ten strokes using a tighter-fitting pestle.
5. Centrifuge homogenate at 1330 *g* for 3 min. After centrifugation decant and keep the supernatant fluid, resuspending the pellet in half the original volume of isolation buffer using a Dounce hand homogenizer.
6. Centrifuge the resuspended material at 1330 *g* for 3 min. Keep the supernatant fraction and discard the pellet.
7. Pool the supernatant fractions from steps 5 and 6 and centrifuge at 21 200 *g* for 10 min.
8. Discard the supernatant and resuspend the pellet in 15% (v/v) of Percoll™, prepared in isolation buffer, such that there are 10 ml/g of original brain homogenate.
9. Prepare a discontinuous Percoll density gradient consisting of 3.5 ml 23% (v/v) Percoll layered above 3.5 ml 40% (v/v) Percoll: add the 40% (v/v) Percoll below the 23% Percoll (v/v) using a 23 gauge needle and a disposable syringe. Use a swinging-bucket rotor.
10. Layer 3 ml of the mitochondrial pellet in 15% (v/v) Percoll on to the discontinuous density gradient and centrifuge at 30 700 *g* for 5 min.
11. Three major bands are obtained: take the material which bands near the interface of the lower two Percoll layers and dilute 1 : 4 by gently mixing with isolation medium. Centrifuge at 16 700 *g* for 10 min.

12. Remove the supernatant fluid using a Pasteur pipette, leaving a few millimetres of medium above the pellet and gently resuspend the pellet. Take care when removing the supernatant fraction as the pellet may not be firm owing to the presence of Percoll which will have been carried over from the gradient.
13. Add fatty acid-free bovine serum albumin (10 mg/ml) equivalent to 0.5 ml/forebrain originally used. Dilute suspension with isolation medium (3 ml/forebrain) and centrifuge at 6900 *g* for 10 min.
14. Rapidly decant the supernatant and gently resuspend the pellet in 0.3 ml of original isolation buffer.

2.2.4 Mitochondrial isolation from tissue culture cells

This method is adapted from that of Attardi and Ching (10) to obtain crude mitochondrial and post-nuclear supernatant fractions, starting from approx. 2×10^8 cells.

Protocol 4. Purification of mitochondria from tissue culture cells

Equipment and reagents

- Plastic Roux flasks
- Rubber policeman
- High-speed centrifuge and rotors
- Potter–Elvehjem homogenizer
- Phosphate-buffered saline
- 0.15 mM $MgCl_2$, 10 mM KCl, 10 mM Tris–HCl pH 6.7
- Sucrose
- 0.15 mM $MgCl_2$, 0.25 M sucrose, 10 mM Tris–HCl pH 6.7
- 0.25 M sucrose, 10 mM Tris–acetate pH 7.0

Method

1. Grow cell monolayers in plastic Roux flasks until they reach confluence and wash twice with 2 × 20 ml of phosphate-buffered saline (0.15 M NaCl, 20 mM KH_2PO_4 adjusted to pH 7.4 using KOH).
2. Scrape the cells from each flask using a rubber 'policeman' into 20 ml (per flask) of phosphate-buffered saline (see above).
3. Pool the cells from three to four flasks and pellet by centrifugation at 1500 *g* for 15 min.
4. Resuspend the pellet in approx. 6 vol. of 0.15 mM $MgCl_2$, 10 mM KCl, 10 mM Tris–HCl pH 6.7.
5. Place the resuspended cells on ice and incubate for 2 min. Homogenize using a Potter–Elvehjem Teflon/glass homogenizer (approx. 500 r.p.m.). Perform five up-and-down passes using the homogenizer: cell breakage can be confirmed by examining the suspension under a light microscope.

6. Add sucrose to a final concentration of 0.25 M and centrifuge at 1500 *g* for 3 min to remove nuclei.
7. Take the supernatant fraction from step 7 and pellet the mitochondria by centrifugation at 5000 *g* for 10 min. Resuspend the resultant pellet in 2 ml of 0.15 mM $MgCl_2$, 0.25 M sucrose, 10 mM Tris–HCl pH 6.7.
8. Repeat step 8 before resuspending the final mitochondrial pellet in approx. 1 ml of 0.25 M sucrose, 10 mM Tris–acetate pH 7.0.

2.2.5 Mitochondrial isolation from *Saccharomyces cerevisiae*

Mitochondria were isolated by adapting slightly the method of Rickwood *et al.* (11).

Protocol 5. Purification of mitochondria from yeast

Equipment and reagents

- High-speed centrifuge and rotors
- Dounce homogenizer
- Distilled water
- 10 mM DTT, 0.1 M Tris–HCl pH 9.3
- Zymolase
- 1.2 M sorbitol, 20 mM KH_2PO_4 pH 7.4 (with KOH)
- Breaking buffer: 0.6 M mannitol, 1 mM PMSF, 20 mM Hepes–KOH pH 7.4
- 0.6 M mannitol, 20 mM Hepes–KOH pH 7.4

Method

1. Grow wild-type yeast strains (e.g. D273-10B or KL14-4A) to early or mid-log phase. Take care not to grow the yeasts to late log or near stationary phase since the development of a thick cell wall inhibits their disruption.
2. Harvest yeast cells by centrifugation at 1000 *g* for 10 min at 4 °C.
3. Wash the cells once in distilled water and resuspend in 2 vol. of buffer containing 10 mM dithiothreitol, 0.1 M Tris–HCl pH 9.3.
4. Incubate the cells for 30 min at 30 °C, then centrifuge the cells at 1000 *g* for 10 min.
5. Wash the pelleted cells once in buffer containing 1.2 M sorbitol, 20 mM KH_2PO_4 adjusted to pH 7.4 using KOH.
6. Resuspend the washed cells to 0.15 g wet weight/ml of the sorbitol-containing buffer used in step 5. Digest the cell walls of the yeast cells using yeast lytic enzyme (0.25 mg zymolase[a]/100 g wet cell weight). Extent of spheroplast formation can be checked by addition of SDS (0.1%, w/v) and counting the remaining intact yeast cells under the light microscope.
7. After a 1 h incubation at 30 °C, pellet the spheroplasts by centrifugation at 1000 *g* for 10 min. Wash the pellet twice in 1.2 M sorbitol, 20 mM KH_2PO_4 adjusted to pH 7.4 using KOH.

Note: It is important that all subsequent steps are carried out at 4°C.

8. Resuspend spheroplasts in 2 vol. of breaking buffer containing 0.6 M mannitol, 1 mM PMSF, 20 mM Hepes–KOH pH 7.4. Homogenize (by hand) using a Dounce homogenizer for a maximum of ten up-and-down strokes.

9 Add 1 vol. of breaking buffer to the homogenate and centrifuge at 1000 *g* for 10 min to remove nuclear and cellular contaminants.

10. Take the supernatant fraction from the previous step and centrifuge at 6500 *g* for 15 min to pellet the mitochondria. Wash the resultant pellet once in buffer containing 0.6 M mannitol, 20 mM Hepes–KOH pH 7.4 to remove PMSF before gently resuspending in the same buffer.

[a] Zymolase-100T from *Arthrobacter luteus* 100 000 U/g.

2.3 Further purification of mitochondrial fractions

The procedures outlined above normally provide a mitochondrial pellet of sufficient purity for most experimental applications. However, in certain circumstances it may be desirable to obtain mitochondria of higher purity. These are often obtained by using the technique of buoyant density centrifugation using either sucrose, Ficoll, or Percoll for gradient formation. In general, sucrose gradients should not be used since high concentrations are required exposing the mitochondria to hyperosmotic conditions capable of damaging the organelle. In addition, the increase in viscosity necessitates much longer centrifugation times. In such situations, the use of Ficoll a non-ionic sucrose polymer is recommended as it is capable of high densities with small changes in osmolarity and viscosity. Percoll is gradually superceding the use of Ficoll in mitochondrial purification schemes as its chemical properties appear to create favourable conditions for allowing the purification of pure, actively-phosphorylating mitochondria from a number of different tissues.

3. Determination of mitochondrial purity

3.1 Introduction

Once mitochondria have been prepared from a particular source, it is often important to make an assessment of their relative purity and functionality. Several methods exist such as the use of electron microscopy, purity of mitochondrial DNA, and the level of marker enzymatic activities from contaminating subcellular fractions. Electron microscopy is an involved process which requires skilled operators and access to sophisticated equipment. Alternatively, DNA purified from mitochondria can be digested using certain restriction endonucleases to produce characteristic restriction maps. Any contamination by nuclear DNA is clearly evident (12). A simpler approach to determining purity is to assay mitochondrial preparations for enzyme activities such as catalase or uricase, acid phosphatase, glucose-6-phosphatase or NADPH-cytochrome *c* reductase and 5′-nucleotidase which indicate the degree of contamination of the preparation by peroxisomes, lysosomes, endoplasmic reticulum, and plasma membrane respectively. The following section details some of the routine techniques commonly used for assessing the state of a purified mitochondrial suspension.

3.2 Use of the oxygen electrode to determine mitochondrial integrity

Oxidative phosphorylation is the name given to the process by which ATP is formed as reducing equivalents are transferred from the reduced cofactors NADH and FADH to molecular oxygen via a series of electron carriers (see Section 5.2). It follows that in mitochondria the processes of respiration and ATP production are very tightly coupled under normal conditions.

Respiratory control refers to the phenomenon whereby the rate of respiration is controlled by the availability of ADP in the cell or more specifically by the overall phosphate potential, defined as the ratio of [ATP]/[ADP]+[P_i]. At high phosphate potential, the availability of ADP and P_i is extremely limited, so that in well coupled mitochondria the respiratory rate of any given substrate is very low under these conditions: this state of respiration is often referred to as state four respiration. On the addition of ADP (plus P_i), the mitochondria rapidly enter active state three respiration, typified by a marked increase in the rate of oxygen uptake. If similar experiments are performed using 'uncoupled' mitochondria where the integrity of the inner membrane to protons has been compromised by physical damage or chemical agents (uncouplers) affecting its permeability to protons, respiration proceeds at maximal uncontrolled rates which are unresponsive to the presence of adenine nucleotides. These aspects of mitochondrial function can be assessed by measuring the uptake of dissolved oxygen from a solution by an isolated mitochondrial suspension using an oxygen electrode apparatus. A schematic representation of the standard model of apparatus employed is illustrated in *Figure 3*.

The apparatus consists of a platinum electrode which is in electrical contact with a silver wire immersed in a saturated solution of potassium chloride. A potential difference (−0.6 V) is applied from an external source across these electrodes such that the platinum electrode becomes negative with respect to the silver electrode. In this situation, a current will flow only if oxygen is present in the surrounding medium as electrolytic reduction of O_2 to water occurs on the surface of the platinum electrode. Under defined temperature and pressure, it can be shown that the current flowing in the circuit is directly

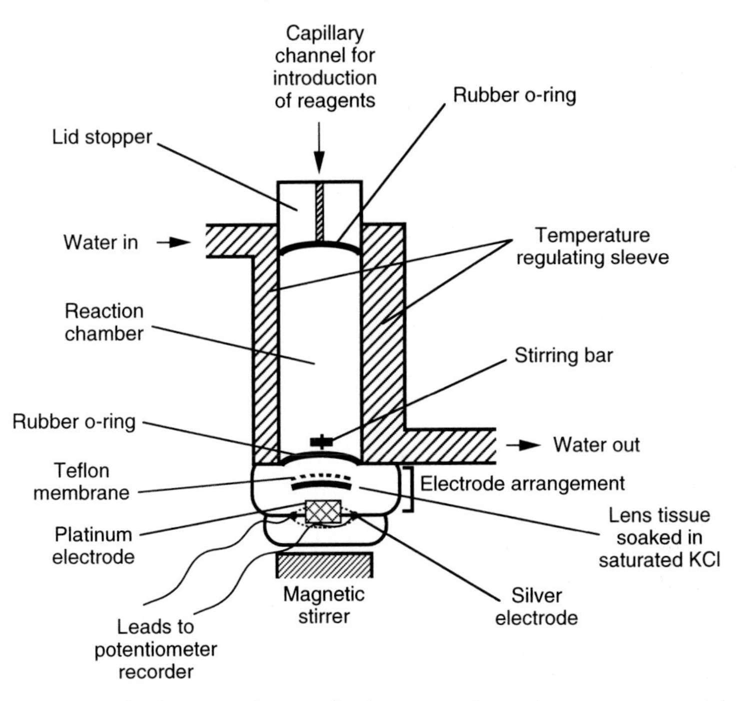

Figure 3. A schematic diagram of a standard oxygen electrode apparatus used for routine measurement of mitochondrial electron transport and calculation of P:O ratios. Substrates are added using a syringe via a capillary channel into the thermostatted reaction chamber. An electrode arrangement at the base of the apparatus measures the decline in current flow which is proportional to the concentration of dissolved O_2 during mitochondrial consumption.

proportional to the oxygen content of the solution. It is also necessary to protect the surface of the platinum electrode with a thin Teflon gas-permeable membrane which prevents its contamination by mitochondrial protein and assay reagents but provides ready access to molecular O_2.

3.2.1 Mitochondrial P:O ratios

The respiratory chain has three coupling sites corresponding to a region where H^+ ions are translocated across the inner membrane. These proton pumping or energy coupling sites provide the driving force for ATP production by generating a proton electrochemical gradient across the inner membrane. Thus the passage of protons through the ATP synthase complex is closely linked to ATP formation. The P:O ratio represents the number of moles of ATP generated per *atom* of oxygen consumed during respiration, i.e. the number of ATP molecules produced as a pair of electrons passes through the respiratory chain to oxygen, the terminal electron acceptor. High respiratory control ratios (ratio of state three to state four respiration) and

P:O ratios close to three and two for NAD- and FAD-linked substrates respectively, are indicative of high quality mitochondria. It follows that in cases where the mitochondrial membrane has been damaged, it is not possible to maintain the proton gradient across the inner membrane, state four respiration rates are increased, and there is a corresponding decline in respiratory control ratios and observed P:O ratios.

A practical guide to the use of the oxygen electrode in the calculation of P:O ratios is given below.

Protocol 6. Calculation of P:O ratios using the oxygen electrode

Equipment and reagents

- Oxygen electrode
- Glass pipette
- Hamilton syringe
- Distilled water
- Fatty acid-free bovine serum albumin
- Assay buffer: 0.22 M mannitol, 0.05 M sucrose, 10 mM NaH_2PO_4, 20 mM MOPS pH 7.2
- 1 M succinate
- 0.1 M ADP

Method

1. Wash the oxygen electrode chamber thoroughly with distilled water before use. Using a glass pipette, add 3.2 ml of mitochondrial assay buffer containing 0.22 M mannitol, 0.05 M sucrose, 10 mM NaH_2PO_4, 20 mM MOPS buffer pH 7.2. The buffer should be heated to 25°C and pre-equilibrated with air prior to use.
2. Add 0.3 ml of fatty acid-free bovine serum albumin (5 mg/ml) and establish a level baseline on the chart recorder prior to the addition of 0.1–0.2 ml of mitochondrial suspension (approx. 20 mg protein/ml). The amount of mitochondria required is dependent on their source, degree of purity, and the substrate under investigation so this volume can be adjusted as necessary. Insert the sealing lid on to the electrode chamber ensuring that no air bubbles remain above the mitochondrial suspension. A small capillary opening in the lid permits the introduction of substrates, inhibitors, and other reagents by means of a Hamilton syringe. Once mitochondria have been added to the electrode, it is essential to work rapidly as they will deteriorate if left in the electrode for prolonged periods.
3. Using a Hamilton syringe, add 50 μl of succinate (1 M stock). A small amount of oxygen consumption may be observed prior to substrate addition, owing to the presence of endogenous substrates within the mitochondrion. Allow a linear rate of respiration to be established before adding 10 μl of ADP (0.1 M). If the mitochondria are of good quality a dramatic increase in O_2 consumption will be observed. Allow trace to return near the original respiration rate before adding an identical aliquot of ADP (see *Figure 4*).

4. Repeat additions of ADP at appropriate intervals until solution becomes anaerobic, i.e. when the rate of respiration drops to zero and further ADP additions are ineffective.

A trace obtained for an experiment of this type is shown in *Figure 4*.

The oxygen content of the assay buffer used in this experiment has previously been calculated and is generally accepted to contain 250–270 nmoles O_2/ml. If the amount of assay buffer in the electrode is known, then by measuring the amount of oxygen consumed in phosphorylating a known amount of ADP, the P:O ratio can be calculated.

$$\text{P:O ratio} = \frac{\text{moles ADP}}{\text{moles atomic oxygen consumed}} = \frac{\text{P}}{\text{O}}$$

where P refers to the number of moles of P_i esterified to ATP.

Note:

(a) The full scale deflection on the oxygen electrode trace represents the total amount of O_2 in 3.5 ml of assay buffer.

(b) It is assumed that all ADP added is converted to ATP when state three respiration recovers to the original state four rate.

(c) When calculating moles of atomic oxygen consumed it is necessary to double the amount of molecular O_2.

Figure 4A illustrates a typical oxygen electrode trace for an NAD-linked substrate while *Figure 4B* represents a similar trace for an FAD-linked

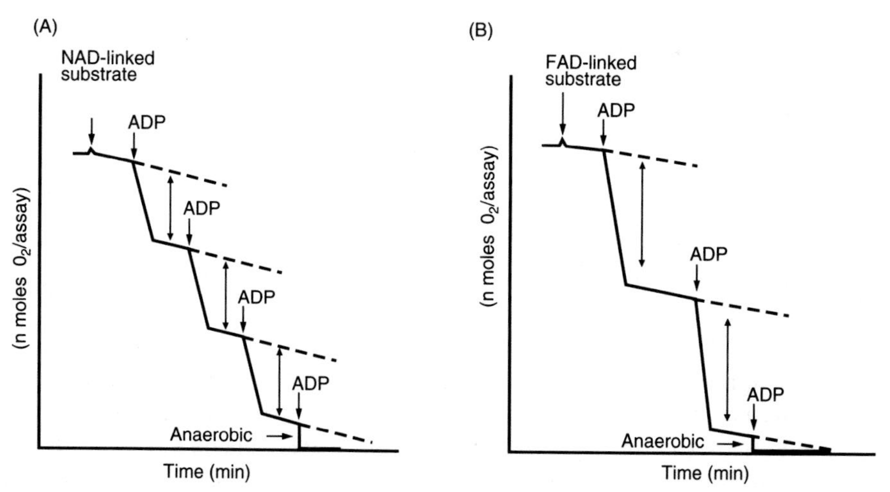

Figure 4. Oxygen electrode traces for calculation of P:O ratios where panels (A) and (B) represent traces for NAD-linked and FAD-linked substrates respectively. On addition of substrate to mitochondrial suspension a slight increase in respiration may be observed owing to the presence of endogenous substrates. Each cycle of state four to state three is initiated by addition of equivalent amounts of ADP until the solution becomes anaerobic.

substrate. Examples of NAD-linked substrates include pyruvate, malate, 2-oxoglutarate, and β-hydroxybutyrate (an intermediate of the fatty acid oxidation pathway): electrons derived from these substrates enter the respiratory chain at the level of complex I (NADH : ubiquinone oxidoreductase) (see *Figure 8*) and pass along the entire length of the respiratory chain to give a P:O ratio of approximately three. Succinate oxidation to fumarate is catalysed by succinic dehydrogenase, an FAD-linked enzyme which is embedded in the inner membrane such that electrons can enter the respiratory chain directly via ubiquinone, passing through the final two energy conserving sites of the electron transport pathway, resulting in the production of approximately two ATP molecules per oxygen atom reduced to water.

Graphical determination of P:O ratios as illustrated in *Figure 4* for any given substrate by the oxygen electrode technique is subject to a number of errors and limitations. The presence of significant amounts of respiration driven by endogenous substrates, normally products of fatty acid β-oxidation, can influence the final value. It is often the case that calculated P:O ratios are lower than the theoretical value. This can be caused by partial uncoupling which activates the latent ATPase activity of intact organelles or result from contamination of isolated mitochondria with other subcellular fractions with significant levels of ATPase which result in significant turnover of ATP.

3.2.2 Inhibitors of mitochondrial function

i. Inhibitors of energy transduction

In actively phosphorylating mitochondria, the ADP-stimulated respiration following substrate addition is accompanied by rapid generation of ATP. The antibiotic oligomycin, interacts directly with mitochondria by binding to the proton channel of the ATP synthase complex, resulting in inhibition of the ADP response and ATP production. Oligomycin-induced inhibition of state three respiration occurs indirectly as a result of the close links between respiration and phosphorylation. Since H^+ ions are unable to flow back through the ATP synthase complex, respiration decreases dramatically on account of a build up of the proton electrochemical gradient on the cytosolic side of the membrane. Routinely oligomycin (5 mg/ml) is added in 5 μl or 10 μl aliquots until an effect is observed.

Inhibition of respiration caused by oligomycin can be overcome by treatment of mitochondria with a class of compounds termed uncouplers which act to abolish respiratory control, resulting in uncontrolled maximal rates of respiration in the absence of ATP synthesis. The uncoupler depicted in *Figure 5A* is 2,4-dinitrophenol (DNP) a weak lipophilic anion capable of transporting H^+ ions across the inner mitochondrial membrane in its uncharged state, thereby increasing the proton conductance of the membrane and dissipating the electrochemical potential gradient. Rapid respiration ensues as mitochondria respond by attempting to maintain this gradient.

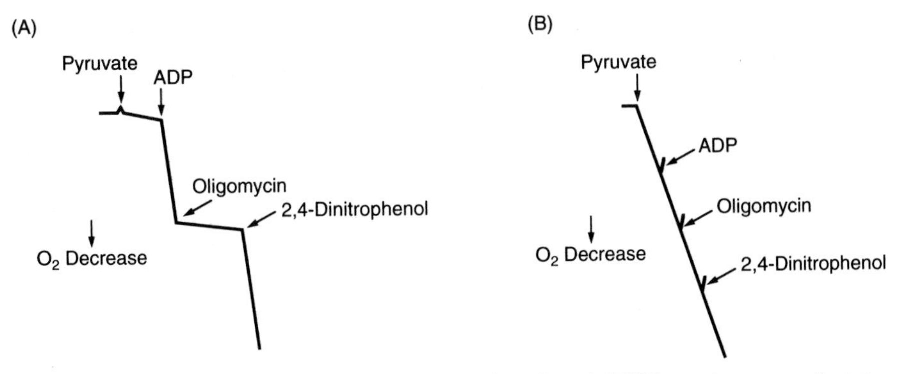

Figure 5. The effects of oligomycin and 2,4-dinitrophenol (DNP) on the rates of state three respiration in (A) coupled mitochondria, and (B) uncoupled mitochondria.

Routinely DNP (5 mM) in ethanol is added in 10 μl or 15 μl aliquots until an effect is observed.

Figure 5B illustrates the result of a similar study in non-phosphorylating (uncoupled) mitochondria which are unable to maintain a proton gradient. In this case, addition of ADP, oligomycin, or 2,4-dinitrophenol elicit no alterations in the rate of respiration, confirming that their mechanism of action is not directly on the electron transport pathway *per se*.

In summary, oligomycin and 2,4-dinitrophenol are classical inhibitors of energy transduction which can be readily distinguished from respiratory inhibitors by virtue of their effects on respiration only in coupled mitochondria where close linkage exists between the processes of respiration and ATP production.

ii. Inhibitors of electron transfer

Compounds belonging to this class of inhibitors are characterized by their direct inhibitory effect on the components of the electron transport chain. As a consequence, compounds in this class exert their effects on respiration either in actively phosphorylating or non-phosphorylating mitochondria. Inhibitors of respiration block electron flow at specific sites along the electron transport chain such that individual substrates which enter the respiratory pathway at different points may bypass the site of inhibition in some cases. This is seen in *Figure 6* where inhibition of oxygen consumption by the appropriate inhibitor is restored each time by the addition of a substrate with a lower P:O value (see Section 3.2.1) providing reducing equivalents to a site 'downstream' from the block in electron flow.

From *Figure 6* it is apparent that oxygen consumption as a result of pyruvate addition to the mitochondria (an NAD-linked substrate), is inhibited by rotenone or amobarbital which interrupt electron transfer within the site I region of the respiratory chain. However addition of succinate (an FAD-

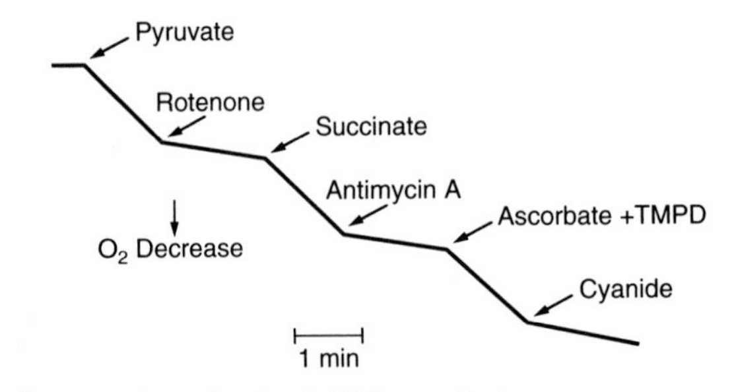

Figure 6. Effect of energy transduction inhibitors of mitochondrial respiration. Inhibition at various points in the electron transport chain can be overcome by addition of substrates which enter further down the sequence of electron carriers. Since cyanide inhibits at the terminal point of the sequence, respiration cannot be restimulated by any further additions.

linked substrate), which enters the chain at complex III is capable of bypassing the induced block. A similar situation exists further down the pathway where inhibition caused as a result of antimycin A addition (a complex III inhibitor), can be overcome by the inclusion of ascorbate and TMPD (*N,N,N′,N′*-tetramethylphenylenediamine) in the reaction medium. Ascorbate is a reducing agent which donates reducing equivalents directly to cytochrome *c* via the dye, TMPD as mediator. Cyanide, CO, and N_3^- inhibit electron flow at the level of cytochrome *c* oxidase (complex IV), the terminal region of the chain which also contains the site of oxygen interaction and as such its inhibitory effects cannot be overcome regardless of the type of substrate. Antimycin A, rotenone and amobarbital are all prepared at 1 mg/ml in ethanol and added to the oxygen electrode system in 10 μl aliquots until an effect is observed. Rapid response to the inhibitors is normally observed within one or two additions.

iii. Inhibition of substrate uptake

The uptake of various metabolites into mitochondria is promoted by the action of specific carrier proteins located in the mitochondrial membrane. Most of these carriers operate by exchange–diffusion mechanisms promoting the simultaneous electroneutral transfer of substrates of equal charge in opposite directions across the inner membrane by facilitated diffusion. In one or two cases, there is active, i.e. energy-dependent transport in which the membrane potential or proton gradient are utilized directly for the transport of specific substrates and ions, e.g. ATP/ADP exchange, Ca^{2+} and P_i uptake. Pyruvate which is a monocarboxylic acid also needs to be transported across the mitochondrial inner membrane, probably in exchange for hydroxyl ions.

As shown in *Figure 7*, α-cyano-4-hydroxycinnamic acid inhibits the respiration of coupled mitochondria by pyruvate. It is ineffective against other substrates suggesting either that a specific carrier protein is involved in the

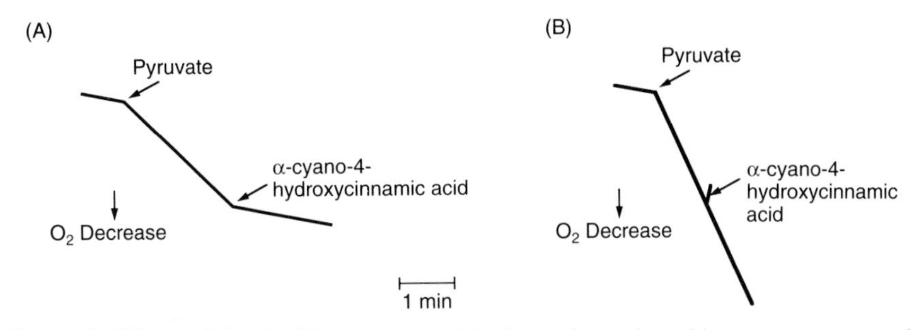

Figure 7. Effect of the inhibitor α-cyano-4-hydroxycinnamic acid on pyruvate uptake using (A) intact mitochondria, and (B) submitochondrial particles.

transport of pyruvate or that pyruvate oxidation by the pyruvate dehydrogenase multienzyme complex is affected directly. Identification of its site of action at the level of pyruvate uptake stems from its inability to block pyruvate-driven respiration in submitochondrial particles (mitochondrial inner membrane fragments), where the inverted orientation of the inner membrane provides direct access for the substrate to its site of oxidation. α-Cyano-4-hydroxycinnamic acid (a derivative of α-cyanocinnamate) is reported to inhibit respiration at 2.5 mM.

3.3 Determination of the integrity of the mitochondrial outer membrane

Damage to the outer membrane often occurs during their purification. Disruption of the mitochondrial ultrastructure can be assessed directly by using electron microscopy, although this technique has the disadvantage of being qualitative, time-consuming, and not particularly convenient. In many cases, it is necessary to assess the degree of damage to this membrane barrier by a rapid and convenient assay, for example, in investigation of the interaction of mitochondrial precursor proteins with outer membrane receptors during the import of nuclear encoded, cytoplasmic precursors into the organelle.

Convenient monitoring of the extent of damage to the mitochondrial outer membrane can be achieved by assaying the rate of oxidation of exogenous, reduced cytochrome *c*. Respiration via endogenous substrates is first inhibited by the addition of rotenone, such that oxygen uptake is dependent on provision of exogenous substrate, in this case reduced cytochrome *c* which is unable to gain access to its binding site on the surface of the inner membrane in intact mitochondria.

The following assay provides a good indication of the degree of damage to the outer membrane. It is based on the premise that exogenous reduced cytochrome *c* is an ineffective substrate in intact organelles as its binding site is rendered inaccessible by the presence of intact outer membrane. Any significant oxidation of added cytochrome *c* is indicative of major membrane damage, permitting cytochrome *c* to penetrate to the outer aspect of the inner membrane. Disruption of mitochondria with suitable detergents unmasks cytochrome *c* oxidase activity in the presence of reduced cytochrome *c* providing a reference standard equivalent to 100% accessibility of the substrate binding site. By comparing this value with that obtained for isolated 'intact' mitochondria, it is possible to estimate the extent of exposure of the inner membrane to externally added cytochrome *c*.

Several detergents are normally employed to strip the outer membrane selectively or solubilize the mitochondria, e.g. digitonin, Lubrol WX, and Triton X-100. This ensures that the 100% reference value is not merely a result of activation of cytochrome *c* oxidase by a specific detergent but is indicative of increased access to the cytochrome *c* binding site on mitochondrial disruption.

The assay is normally performed in a final volume of 3 ml which includes: 120 mM KCl, 0.1 mM 2,4-dinitrophenol (DNP), 1 μM rotenone, 30 μM cytochrome *c* (as ferrocytochrome *c*), 10 mM Tris–HCl pH 7.4, at 25 °C. DNP is normally added to the assay medium to uncouple the mitochondria and prevent limitation of the rate of reaction by phosphorylation. Rotenone is also included to exclude any interference from endogenous substrates.

To obtain a 100% activity value for the enzyme, treat a mitochondrial suspension with 0.2–1.0% Lubrol WX (or any of the other detergents listed above). Place on ice for 15 min and then dilute sample into assay medium such that the final concentration of Lubrol WX is less than 0.02%. Measure the decrease in absorbance at 550 nm and allow the reaction to proceed until an initial linear rate has been established.

Now repeat the same procedure using an untreated mitochondrial suspension ensuring that comparable amounts of mitochondrial protein are added to the assay. In order to determine the percentage damage caused to the mitochondrial outer membrane, compare the initial linear reaction rate obtained in both cases.

From previous studies it has been shown that rat liver mitochondria with 5–10% breakage of the outer membrane are routinely obtained. Assay of isolated heart mitochondria reveals that they tend to be more susceptible to damage with 30–50% of the outer membrane depending on the isolation procedure. It should be noted that an assessment of mitochondrial damage should be performed with each new batch of mitochondria as there is often considerable variation between preparations.

3.4 Glucose hexokinase trap method for estimation of P:O ratios

The amount of phosphate which is esterified during ATP synthesis can be measured directly by estimating the incorporation of $^{32}P_i$ into ATP. This method increases the sensitivity of the P:O assay performed using the oxygen electrode. It is a more general technique for measurement of ATP synthesis in impure preparations, partially uncoupled mitochondria, or submitochondrial particles where there are high background levels of ATPase. It tends to maximize ATP yields in contrast to the graphical method described in Section 3.2.1 where observed P:O ratios are usually underestimates caused by limited recycling of ATP. The reaction being followed is shown below:

$$\text{ADP} + [^{32}\text{P}]\text{P}_i \longrightarrow [^{32}\text{P}]\text{ATP} \xrightarrow[\text{glucose}]{\text{hexokinase}} [^{32}\text{P}]\text{G-6-P}$$

Excess hexokinase and glucose are included in the assay so that [γ-^{32}P]ATP is immediately converted to the more stable radiolabelled glucose-6-phosphate (G-6-P) which can be easily extracted and measured by scintillation counting. The reaction is normally terminated as the solution becomes anaerobic, so the amount of O_2 consumed is also known.

Protocol 7. Use of the oxygen electrode with the glucose hexokinase trap

Equipment and reagents

- Oxygen electrode
- Assay buffer: 0.22 M mannitol, 0.05 M sucrose, 10 mM NaH_2PO_4, 20 mM MOPS pH 7.2
- Fatty acid-free bovine serum albumin
- EDTA–Na^+
- ATP
- Glucose
- Hexokinase
- [^{32}P]P_i
- 1 M succinate
- 35% (w/v) perchloric acid
- 5 M H_2SO_4
- Silicotungstic acid
- Benzene/isobutanol (1:1)
- 10% (w/v) ammonium molybdate
- Water saturated isobutanol

Method

1. To the assay chamber of the oxygen electrode add 3.2 ml of assay buffer which includes 0.22 M mannitol, 0.05 M sucrose, 10 mM NaH_2PO_4, 20 mM MOPS pH 7.2.
2. Add 0.3 ml of fatty acid-free bovine serum albumin (5 mg/ml). In addition, 100 µM EDTA (sodium salt), 1 mM ATP, 30 mM glucose, 75 U of hexokinase, and 15 mM [^{32}P]P_i (approx. 1 mCi/mmol) are present in the assay.
3. Add between 0.1–0.3 ml of mitochondrial suspension and initiate the reaction by adding 50 µl of succinate or required substrate (1 M stock).
4. Allow reaction to proceed terminating the reaction immediately on anaerobiosis by addition of 50 µl of 35% (w/v) perchloric acid per 0.5 ml assay solution.

Protocol 7. *Continued*

5. Mix the deproteinized reaction mix in a tube containing 0.4 ml 5 M H_2SO_4, 1 ml of silicotungstic acid,[a] and 5 ml of benzene/isobutanol (1:1). Add water to bring the final volume of the water phase (including the mitochondrial sample) to 4.5 ml. Add 0.5 ml of 10% (w/v) ammonium molybdate immediately and shake for 15 sec.
6. Remove the aqueous layer and re-extract by adding 5 ml of isobutanol saturated with water to remove the last traces of $^{32}P_i$. Measure the radioactivity incorporated into G-6-P by scintillation counting and calculate the moles of G-6-P (equivalent to ATP synthesized) per moles of O_2 consumed to obtain P:O ratios.

[a] Silicotungstate reagent is prepared as follows: add 5.7 g sodium silicate nonahydrate and 79.4 g sodium tungstate dihydrate to 500 ml water. Add 15 ml of concentrated H_2SO_4 and dilute to a final volume of 1 litre.

4. Subfractionation of mitochondria

The development of subfractionation techniques for physical separation of the individual mitochondrial compartments, namely outer and inner membranes, matrix and intermembrane space has facilitated the accumulation of much of our current knowledge relating to the enzymatic distribution, metabolic activities, and specialized functions of individual mitochondrial sublocations.

Two general approaches can be adopted here. It is feasible to achieve specific release of the outer membrane by hypotonic treatment of intact organelles or by exposure to selective detergents such as digitonin or Lubrol WX which preferentially disrupt the outer membrane. This permits the isolation of relatively pure outer membrane, a soluble intermembrane space fraction, and mitoplasts (mitochondria minus outer membrane). Care is required to ensure that there is minimal damage to the inner membrane with consequent leakage of matrix enzymes contaminating the intermembrane space fraction.

In the preparation of inner membrane fragments (submitochondrial particles), mitochondria can be totally disrupted by sonication or high pressure extrusion from a French press. In this case, inner and outer membranes may be separated by virtue of their differing densities on sucrose gradients (13); however, the soluble fraction is clearly a mixture of the contents of the intermembrane space and mitochondrial matrix.

4.1 Preparation of submitochondrial particles by sonication

During rupture of the organelle, the inner membrane tends to invert, assuming an orientation opposite to that normally seen in the intact mitochondrion. Preparations of this type are normally contaminated with outer membrane vesicles, although this is less apparent in preparations using beef heart where the mitochondria tend to be richer in inner membrane.

Various protocols exist in the literature for the preparation of submitochondrial particles by sonication, where the phosphorylation capacity of a particular preparation is largely dependent on the composition of the sonicating medium. Media which contain Mg^{2+} ions and ATP result in the production of particles with a high phosphorylation efficiency. The following protocol generally leads to the production of non-phosphorylating membrane fragments with an inverted orientation, i.e. 'inside out' with respect to intact mitochondria.

Protocol 8. Use of sonication to prepare submitochondrial particles

Equipment and reagents

- Probe sonicator
- 20 mM KH_2PO_4 pH 7.4 (with KOH)
- High-speed and ultracentrifuges with fixed-angle rotors

Method

1. Resuspend 200 mg of mitochondrial protein in 5 ml of 20 mM KH_2PO_4 adjusted to pH 7.4 using KOH.
2. Put tube on ice and place sonicator probe into the solution ensuring that it does not touch the bottom or sides of the tube.
3. Sonicate at a setting of 70–80 watts for 15 sec, then rest for a period of 15 sec to allow cooling. Repeat this step twice.
4. Spin sonicated solution at 6500 *g* for 7 min. Keep the supernatant fraction and resuspend the pellet in 5 ml of 20 mM KH_2PO_4 adjusted to pH 7.4 using KOH.
5. Repeat the sonication described in step 4, then spin the sonicated suspension at 6500 *g* for 7 min.
6. Take the supernatant from this second centrifugation and pool with the corresponding fraction from step 4. Spin the combined supernatants at 105 000 *g* for 1 h.
7. Retain the pellet and resuspend in 5 ml of 20 mM KH_2PO_4 adjusted to pH 7.4 using KOH. Centrifuge at 105 000 *g* for 1 h.
8. Resultant pellet contains the SMPs—resuspend by hand homogenization in 3–5 ml of 20 mM KH_2PO_4 adjusted to pH 7.4 using KOH.

4.2 Preparation of mitoplasts using digitonin

Digitonin is a surface active plant steroid which causes fragmentation of biological membranes by forming complexes with cholesterol. In general, it is the outer membrane which contains significant amounts of cholesterol compared to the inner membrane and is more susceptible to the action of this detergent compared to the inner membrane. At concentrations of 0.1 mg/mg protein the outer membrane breaks down into smaller fragments and reseals into vesicles, the inner membrane remaining intact and conserving its matrix to form particles called mitoplasts. By increasing the concentration of digitonin it is possible to rupture the inner membrane itself to release the matrix enzymes. When this happens the inner membrane reseals to form sealed inner membrane ghosts.

Protocol 9. Preparation of mitoplasts using digitonin

Equipment and reagents

- Isolation buffer: 0.225 M mannitol, 0.075 M sucrose, 0.5 mM EGTA, 2 mM MOPS pH 7.4
- High-speed and ultracentrifuge
- Recrystallized digitonin
- Lubrol WX

Method

1. Solubilize recrystallized digitonin at 1.2% (w/v) in mitochondrial isolation buffer containing 0.225 M mannitol, 0.075 M sucrose, 0.5 mM EGTA, 2 mM MOPS pH 7.4. If necessary the solution can be heated to dissolve the digitonin.
2. Take isolated rat liver mitochondria and stirring gently on ice, treat with digitonin at 0.12 mg digitonin/mg mitochondrial protein. Leave for 15 min.
3. Dilute the digitonin treated sample with 3 vol. of isolation medium. Centrifuge at 9000 *g* for 10 min.
4. (a) Take the pellet from step 3 which contains the inner membrane vesicles and treat with 0.16 mg of Lubrol WX per mg of mitochondrial protein (i.e. mitoplasts). Leave on ice for 10 min and dilute with 3 vol. of isolation buffer before centrifugation at 144 000 *g* for 60 min to sediment the inner membrane vesicles. The matrix enzymes can be recovered from the supernatant fraction.

   (b) Take the supernatant fluid from step 3 and centrifuge at 144 000 *g* for 60 min: outer membrane vesicles can be recovered in the pellet with intermembrane space enzymes in the supernatant fluid.

Another use for digitonin is for the selective solubilization of tissue culture cells to obtain cytosolic and particulate fractions. The success of the fractionation can subsequently be assessed by assaying the fractions for the appropriate marker enzyme and is convenient for analysing patterns of enzyme distribution in small numbers of cells where standard techniques of mitochondrial preparation may be impractical or unnecessary.

Take a 0.5–1.0 ml aliquot of cells (approx. 5×10^6 cells) and mix with digitonin ranging in concentration between 0.05–2.00 mg/ml. Leave suspensions on ice for 2 min, then spin at 14000 *g* for 10 min in a microcentrifuge. Remove the soluble supernatant which represents the cytosolic fraction, and assay for the differential release of plasma membrane, cytosolic, and mitochondrial markers such as 5′-nucleotidase, lactate dehydrogenase, and citrate synthase respectively.

Digitonin may also be employed to increase the overall purity of mitochondrial suspensions. By treating mitochondria with low concentrations of digitonin, it is possible to cause selective fragmentation of any lysosomal or peroxisomal membranes which may be contaminating the preparation. After detergent treatment it is a simple manner to remove any released enzymes by centrifugation (14).

4.3 Separation of 'right side out' and 'inside out' submitochondrial particles by affinity chromatography on CNBr–cytochrome *c* coupled Sepharose

When the mitochondria and inner membrane are disrupted they disintegrate into small vesicles to produce submitochondrial particles. The membrane can assume two different orientations such that mixed populations of vesicles are formed with opposing polarities. In right side out or non-sealed vesicles, the binding site for cytochrome *c* is exposed on the surface of the membrane, whereas in sealed inside out vesicles it is inaccessible. For analysis of inner membrane transport activities, well-defined populations of vesicles are required as is also the case for surface-specific probing of the transmembrane organization of mitochondrial carriers with proteases, antibodies, or non-permeant covalent chemical, or radiochemical 'tags'. Cytochrome *c* is readily released from the inner membrane on disruption of mitochondria; hence on applying these cytochrome *c*-depleted membrane fragments to a cytochrome *c*–Sepharose affinity column the right side out and non-sealed vesicles with exposed binding sites are selectively absorbed to the immobilized cytochrome *c*, whereas the inside out particles are not retained and will elute in the void volume.

From mitochondria with abundant cristae, the bulk of inner membrane vesicles tend naturally to form an 'inside out' formation although this is dependent on the source of the mitochondria. The quality of preparation can be monitored routinely, once again with exogenous cytochrome *c* as substrate in the presence and absence of suitable detergents. As with 'intact' mitochondria the binding site for cytochrome *c* should be totally 'inaccessible' in homogeneous populations of sealed 'inside out' vesicles.

Protocol 10. Affinity chromatography using CNBr-activated Sepharose

Equipment and reagents

- Buchner funnel
- 10 ml disposabe syringe plugged with glass wool
- CNBr-activated Sepharose
- Distilled water
- 1 mM HCl
- 0.1 M Na_2CO_3 pH 8.0
- Cytochrome *c*
- 50 mM ethanolamine–HCl pH 8.0
- 0.1 M CH_3COONa pH 5.0
- 1 M KCl, 1% (v/v) Lubrol WX
- 0.25 M sucrose, 1 mM EDTA, 10 mM Tris–HCl pH 7.4
- 0.15 M KCl, 10 mM Tris–HCl, pH 7.4
- 0.25 M sucrose, 1 mM EDTA, 1 M KCl, 10 mM Tris–HCl pH 7.4

Method

1. Swell 5 g of CNBr-activated Sepharose[a] in 500 ml of distilled water for 2 h. Wash on a Buchner funnel with 1 litre of distilled water, followed by 1 litre of 1 mM HCl, and again with 1 litre of distilled water.
2. Suspend the washed CNBr-activated Sepharose in 10 ml of 0.1 M Na_2CO_3 pH 8.0 containing 100 mg of cytochrome *c*. Incubate at room temperature for 90 min stirring occasionally. Wash the gel on a Buchner funnel with a further 100 ml of 0.1 M Na_2CO_3 pH 8.0.
3. After washing, resuspend the gel in 50 mM ethanolamine–HCl pH 8.0 and incubate for 1 h at room temperature.
4. Carry out successive washings of the gel on a Buchner funnel as follows: 200 ml of 0.1 M CH_3COONa pH 5.0; 200 ml of 1 M KCl containing 1% (v/v) Lubrol WX; 200 ml of 0.1 M Na_2CO_3 pH 8.0. Repeat this series of washes a further three times.
5. Equilibrate the cytochrome *c*–Sepharose with buffer containing 0.25 M sucrose, 1 mM EDTA, 10 mM Tris–HCl pH 7.4. Pour the gel matrix into a 10 ml disposable plastic syringe plugged with glass wool at one end to prevent run-through.
6. Equilibrate the column with approx. five column volumes (50 ml) of buffer containing 0.25 M sucrose, 1 mM EDTA 10 mM Tris–HCl pH 7.4.
7. Take sample of prepared submitochondrial particles and remove cytochrome *c* by washing twice in 0.15 M KCl, 10 mM Tris–HCl, pH 7.4 (centrifuge at 105 000 *g* for 1 h to harvest the membranes).
8. Take the membranes from step 7 (now stripped of cytochrome *c*), and suspend in 0.25 M sucrose, 1 mM EDTA, 10 mM Tris–HCl pH 7.4 at a concentration of 2 mg/ml. Apply 1 ml of resuspended membranes to the top of the column allowing solution to run on to the gel bed before washing the column with approx. 20 ml of suspension buffer, collecting 1 ml fractions.

9. The inside out particles do not bind to the column and will be eluted. Read the absorbance of the collected 1 ml fractions at 280 nm to detect the protein peak.
10. Once the protein peak corresponding to the inside out particles has been detected, wash the column with buffer containing 0.25 M sucrose, 1 mM EDTA, 1 M KCl, 10 mM Tris–HCl pH 7.4: this will elute the right side out particles which will have bound to the column. Again, collect 1 ml fractions and detect protein at 280 nm.

[a] Pharmacia LKB, Uppsala, Sweden.

4.4 Assaying of mitochondrial marker enzymes

The following section details the more commonly used assays which are used during the systematic fractionation of mitochondria.

4.4.1 Citrate synthase (citrate-oxaloacetate-lyase EC 4.1.3.7)

This enzyme is a marker used exclusively for the mitochondrial matrix, therefore, its activity is not readily measurable in the intact mitochondrion owing to the impermeability of the substrates.

Reaction catalysed:

$$\text{acetyl CoA} + \text{oxaloacetate} + H_2O \longrightarrow \text{citrate} + \text{CoASH} + H^+$$

The activity of this enzyme is measured by following the deacetylation of acetyl CoA at 412 nm. The production of acetyl CoA is measured by the formation of the mercaptide with DTNB (5,5′-dithio-bis-2-nitrobenzoic acid). The assay is performed in a final volume of 1 ml and includes 200 μl of 0.5 M Tris–HCl pH 8.0, 200 μl of 0.5 mM DTNB, and 100 μl of 3 mM acetyl CoA (all freshly prepared) and 0.05–0.20 mg of mitochondrial protein.

(a) Warm to 25 °C before initiating the reaction and follow the absorbance for 1–2 min to measure endogenous levels of thiol or deacylase activity.
(b) Add 100 μl 5 mM sodium oxaloacetate (freshly prepared) to initiate the reaction. Enzyme units are expressed as amount of enzyme that produces 1 μmol of CoA/min at 25 °C (ε_{412} 13 000 M^{-1} cm^{-1}).

4.4.2 Cytochrome *c* oxidase activity (EC 1.9.3.1)

This enzyme is used as a marker for the inner mitochondrial membrane. It involves following the decrease in absorbance of cytochrome *c* (α band) at 550 nm coupled to the oxidation of cytochrome *c*. The assay is performed in a final volume of 3 ml.

(a) Prepare the cytochrome *c* solution by dissolving 1 mg cytochrome *c* in 30 ml of 30 mM NaH_2PO_4 adjusted to pH 7.4 with NaOH.

(b) Add 100 μl sodium dithionite to this solution and shake vigorously for 2 min to remove excess dithionite.
(c) To 2.98 ml of this solution add a sample of mitochondrial protein in a volume of 200 μl.

Enzyme units are expressed as amount of enzyme that oxidizes 1 μmol of cytochrome *c*/min at 25 °C (ε_{550} 28 500 M^{-1} cm^{-1}).

4.4.3 Adenylate kinase (ATP : AMP phosphotransferase EC 2.7.4.11)

This enzyme is used as a marker for the mitochondrial intermembrane space

Reaction catalysed:

$$AMP + ATP \rightleftharpoons 2\,ADP$$

This reaction involves monitoring the reduction of NADP at 340 nm. The assay is performed in a final volume of 1 ml.

(a) Prepare the assay mixture of 3 mM ADP, 15 mM glucose, 10 U of hexokinase, 0.4 U of glucose-6-phosphate dehydrogenase, 5 mM $MgCl_2$, 0.75 mM NAPD, 0.45 mM KCN, 70 mM glycyl-glycine buffer, pH 8.0.
(b) Incubate mixture at 25 °C for approx 5 min before initiating the reaction.

Units of enzyme activity are defined as the amount of enzyme which produces 1 μmol NADPH/min at 25 °C (ε_{340} 6200 M^{-1} cm^{-1}).

4.4.4 Monoamine oxidase (monoamine oxygen oxidoreductase deaminating EC 1.4.3.1)

This enzyme is used as a marker for the mitochondrial outer membrane where activity is measured by following the oxidation of leuco 2′, 7′-dichlorofluorescein at 502 nm coupled to the monoamine oxidase-dependent production of H_2O_2 in the presence of benzamidine. The assay is performed in a final volume of 3 ml.

(a) Prepare the assay mixture from 600 μl of 0.25 mM leuco 2′,7′-dichlorofluorescein diacetate (freshly prepared in 10 mM NaOH), 1.8 ml of horse-radish peroxidase (dissolve 0.83 mg in 10 ml sodium phosphate buffer pH 7.15).
(b) Add mitochondrial protein and water to give a final volume of 2.95 ml.
(c) Pre-warm the cuvette to 25 °C and monitor the absorbance at 502 nm for 2–3 min until a baseline is obtained.
(d) Initiate the reaction by adding 50 μl of 0.1 M benzamidine.

5. Medical aspects of mitochondrial research

5.1 Mitochondrial biogenesis

The biogenesis of mitochondrial ultrastructure represents a complex feat in cell growth and development which must be closely organized and regulated in parallel with other programmed cellular activities. This organelle is the product of proteins, encoded not only by nuclear genes, but by DNA present within the mitochondrial matrix itself. It is now well established that certain proteins of the mitochondrial inner membrane, including cytochrome *c* and ATPase, are synthesized through the joint co-operation of the mitochondrial and cytoplasmic systems of protein synthesis. The protein synthesizing machinery of the mitochondrion is limited, coding only for 13 proteins in mammals, with the vast majority of proteins being synthesized outside the mitochondrion on cytoplasmic ribosomes. These proteins imported into the organelle are often synthesized with an additional 10–70 amino acids at their N termini, this leader sequence targetting the protein to its correct location (15). In the case of the mitochondrion any imported protein must also be directed to specific sublocations within the organelle, i.e. the inner or outer membranes, matrix, or intermembrane space.

5.2 Oxidative phosphorylation and the electron transport chain

As discussed in earlier sections, oxidative phosphorylation describes the process in which electrons are transferred from reduced cofactors to molecular oxygen. As electrons are passed along the respiratory chain a H^+ ion gradient is generated with the subsequent re-entry of these ions through specific proton channels in the ATP synthase used to generate the formation of ATP.

The electron transport chain is composed of four complexes located in the inner mitochondrial membrane. A schematic diagram of their arrangement in the membrane is shown in *Figure 8*. It represents an assembly of approximately 20 discrete oxidation–reduction carriers organized into four multimeric complexes termed complexes I, II, III, IV which exist as discrete entities in the plane of the membrane, interacting with each other by random collision. Complex II (succinate ubiquinone oxidoreductase) is largely embedded on the inside of the inner membrane, channelling electrons directly into complex III. Cytochrome *c* (a peripheral hydrophilic protein), and ubiquinone (a lipid soluble molecule which transfers electrons from complexes I and II to complex III), constitute the 'mobile carriers' of the chain functioning as electron mediators in transferring reducing equivalents between the discrete, freely-diffusing complexes

The sequence of the electron carriers in the membrane was established as a result of oxygen electrode experiments (see Section 3.2.1) and spectroscopic techniques. It is possible to introduce and extract electrons at a number of locations along the respiratory chain. For example, NADH and associated substrates such as pyruvate and β-hydroxybutyrate act to reduce complex I, succinate introduces its electrons at complex III, with TMPD entering at complex IV (cytochrome *c* oxidase) via cytochrome *c*. (Note: it is necessary to add ascorbate in conjunction with TMPD in order to regenerate it from

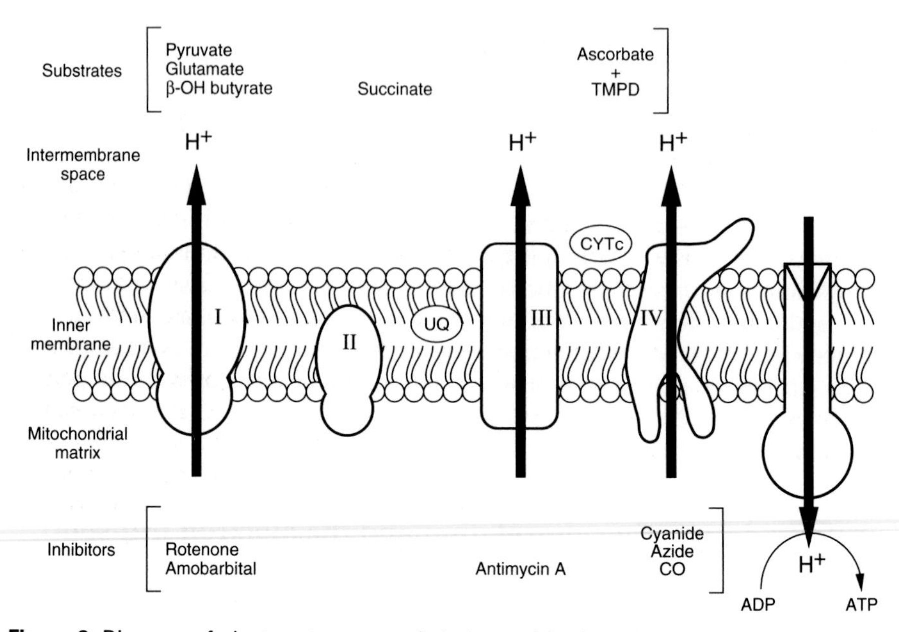

Figure 8. Diagram of electron transport chain located in the mitochondrial inner membrane. Substrates and inhibitors affecting the individual complexes are shown above and below respectively.

the oxidized form.) Also shown in *Figure 8* are the sites of action of various inhibitory compounds whose discovery enabled the relative positions of the sites of electron entry to be determined. By adding substrates which enter at points downstream of these inhibitors it is possible to bypass such blocks.

5.3 Mitochondrial myopathies

Mitochondrial myopathies represent a group of human diseases involving abnormalities in the mitochondria, mainly from skeletal muscle and brain, since these tissues possess an abundance of mitochondria on account of their high energy demands. Such defects are relatively rare and may be caused by rearrangements, deletions, insertions, or point mutations in the nuclear or mitochondrial genome. Primary mitochondrial defects are apparent in the cellular DNA and the application of various molecular biological techniques can aid in their diagnosis. This contrasts with secondary defects which are often associated with the respiratory chain; however, other classes of mitochondrial dysfunction do exist and for an in depth treatment of this topic the reader is directed to papers dealing specifically with this subject area.

5.4 Techniques applied in the investigation of mitochondrial abnormalities

Various approaches can be employed in the study of the mitochondrial disease state: these include immunoblotting of the complexes using antibodies raised to the holocomplexes and individual subunits (16). This method provides a very sensitive means of determining deficiencies in the amount of complex or individual protein subunits present. Determination of cytochrome concentrations can also be examined using the technique of low temperature difference spectroscopy where low levels of cytochrome components in the respiratory chain are detrimental to mitochondrial function, a total lack of cytochromes being a lethal condition. Since the vast majority of mitochondrial disorders occur at the level of electron transport or ATP synthesis, it is desirable to be able to measure the relative activities of the individual complexes. Deficiencies in complex I activity represent the most common defects, whereas complex II deficiency is relatively rare. For investigations of this type it is more convenient to isolate mitochondria from freshly obtained skeletal muscle (*Protocol 11*), since the specific lesion may be confined to a specific tissue or cell type, so no apparent abnormality can be detected in cultured fibroblasts usually taken from the skin of the patient.

Protocol 11. Purification of mitochondria from skeletal muscle tissue

Equipment and reagents

- Potter–Elvehjem homogenizer
- Nylon mesh (200 μm)
- High-speed centrifuge and rotors
- Extraction buffer: 0.25 M sucrose, 40 mM KCl, 3 mM EDTA, 20 mM Tris–HCl pH 7.4
- Percoll

Method

1. Wash 100–500 mg of freshly obtained biopsy material in 5 ml of extraction medium which includes 0.25 M sucrose, 40 mM KCl, 3 mM EDTA, 20 mM Tris–HCl pH 7.4.
2. Mince the tissue finely using scissors and using a Potter–Elvehjem motorized homogenizer (500 rev/min), homogenize the tissue in 30 ml of extraction buffer using approx. three up-and-down strokes.
3. Filter the homogenate through a fine nylon mesh (200 μm) and centrifuge at 2000 *g* for 10 min.
4. Remove the supernatant fraction and keep on ice: resuspend the pellet in 5 ml of extraction buffer using approx. three up-and-down passes of the homogenizer at 500 rev/min.
5. Centrifuge at 2000 *g* for 10 min. Remove the supernatant fraction and

Protocol 11. *Continued*

pool with the supernatant fraction obtained from step 4. Centrifuge at 10 000 *g* for 10 min.

6. Take 1 ml of extraction buffer containing 5% (v/v) Percoll and put into a 1.5 ml tube. Layer the mitochondrial pellet on to this gradient and centrifuge at 10 000 *g* for 10 min.
7. Resuspend the resultant mitochondrial pellet in the smallest volume of extraction buffer possible.

5.4.1 Assaying of the individual complexes of the respiratory chain

The following assays are used to measure the activity of individual complexes of freshly isolated mitochondrial membranes *in situ*. Fractionation of the respiratory chain into four discrete complexes can be achieved while retaining the electron transfer activity of each complex (17). However, small scale isolation of individual complexes from cultured human cells or biopsy samples is not readily performed on a routine basis.

i. Complex I (NADH:ubiquinone oxidoreductase)

(a) Prepare the following assay medium: 2 μg/ml antimycinA, 2 mM potassium cyanide, 2.5 mg/ml fatty acid-free bovine serum albumin (BSA), 0.13 mM NADH, 5 mM $MgCl_2$, 35 mM KH_2PO_4 and adjust the pH to 7.4 using KOH.

(b) Perform the assay at 30 °C.

(c) Initiate the reaction by the addition of enzyme (approx. 15–30 mg protein) and follow by measuring the decrease in absorbance at 340 nm.

(d) Perform this assay twice, both in the presence and absence of rotenone (2 μg). Complex I activity is then calculated as the difference between the two rates.

ii. Complex II (succinate:ubiquinone oxidoreductase)

(a) Prepare the following assay medium: 2 μg/ml antimycinA, 50 μM DCPIP (electron acceptor), 2 mM potassium cyanide, 5 mM $MgCl_2$, 35 mM KH_2PO_4 and adjust the pH to 7.2 using KOH.

(b) Incubate the mitochondrial extract (15–30 mg) with 20 mM succinate for 10 min at 30 °C to activate the enzyme fully.

(c) Add the extract to the pre-warmed assay medium and initiate the reaction by adding 65 mM ubiquinone and follow the decrease in absorbance at 600 nm (ε_{600} DCPIP 19.1 mM^{-1} cm^{-1}).

iii. Complex III (ubiquinol:cytochrome c *reductase)*

For assaying of this complex it is necessary to use ubiquinol which is produced by chemical reduction of ubiquinone, see *Protocol 12*.

(a) Prepare an assay medium containing the following: 2 μg/ml rotenone, 2.5 mg/ml fatty acid-free BSA, 50 μM ubiquinol, 15 μM cytochrome *c* (III). 2 mM potassium cyanide, 5 mM $MgCl_2$, 35 mM KH_2PO_4 and adjust to pH 7.2 with KOH.

(b) Prior to starting the reaction, measure the non-enzymic reduction of cytochrome *c* for 1 min at 550 nm (30°C), this rate being subtracted from that in the presence of enzyme to give the true enzymic rate (ε_{550} cyt *c* 20 mM^{-1} cm^{-1}).

(c) Initiate the reaction proper by adding enzyme (10–20 μg).

Protocol 12. Preparation of reduced ubiquinone

Equipment and reagents

- Fume cupboard
- Nitrogen gas
- Ubiquinone
- Ethanol
- 6 M HCl
- Sodium borohydride
- Diethyl ether/cyclohexane (2:1)
- 2 M NaCl

Method

1. Dissolve ubiquinone (10 μM) in 1 ml of ethanol. Reduce pH to 2.0 using 6 M HCl.
2. Add 0.5 g of sodium borohydride dissolved in 1 ml of water.
3. Add 3 ml of diethyl ether/cyclohexane (2:1) and mix vigorously. Perform this procedure in a well ventilated area.
4. Allow phases to separate, then remove upper phase into 1 ml of 2 M NaCl. Mix vigorously.
5. Again, allow the phases to separate and then remove the upper phase. Dry under a stream of N_2 gas in a fume-cupboard.
6. Add 1 ml of ethanol (pH adjusted to 2 using HCl) to dissolve ubiquinol.

6. Concluding remarks

In this chapter, we have attempted to provide some general information on the routine approaches to the analysis of mitochondrial function. There has been a revival in interest in the study of mitochondria in recent years with the realization through the application of more sensitive and sophisticated diagnostic procedures that genetic defects in mitochondrial function occur with much higher frequency than previously suspected. The basic biochemical approaches outlined in this chapter will permit initial assessment of the general nature of the lesion and indicate a possible site of action. Clearly, at this stage more refined investigation is necessary using specific immunological probes to detect whether the particular mitochondrial protein under study is deficient or expressed in an abnormal form. Finally, analysis of the cDNA to the gene in question or even ultimately the overall genomic organization may be required to provide a complete picture of the molecular basis of a given lesion. Genetic analysis of individual patients carrying mutations on either mitochondrial or nuclear encoded polypeptides of the organelle is, however, beginning to produce some fascinating insights into the co-ordinated expression of the two genomes and the molecular mechanisms underlying mitochondrial biogenesis.

References

1. Claude, A. (1946). *J. Exp. Med.*, **84**, 61.
2. Kennedy, E. P. and Lehninger, A. L. (1949). *J. Biol. Chem.*, **179**, 957.
3. Palade, G. E. (1953). *J. Histochem. Cytochem.*, **1**, 188.
4. Tyler, D. D. (1992). In *The mitochondrion in health and disease*, pp. 147–194. VCH, New York, Cambridge.
5. Chance, B. and Hagihara, B. (1963). In *Proc. 5th Intl. Congress Biochem.* (ed. A. N. M. Sissakian), Vol. 5, pp. 3–37. Pergamon Press, NY.
6. Reinhart, P. H., Taylor, W. M., and Bygrave, F. L. (1982). *Biochem. J.*, **204**, 731.
7. Smith, A. L. (1967). In *Methods in enzymology* (ed. R. W. Estabrook and M. E. Pullman), Vol. 10, pp. 81–6. Academic Press, New York and London.
8. Sims, N. R. (1990). *J. Neurochem.*, **55**, 698.
9. Lai, J. C. K., Walsh, J. M., Dennis, S. C., and Clark, J. B. (1977). *J. Neurochem.*, **28**, 626.
10. Attardi, G. and Ching, E. (1979). In *Methods in enzymology* (ed. S. Fleischer and L. Packer), Vol. 56G, pp. 66–79. Academic Press, New York, San Francisco, and London.
11. Rickwood, D., Dujon, B., and Darley-Usmar, V. M. (1988). In *Yeast: a practical approach* (ed. I. Campbell and J. H. Duffins), pp. 185–254. IRL Press, Oxford, Washington, DC.
12 Hauswirth, W. W., Lim, L. O., Dujon, B., and Turner, G. (1987). In *Mitochondria: a practical approach* (ed. V. M. Darley-Usmar, D. Rickwood, and M. T. Wilson), pp. 171–244. IRL Press, Oxford, Washington, DC.
13. Sottocasa, G. L., Kuylenstierna B., Ernster, L., and Bergstrand, A. (1967). In *Methods in enzymology* (ed. R. W. Estabrook and M. E. Pullman), Vol. 10, pp. 448–463. Academic Press, New York and London.
14. Loewenstein, J., Scholte, H. R., and Wit-Peeters, E. M. (1970). *Biochim. Biophys. Acta*, **223**, 432.
15. Glover, L. A. and Lindsay, J. G. (1992). *Biochem. J.*, **284**, 609.
16. Towbin, H., Staehelin, T., and Gordon, J. (1979). *Proc. Natl. Acad. Sci. USA*, **76**, 4350.
17. Ragan, C. I., Wilson, M. T., Darley-Usmar, V. M., and Lowe, P. N. (1987). In *Mitochondria: a practical approach* (ed. V. M. Darley-Usmar, D. Rickwood, and M. T. Wilson), pp. 79–112. IRL Press, Oxford, Washington, DC.

5

Isolation and characterization of peroxisomes

A. VÖLKL, E. BAUMGART, and H. D. FAHIMI

1. Introduction

Peroxisomes (PO) are organelles ubiquitously distributed to most pro- and eukaryotic cells (1, 2). Depending on the cell type and the sets of enzymes which they contain, synonymous designations such as glyoxisomes or glycosomes are used with 'microbodies' as the all-embracing term. Mostly, PO are spherical or spheroidal particles with a diameter between 0.2–1.0 μm. In some tissues, however (3–6), they may be irregularly shaped.

PO are subcompartmentalized with a unit membrane surrounding the so-called matrix compartment which comprises the bulk of peroxisomal enzymes. Matrix proteins are mostly quite soluble apart from urate oxidase and α-hydroxyacid oxidase B which tend to aggregate, thus forming the crystalline 'core' and the 'marginal plate' of hepatic and renal PO respectively (1, 6).

PO are most abundant in mammalian liver and kidney. They are predominantly engaged in cellular lipid metabolism (7), both anabolic as well as catabolic: biosynthesis of ether lipids, cholesterol, and dolichols on the one hand, the degradation of fatty acids and derivatives via β-oxidation on the other. In addition, enzymes, collectively called oxidases which are involved in the catabolism of purines, polyamines, D-amino acids and α-hydroxyacids are present (8). Consistently, these oxidases generate hydrogen peroxide which is destroyed by catalase, the most abundant enzyme of hepatic PO.

It is unique to hepatic and renal PO of rodents to undergo massive proliferation upon treatment of the animals with various xenobiotics (9). Usually, this process is accompanied by the induction of some of their enzymes, particularly those of the fatty acid β-oxidation system. The two phenomena however, are apparently not necessarily interrelated and may occur independently. Moreover, there are marked interspecies differences in the response of PO to various drugs with rodents responding vigorously but guinea-pig and primates being only weakly affected (10).

In order to investigate the unique biochemical, functional, and structural features of PO, the isolation of highly purified fractions of this organelle is indispensable. This however, is hampered by two serious problems: first, the relative paucity of PO even in liver (only about 2% of the total protein); secondly, their considerable fragility. Thus, mild conditions during homogenization as well as the different centrifugation steps have to be maintained to minimize mechanical, hydrostatic, and osmotic stress. Although most tissues and cells, including those of plant origin contain PO (or microbodies), the most widely used source of PO is rat liver and the protocols which follow apply to that tissue. Other tissues or cells will require alternative homogenization methods (see Chapter 1), but the centrifugation and analytical protocols should be broadly applicable.

2. Preparation and purification of rat hepatic PO

2.1 General considerations

In general, preparation and purification of rat hepatic PO is accomplished in three steps:

Step 1: homogenization of the tissue or disruption of the cells.

Step 2: subfractionation of the homogenate by differential contrifugation to obtain a crude PO fraction.

Step 3: isolation of purified PO by density gradient centrifugation of the crude PO fraction (for a comprehensive treatise on centrifugation see ref. 11).

Step 1

(a) Homogenization of tissues is usually carried out by means of a motor-driven Potter–Elvehjem tissue grinder with loose-fitting pestle. Cells are disrupted either by French press or by using a Dounce tissue grinder.

(b) Homogenates are normally prepared in isotonic 0.25 M sucrose adjusted to physiological pH with low concentrations of buffers such as MOPS or Hepes.

(c) Homogenization medium is routinely supplemented with additives:

  i. Ethanol to prevent the formation of inactive catalase compound II (a complex of catalase and H_2O_2 which leads to inactivation of the enzyme).

  ii. Sulfhydryl protecting agents as antioxidants.

  iii. Chelators (EDTA) to reduce the contamination of PO by microsomal fragments.

  iv. A cocktail of protease inhibitors.

Step 2

(a) According to the classical fractionation scheme of De Duve *et al.* (12), five main fractions are routinely obtained: nuclear (N), heavy mitochondrial (M), light mitochondrial (L or D), microsomal (P), and cytosolic fraction (S).

(b) The L or D fraction is the most enriched in PO but also contains lysosomes and substantial amounts of mitochondria and microsomes which have to be separated from PO by density gradient centrifugation.

Step 3

(a) Isopycnic density gradient centrifugation is commonly used to separate PO from contaminants.

(b) Gradients covering a linear density range of about 1.10–1.30 g/ml are either made of sucrose, or a medium such as Percoll™, or one of the iodinated density gradient media (for the criteria a gradient medium should fulfil, see ref. 11).

i. Sucrose is disadvantageous because of the high osmolarity of dense solutions, and the necessity to pre-treat animals with Triton WR-1339 to lower the density of lysosomes which otherwise will contaminate the peroxisomal fraction (13).

ii. Percoll gradients are most convenient for isolating PO under isotonic conditions, yet the purity of the final PO fraction is unsatisfactory (14).

iii. Separation and purification of PO on gradients made of the iodinated media such as Nycodenz, metrizamide, or Iodixanol is excellent and avoids any pre-treatment of animals, but the media are quite expensive.

(c) In any of the above gradients PO band at the rather high density of 1.22–1.25 g/ml due to the permeability of their membranes to low molecular weight substances (15).

(d) So far three approaches have been developed:

i. In the classic procedure (13) sucrose gradients and the specialized Beaufay rotor was employed which is not generally available.

ii. A self-generating Percoll gradient in conjunction with a vertical rotor are used for the isolation of PO under isotonic conditions (16).

iii. In the most straightforward approach Nycodenz or metrizamide gradients are used in combination with a vertical rotor (14, 17–19).

2.2 Experimental design

The approach subsequently outlined in *Protocols 1–4* makes use of an exponentially shaped metrizamide gradient and a vertical rotor. It is a modification of a procedure developed earlier to isolate highly purified (about 98%) PO from normal rat livers (17). In the meantime it has been extended to the isolation of rat hepatic PO subpopulations (20). It has been further applied to the livers of several other mammalian species (10) as well as to the isolation of renal PO (6) and of PO from invertebrate tissues (21). In *Protocols 5* and *6* the basic procedures for Nycodenz and Percoll gradient centrifugation will be also provided.

2.2.1 Equipment and solutions

Equipment:

(a) Perfusion device. Although perfusion is not essential for the isolation of PO from the homogenate, the catalase in erythrocytes may make calculations of rates of recovery of PO difficult if this enzyme is used as a marker.
(b) 30 ml Potter–Elvehjem tissue grinder with loose-fitting pestle (clearance 0.1–0.15 mm) and a motor-driven homogenizer (up to 1500 r.p.m.).
(c) Refrigerated low and high speed centrifuge (e.g. Beckman TJ-6 and J2-21) including a fixed-angle rotor (e.g. JA-20) and capped polycarbonate tubes of 50 ml.
(d) Ultracentrifuge (e.g. Beckman L90) and a vertical-type rotor (e.g. VTi50) as well as corresponding tubes (e.g. Quick-seal polyallomer, Beckman).
(e) Refractometer.
(f) Animals: rats of 220–250 g body weight. Females are preferred as hepatic PO are less fragile compared to those from male rats.

Solutions:

(a) Homogenization medium (HM): 0.25 M sucrose, 5 mM MOPS, 1 mM EDTA (adjusted to pH 7.4 with NaOH), containing 0.1% (v/v) ethanol, 0.2 mM dithiothreitol (DTT), plus protease inhibitors: 0.2 mM phenylmethysulfonyl fluoride (PMSF), 1 μg/ml leupeptin, 1 μg/ml aprotenin, 15 μg/ml phosphoramidon. These inhibitors are normally stored at 100 × concentration (the PMSF in isopropanol or ethanol). See Chapter 1, Section 4 for more information.
(b) Gradient buffer (GB): 5 mM MOPS, 1 mM EDTA (adjusted to pH 7.4 with NaOH), containing ethanol, DTT, and protease inhibitors as in HM.

2.2.2 Preparation of a crude peroxisomal fraction

The following items should be kept in mind:

(a) Starve the animals overnight to minimize a high glycogen concentration which otherwise interferes with subcellular fractionation.
(b) Cool media in advance and keep them on ice, and pre-cool tubes, tissue grinder, rotors, etc.
(c) Store homogenates and subcellular fractions on ice.
(d) g Forces given are calculated at the average radius (in millimetres) of the rotor used according to the equation:

$$g = (\text{r.p.m.}/1000)^2 \times 1.12 \times r_{av}$$

(e) Centrifugation time refers to the entire period from the start till the stopping of the rotor.

(f) The procedures should be carried out as quickly and gently as possible to avoid mechanical or enzymic degradation of the organelles.

Protocol 1. Perfusion and homogenization

1. Anaesthetize the animal, e.g. by i.p. injection of chloralhydrate. This must be carried out by a properly trained operative with the appropriate licence.
2. Weigh the animal and open abdominal cavity.
3. Perfuse liver with 0.9% saline via the portal vein until all blood is drained away.
4. Remove liver and dissect away connective tissue.
5. Weigh, and cut the liver in small pieces, then transfer into a Potter–Elvehjem tissue grinder in ice.
6. Add 3 ml/g (wet weight of liver) of ice-cold HM and homogenize at 1000 r.p.m. with a single stroke of a loose-fitting pestle.
7. Pour the homogenate into a 50 ml centrifuge tube.

Protocol 2. Subcellular fractionation

1. Centrifuge the total homogenate at 100 *g* for 10 min in a refrigerated low speed centrifuge to remove debris, unbroken hepatocytes, and blood cells.
2. Carefully pour off the supernatant from the loose pellet, and resuspend the pellet in 2 ml/g of ice-cold HM.
3. Rehomogenize at 800 r.p.m./1 stroke and centrifuge again as in step 1.
4. Pour off the second supernatant and combine it with the first one; discard the pellet.
5. Centrifuge the combined supernatants at 1950 *g* for 10 min in a refrigerated high speed centrifuge.
6. Decant supernatant from the firm pellet; resuspend pellet manually in 1 ml/g of ice-cold HM using a pestle or a glass rod, and centrifuge again at 1950 *g*.
7. Save the supernatant combining it with the first one, this is the post-mitochondrial supernatant. The pellet contains the majority of mitochondria, large sheets of membrane, and nuclei.
8. Subject the post-mitochondrial supernatant to 25 300 *g* for 20 min in a high speed centrifuge at 4°C.

Protocol 2. *Continued*

9. Aspirate the supernatant *including* the reddish fluffy layer on top of the pellet.
10. Resuspend the pellet carefully in 2 ml/g of ice-cold HM using a glass rod[a] and recentrifuge at 25 300 *g* for 15 min.
11. The final pellet is gently resuspended again by means of a glass rod [a] in 1 ml/g of ice-cold HM and comprises the crude peroxisomal fraction.
12. The corresponding supernatant contains microsomes and soluble proteins (mostly of cytosolic origin).

[a] This is a crucial step. HM should be added dropwise and any aggregates should be resuspended carefully and completely.

The distribution of the major organelles in each fraction of the differential centrifugation procedure is shown in *Figure 1*.

2.2.3 Purification of PO by density gradient centrifugation

The volume of a crude peroxisomal fraction (5 ml) which can be top-loaded on to a gradient corresponds to one liver of approximately 5–6 g. The metrizamide gradient to be used is prepared in advance as a step gradient. It is frozen and transformed to slightly exponential shape by thawing.

Protocol 3. Preparation of a metrizamide gradient

1. Dissolve metrizamide in GB to a concentration of 60% (w/v). This process is time-consuming and should be done the day before.
2. Prepare dilutions of the stock solution adjusting densities to: 1.26, 1.225, 1.19, 1.155, and 1.12 g/ml respectively by means of a refractometer and the formula:

$$\text{density} = 3.350 \times \text{refractive index} - 3.462.$$

3. Layer sequentially 4, 3, 6, 7, and 10 ml of metrizamide solutions (1.26–1.12 g/ml) in a 40 ml centrifuge tube.
4. Immediately freeze the discontinuous gradient in liquid nitrogen and store it at −20°C.
5. Prior to use quickly thaw the gradient at room temperature in a metallic stand.

Protocol 4. Metrizamide density gradient centrifugation

1. Layer 5 ml of the crude peroxisomal fraction on top of the gradient, fill up with GB, and seal the tube.

2. Centrifuge in a vertical rotor at an integrated force of 1.256×10^6 $g \times$ min (g_{max} = 33 000) using slow acceleration/deceleration (total time: about 60 min).
3. Highly purified PO band at a density of 1.23–1.24 g/ml (greenish band) well separated from mitochondria (1.15 g/ml) and microsomes (1.12 g/ml) as is indicated in *Figure 2*.
4. Recover PO fraction by means of a fraction collector or by puncturing the gradient tube and aspiration by means of a syringe.
5. Store fraction at −80 °C.
6. To remove metrizamide, which interferes with the determination of some of the peroxisomal enzymes (urate oxidase) or of protein (Lowry method) dilute the fraction about tenfold with HM and pellet the organelles for 20 min at 25 300 *g*.

Protocol 5. Nycodenz gradient centrifugation (14)

1. Pipette in a 25 ml thick wall polycarbonate tube 2 ml of 56% Nycodenz in GB (ρ = 1.30 g/ml), 4 ml of 45% (ρ = 1.24 g/ml), and 16 ml of 30% (ρ = 1.15 g/ml), and store the tube on ice.
2. Layer 2.5 ml of the crude peroxisomal fraction (corresponding to 5 g liver) on to the step gradient.
3. Centrifuge for 1 h at 35 000 *g* in a fixed-angle rotor (e.g. Beckman 70Ti) using a slow acceleration/deceleration mode.
4. Intact PO will band at the interphase between the upper and the middle layer, while the majority of the contaminating organelles will not enter the first layer.
5. Collect fractions by means of a peristaltic pump (flow rate: 1 ml/min) and a long needle connected via tubing and inserted to the bottom of the tube.

Protocol 6. Isotonic Percoll gradient centrifugation

1. To prepare an isotonic Percoll solution add together 20 g of Percoll (ρ = 1.130), 5 ml of 10 mM EDTA-Na_2 (unbuffered), 3.77 g sucrose, 50 μl of ethanol, and 50 μl of 1 M DTT. Bring solution to pH 8.0 by means of 20 mM MOPS (1.5–3.0 ml) and to volume (50 ml) with water.
2. Layer 2–3 ml of the crude PO fraction (corresponding to one liver) on 36 ml of the cold isotonic Percoll solution in a 40 ml centrifuge tube.
3. Centrifuge for 1 h at 35 000 *g* in a fixed-angle rotor, using slow acceleration/deceleration.

Protocol 6. *Continued*

4. PO band near the top of the gradient (greenish band) just below the microsomal fraction (pinkish) with mitochondria and lysosomes migrating towards the bottom of the gradient.
5. Collect fractions by means of a peristaltic pump and a long needle inserted to the bottom of the gradient. Alternatively, as the PO are in the top of the gradient, the latter might be unloaded by upward displacement.

2.3 Analysis of results

2.3.1 Subcellular fractionation

Marker enzymes of rat hepatic subcellular organelles are summarized in *Table 1* together with the references of their standard assays and that of the Lowry method for the determination of protein concentration. The distribution of the various subcellular organelles in the main fractions is assessed by the percentage distribution (PD) and the enrichment over the homogenate—relative specific activity (RSA)—of the marker enzymes (11). PD = enzyme activity in a fraction divided by enzyme activity in the homogenate (i.e. $U/\Sigma U) \times 100$. PD and RSA values may be tabulated, or more clearly displayed together in the form of a histogram (12) as shown in *Figure 1*, which is based on the data obtained on the differential centrifugation fractions from the liver homogenate (17). Since the distribution of an enzyme is equivalent to the specific activity × protein in the fraction, plotting RSA values of each fraction versus the protein in each fraction leads to a fraction-specific rectangle with its area related to the PD of the marker enzyme, and hence the relative amount of the organelle recovered in that fraction.

According to *Figure 1*, PO are indeed mostly enriched in the light mito-

Table 1. Marker enzymes of rat hepatic subcellular organelles

Organelle	Enzyme[a]	Reference
Peroxisomes	Catalase	22
	Urate oxidase	23
Microsomes	Esterase	24
	NADPH–Cyt *c* red	25
Mitochondria	Cyt *c* ox	26
	L-GluDH	27
Lysosomes	Acid phosphatase	28
	β-Glucuronidase	29
	Protein	30

[a] NADPH–Cyt *c* red, NADPH–cytochrome *c* reductase; Cyt *c* ox, cytochrome *c* oxidase; L-GluDH, L-glutamate dehydrogenase.

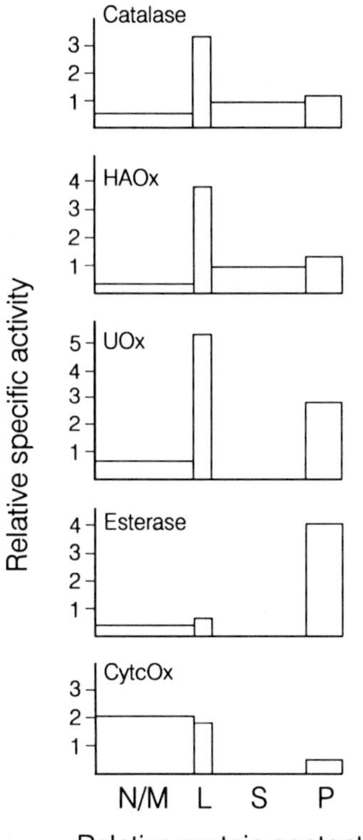

Figure 1. Distribution of marker enzyme activities in main fractions obtained by differential centrifugation of a rat liver homogenate. Abbreviations: HAOx, α-hydroxyacid oxidase; UOx, urate oxidase; Cyt*c*Ox, cytochrome *c* oxidase. Relative protein content is mg/Σmg; relative specific activity = specific activity in the fraction over specific activity in the homogenate. For the meaning of N–S see Section 2.1. Note the four- to fivefold enrichment of PO in the L fraction.

chondrial fraction (L) justifying it to be called an enriched peroxisomal fraction (four- to fivefold enrichment). It is, however, a crude preparation as it is obviously contaminated by microsomes and mitochondria.

2.3.2 Metrizamide density gradient centrifugation

Plotting the relative concentrations of marker enzymes in each collected fraction of a gradient versus the relative volume of each fraction produces a histogram which reflects the distribution of organelles within the gradient. Such a histogram is shown in *Figure 2* which demonstrates the distribution and enrichment of cell organelles after density gradient centrifugation of a crude peroxisomal fraction prepared from rat liver (17). Apparently, PO are mainly recovered in fractions two to four with the mean equilibrium density of 1.245 g/ml. Mitochondria mostly band in a density range of about 1.153 g/ml corre-

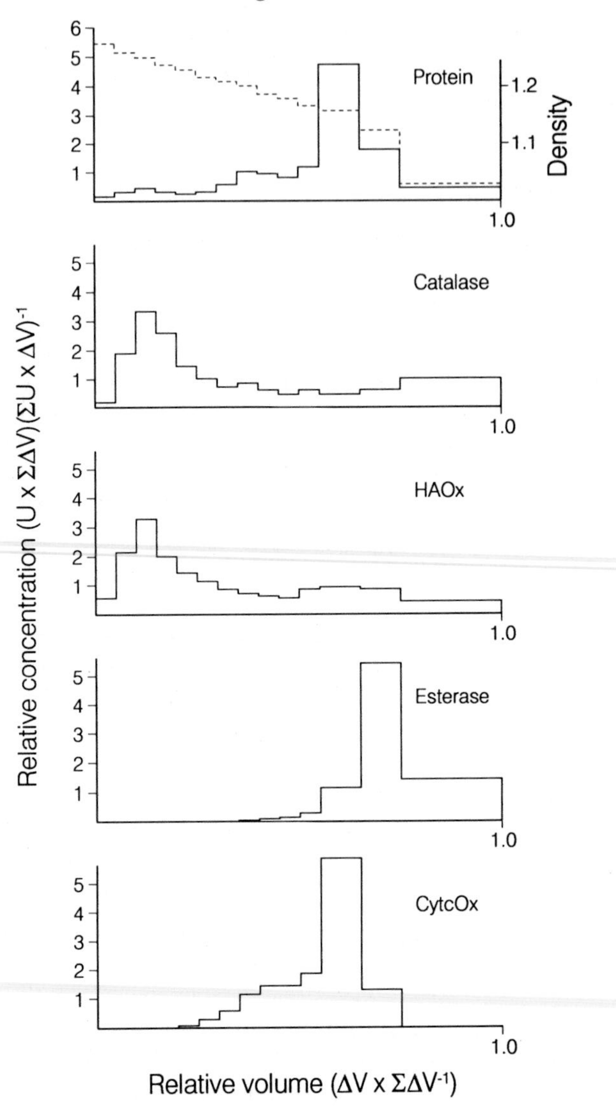

Figure 2. Distribution of protein and enzyme activities following metrizamide density gradient centrifugation of a light mitochondrial fraction. ΔV represents the volume of a single fraction, U its corresponding enzyme activity, $\Sigma\Delta V$ the total gradient volume, and ΣU the total units recovered from all gradient fractions. For abbreviations used see *Figure 1.* Note that PO are mainly concentrated at a mean density of 1.245 g/ml, well separated from all other cell organelles.

sponding to the bulk of protein. Microsomes equilibrate predominantly in the penultimate top fraction (ρ = 1.115 g/ml). Some activities of the peroxisomal enzymes are also found in the uppermost gradient fractions reflecting the fragmentation of the fragile PO during centrifugation.

According to *Figure 2*, the PO fraction is well separated from other cell organelles with a purity of approximately 95–98%, and an enrichment over the homogenate of about 40-fold.

3. Subfractionation of purified rat hepatic peroxisomes

3.1 Extraction of soluble matrix proteins

PO are fragile organelles and isolated PO, albeit seemingly intact are permeable to small molecules (15), although the leakiness of the membrane most probably represents an isolation artefact. Consequently, matrix proteins are supposed to be extracted quite easily. Indeed, simply diluting isolated PO fourfold with ice-cold water and recentrifuging them at about 100 000 *g* causes the release of about 35% of matrix proteins (31) without any substantial loss of enzyme activities. This portion may be increased to 50% by repeatedly freezing and thawing the PO, and nearly brought to 100% by the pyrophosphate pre-treatment according to Leighton *et al.* (32). It has to be pointed out however, that the matrix enzymes are apparently not released completely, and more important, not in parallel; apparently this is due to their divergent solubility or to intermolecular associations which attach these proteins to each other or to the membrane (31). Thiolase was found to be the most soluble peroxisomal enzyme, and the so-called trifunctional protein the least, with catalase and acyl CoA oxidase in between.

3.2 Extraction of core-bound urate oxidase

Disassembly of PO by any of the procedures mentioned above does not release urate oxidase. This is only accomplished by the carbonate treatment of PO which not only solubilizes all matrix proteins including urate oxidase but also removes peripheral membrane proteins (for details see Section 3.3). A highly purified preparation of free crystalloid cores however, is obtained by subjecting a light mitochondrial fraction to isopycnic density gradient centrifugation at 130 000 *g* for 60 min on a linear (20–50%, w/v) metrizamide gradient (33,34). Under these conditions free cores band well separated from the PO fraction in the density range of 1.28–1.30 g/ml. The core fraction thus obtained consists mainly if not exclusively of urate oxidase as is demonstrated by SDS–PAGE (see *Figure 3*).

3.3 Isolation of integral membrane proteins of PO

Peroxisomal membrane proteins (PMP) constitute about 10% of total peroxisomal proteins (35). Though the functions of a few of them have been unraveled, most PMPs still have to be characterized biochemically as well as metabolically.

In subfractionating isolated PO to prepare integral PMPs procedures such as osmotic shock, freezing and thawing, sonication, or even pyrophosphate treatment are not sufficient. The most straightforward alternative is the carbonate treatment introduced by Fujiki *et al.* (35). It completely removes matrix enzymes, urate oxidase included, as well as peripheral membrane proteins. It is non-destructive (inasmuch as there is no hydrolytic cleavage of polypeptides), though membranes are fragmented to sheets (35), yet it abolishes the activity of most of the enzymes.

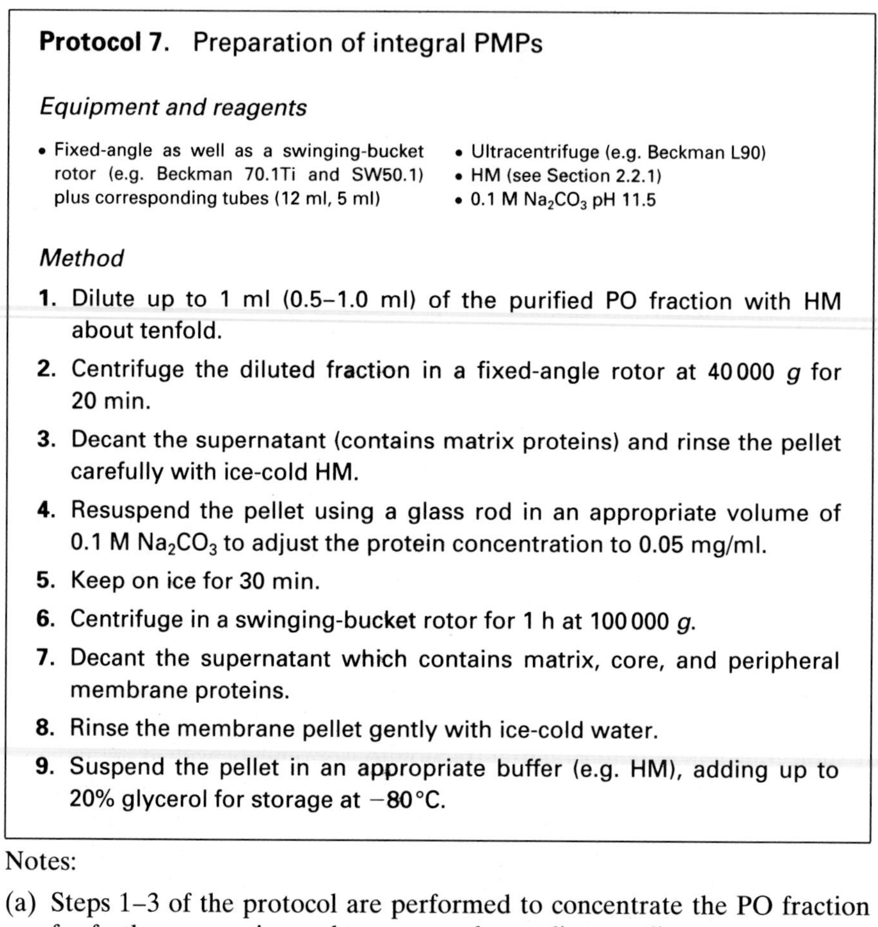

Protocol 7. Preparation of integral PMPs

Equipment and reagents

- Fixed-angle as well as a swinging-bucket rotor (e.g. Beckman 70.1Ti and SW50.1) plus corresponding tubes (12 ml, 5 ml)
- Ultracentrifuge (e.g. Beckman L90)
- HM (see Section 2.2.1)
- 0.1 M Na_2CO_3 pH 11.5

Method

1. Dilute up to 1 ml (0.5–1.0 ml) of the purified PO fraction with HM about tenfold.
2. Centrifuge the diluted fraction in a fixed-angle rotor at 40 000 *g* for 20 min.
3. Decant the supernatant (contains matrix proteins) and rinse the pellet carefully with ice-cold HM.
4. Resuspend the pellet using a glass rod in an appropriate volume of 0.1 M Na_2CO_3 to adjust the protein concentration to 0.05 mg/ml.
5. Keep on ice for 30 min.
6. Centrifuge in a swinging-bucket rotor for 1 h at 100 000 *g*.
7. Decant the supernatant which contains matrix, core, and peripheral membrane proteins.
8. Rinse the membrane pellet gently with ice-cold water.
9. Suspend the pellet in an appropriate buffer (e.g. HM), adding up to 20% glycerol for storage at −80 °C.

Notes:

(a) Steps 1–3 of the protocol are performed to concentrate the PO fraction for further processing and to remove the gradient medium.

(b) The pH of the carbonate solution is of crucial importance and should be kept at 11.0–11.5. A lower pH is less effective or even ineffective (35).

(c) $K_2B_4O_7$ or K_2CO_3 may replace Na_2CO_3.

(d) PO obtained by metrizamide gradient centrifugation should be subsequently treated with carbonate immediately. When frozen prior to the treatment, isolated membranes prove to be contaminated by urate oxidase which cannot be removed.

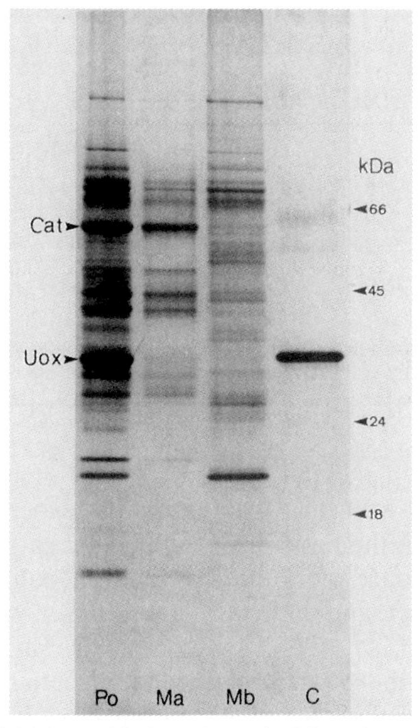

Figure 3. SDS–PAGE of highly purified rat hepatic PO and corresponding subfractions. Amounts of protein loaded are: highly purified PO (Po), matrix fraction (Ma), and membrane fraction (Mb), 5 μg each; core fraction (C), 1 μg. Standards employed are: 66 kDa, bovine serum albumin; 45 kDa, ovalbumin; 24 kDa, trypsinogen; 18.5 kDa, bovine β-lactoglobulin. Note that catalase (Cat) and urate oxidase (Uox) are exclusively confined to the matrix and core fraction respectively.

In *Figure 3* the polypeptide pattern of highly purified PO and corresponding subfractions as revealed by SDS–PAGE are compared. The subfractions were prepared as outlined in Sections 3.1–3.3. SDS–PAGE was performed according to Laemmli (36) using a 10–20% gradient gel (18 × 40 cm) for resolving, and Ag staining (37) to visualize the polypeptide bands. Comparing the polypeptide pattern, it is evident from lanes 1 (Po) and 2 (Ma), that catalase, the most abundant PO protein, is clearly a matrix protein, as are the bulk of peroxisomal proteins (not specified). In contrast, urate oxidase is confined exclusively to the core fraction (lane C) and is neither soluble nor associated with the membrane (lane Mb).

4. Comparative characterization of isolated hepatic peroxisomes from different species

4.1 Enzyme activities of PO fractions

The approach described for the preparation of highly purified rat hepatic PO has been applied to the isolation of PO fractions from the livers of monkey

Table 2. Properties of purified peroxisomal fractions[a] from livers of different species

	Protein (mg/ml	Cat	AOx	HAOx	UOx	PH	Esterase	CytcOx
					U/mg			
Maccaca	0.39	5.74	25.36	262.1	–	2.1	0.56	0.003
Marmoset	0.41	0.36	15.96	127.7	0.003	–	1.31	0.019
Guinea-pig	3.01	4.24	33.83	61.6	0.006	0.6	0.46	0.017
Rat	0.26	7.65	55.82	193.2	0.16	4.3	0.11	0.010
Mouse	0.16	12.17	77.99	35.5	0.08	0.8	0.57	0.030

[a] Cat, catalase; AOx, acyl CoA oxidase; HAOx, L-α-hydroxyacid oxidase; UOx, urate oxidase; PH, peroxisomal trifunctional protein; CytcOx, cytochrome *c* oxidase.

(*Maccaca arctoides*), marmoset, guinea-pig, and mouse. Amounts of 8–9 g liver were processed each in a representative experiment and only minor changes were made in the basic protocol: the guinea-pig was starved for 48 h instead of 24 h to reduce more effectively the hepatic fat load; the livers of the monkey, marmoset, and mice were not perfused with physiological saline to wash out blood.

The properties of the peroxisomal fractions obtained are summarized in *Table 2* compared to those of highly purified rat hepatic PO. Mostly, the respective enzyme activities measured are approximately in the same range, there are however some notable discrepancies.

(a) Catalase activity of marmoset PO is remarkably low, supposedly not due to an unusually high fragility of the particles but to an inherently extremely low level of the enzyme in the marmoset liver (1.5 U/g liver versus 52 U/g liver in the rat).

(b) Urate oxidase is lacking in Maccaca PO, as is common in hepatic PO of hominoids and humans, but this enzyme is also of quite low concentration in PO of the guinea-pig.

(c) Hydroxyacid oxidase activities vary considerably, with mouse PO containing the lowest and Maccaca PO the highest level.

(d) Whereas activities of acyl CoA oxidase, the initial enzyme of the peroxisomal β-oxidation system are widely consistent, those of the second one, the trifunctional protein are more discrepant, being highest in rat PO. This might indicate that the enzyme is not exclusively involved in the degradation of fatty acids via β-oxidation but may have other functions.

Comparing the individual PO preparations it has to be considered however, that they are differentially contaminated. Thus, only the rat peroxisomal fraction may be designated 'highly purified' consistent with the predictable 40-fold enrichment over the homogenate and a rate of contamination by microsomes and mitochondria of about 2%.

4.2 Immunoblotting of PO fractions

For further characterization, PO fractions have been subjected to SDS–PAGE followed by immunoblotting. Resolving gels (8.5 × 5 × 0.1 cm) of 10% (catalase, urate oxidase, peroxisomal *thiolase*, peroxisomal trifunctional protein) and 12.5% (acyl CoA oxidase) were used, and electrotransfer was accomplished by semi-dry blotting (38). Immunocomplexing and visualization of the immune complexes were performed as described previously (39). Enzyme proteins were purified from rat liver according to standard procedures (for details see ref. 39), and employed for immunization of New Zealand rabbits. Antisera obtained were processed to prepare total IgG fractions (40), and the respective polyclonal antibodies were assessed for monospecificity.

The results of the immunoblotting experiments are compiled in *Figure 4*.

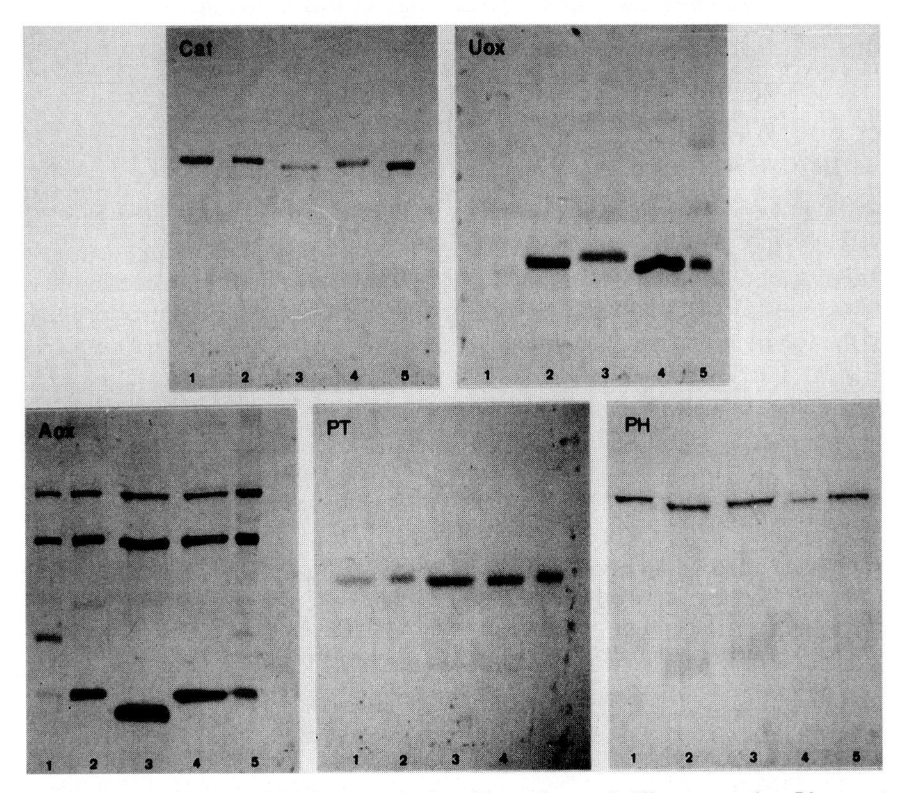

Figure 4. Immunoblots of PO fractions isolated from livers of different species. Divergent amounts of protein (1–5 μg) were loaded per lane to obtain signals of comparable staining intensity. Lanes 1–5: monkey, rat, guinea-pig, marmoset, mouse (except Uox: lane 4, mouse; lane 5, marmoset). Cat, catalase; Uox, urate oxidase; Aox, acyl CoA oxidase; PT, peroxisomal thiolase; PH, peroxisomal trifunctional protein. Note the lack of Uox as well as the modified Aox polypeptide pattern in monkey PO (lane 1), and the minor differences in the molecular weights of the individual enzyme proteins.

Each antibody used cross-reacted with the corresponding antigenic protein of the PO fractions studied, and only one polypeptide was recognized. The main findings disclosed by the blots are:

(a) No urate oxidase protein is detectable in PO of the Maccaca monkey which is in line with the lack of enzymatic activity (see *Table 2*).

(b) Of the typical polypeptide pattern of acyl CoA oxidase, the fragment commonly appearing at about 21 kDa in Western blots is scarcely visible in Maccaca PO. Instead, a band at a molecular weight of about 30 kDa is observed. This points to a unique processing of peroxisomal acyl CoA oxidase in the liver of Maccaca monkeys.

(c) There are only minor species-dependent differences in the apparent molecular weights of the individual enzyme proteins. This might indicate a high degree of interspecies homology in the amino acid sequences of those proteins.

5. Embedding of PO fractions for electron microscopy (EM)

Morphological characterization of PO fractions is an essential procedure to provide crucial information on the integrity, heterogeneity, and the form and size distribution of the isolated organelles; but also their enzyme content may be demonstrated using immuno-EM or cytochemistry. Since matrix proteins of the fragile PO may be easily extracted, great care must be taken to process the PO fractions immediately after their isolation. Filtration of the fixed organelles on to a filter support, post-fixation with reduced osmium, and embedding in Epon 812 usually provide an excellent morphology (see *Figure 5A*), whereas centrifugation steps after fixation will impair their morphological appearance (*Figure 5C*). To identify core-free organelles as PO, histochemical staining for the peroxidatic activity of catalase using the alkaline DAB method (41,42) may be applied. Additionally, post-embedding immunocytochemistry of LR White embedded material may be used to localize proteins in individual PO.

In the following protocols the fixation and filtration of isolated PO on to a filter support, and their embedding by different approaches will be described.

5.1 Equipment and solutions

Equipment

- Filtration apparatus according to Baudhuin *et al.* (43)
- Gas tank with dry nitrogen and reduction valves
- Eppendorf centrifuge
- Water-bath with shaker (37 °C)

- Carousel for incubating glass vials with filters
- Polymerization oven (50 °C and 60 °C)
- Ultramicrotome and a knifemaker with glass strips
- Transmission electron microscope
- Millipore filters (13 mm diameter; 0.05 μm pore size)
- Knives for dissection or razor blades
- Plastic beakers with cover for preparing Epon 812
- BEEM capsules (size 3) and capsule holders
- Gelatine capsules
- Glass vial with cover (< 5 ml)
- Desiccation chamber with airtight cover
- Nickel grids (200 mesh)
- Fine forceps (watchmaker tweezers)

Reagents

(a) Fixatives:
- 25% (w/w) aqueous glutaraldehyde (GA), ultrapure stock
- 0.5% GA in aqueous metrizamide ($\rho = 1.24$ g/ml)
- 0.25% GA in GB (see Section 2.2.1)
- 1.25% GA in cacodylate buffer (see below)
- 1.25% GA in TBS buffer (see below)

(b) Buffers:
- 0.2 M sodium cacodylate stock buffer pH 7.4
- 0.05 M Teorell–Stenhagen stock buffer pH 10.5, containing 0.05 M phosphoric acid, 0.075 M boric acid, 0.035 M citric acid, and 0.345 M NaOH
- TBS buffer: 20 mM Tris, 150 mM NaCl pH 7.4
- TBSA buffer: 0.1% and 1% (w/v) bovine serum albumin in TBS

(c) Agar–agar: 7.5% (w/v). Stir on a hot plate until solution boils and cool down to 37 °C with continuous stirring.

(d) Graded series of ethanol solutions: 70%, 80%, 90% (v/v), and absolute.

(e) 2% (w/v) osmium tetroxide.

(f) Epon 812 mixture. In a 25 ml beaker mix: 8.1 g methyl-5-norbornene-2,3-dicarboxylic anhydride (MNA), 6.1 g (2-dodecen-1-yl) succinic anhydride (DDSA), 13 g Epon 812, and stir for 30 min. Add 0.275 ml 2,4,6-Tris(dimethylaminomethyl)phenol (DMP-30) and stir for another 30 min prior to use.

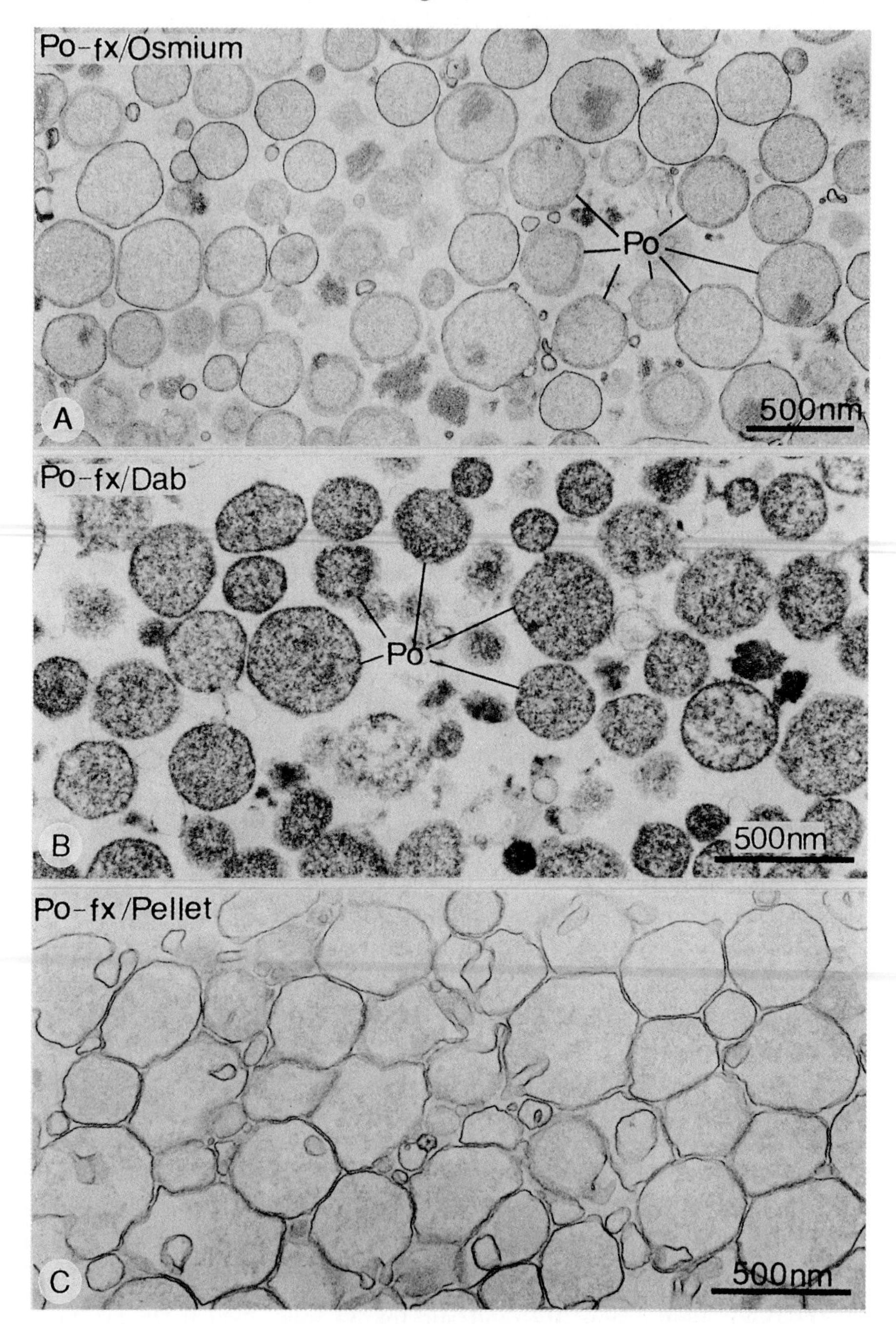

Figure 5. Electron micrographs of PO. (A) Filtration technique—OsO_4 fixed. (B) Filtration technique—DAB treated PO. (C) Pelleting technique. For further details see text.

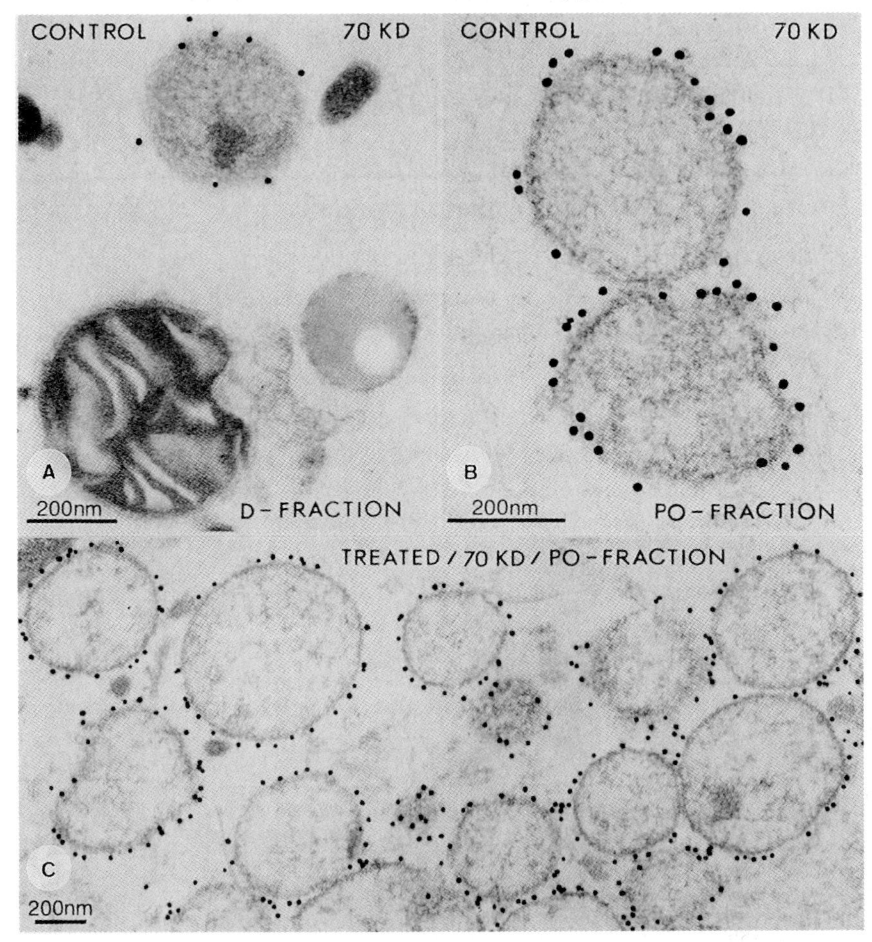

Figure 6. Electron micrographs of isolated PO processed by pre-embedding immuno-electron microscopy using the protein A–gold technique and an antibody to the 70 kDa PO membrane protein. (A) and (B) show PO from a control animal. (C) PO from an animal treated with PO proliferator showing more extraction of the matrix than in control PO. Fraction D corresponds to fraction L in *Figure 1*.

(g) DBA medium: 0.2% (w/v) 3,3′-diaminobenzidine, 0.15% H_2O_2, 0.01 M TS buffer. Always prepare this medium fresh and keep in the dark. DAB does not dissolve completely in TS, therefore dissolve in distilled water, before adjusting to 0.01 M TS. The solution is slightly yellow-rose coloured; if DAB is oxidized it turns brown and precipitates.

5.2 Processing of PO fractions for EM

Fixation of the cell organelles immediately after isolation is a prerequisite for high quality EM morphology. To avoid extraction of PO during fixation and to preserve the antigenicity of the organellar proteins fixatives should contain the density gradient medium (e.g. metrizamide) with the same density used for the isolation of the PO fraction, i.e. 1.24 g/ml, and low concentrations of GA (0.25%).

Protocol 8. Fixation of PO during filtration

1. Keep isolated PO prior to fixation only briefly on ice.
2. Use 400 μg PO proteins per Millipore filter.
3. Add the same volume of 0.5% GA fixative.
4. Adjust the volume to 800–1000 μl, adding 0.25% GA in GB.
5. Keep on ice for 15 min, while setting-up the filtration device, connecting the nitrogen tank to the filter system, and inserting the Millipore filters.
6. Overlay the filters with the pre-fixed PO solution, close the cover of the device, and apply 1 atm with N_2 gas to the filters.
7. After filtering all the solution, release the pressure, open the filtration device, and take out the filters with forceps.[a]
8. Cut the filters with a razor blade in pieces.
9. Make sure that pieces will not dry out during sectioning by placing them into the appropriate fixative.

[a] Only the rim of the filters should be touched; they will stay white, and PO can be seen as a slightly beige overlay on the filters.

Protocol 9. Embedding of PO filters for routine EM

1. Fix the PO on the filter for another 10 min with 1.25% GA in cacodylate buffer.
2. Remove the fixative with a Pasteur pipette without touching the PO.
3. Wash the filters immediately three times with 0.1 M cacodylate buffer to remove all traces of fixative.
4. Post-fix PO for 5 min adding 1 ml of 2% aqueous OsO_4 solution.
5. Add 1 ml of potassium ferricyanide solution to reduce the osmium: final concentration 1% OsO_4, 150 mg/ml $K_4[Fe(CN)_6]$.
6. Close the top of the vials and continue post-fixation for 1 h.
7. Decant the solution of reduced OsO_4 and rinse filters briefly with distilled water.
8. Cut the filter into pieces of 2 mm width and 5 mm length and insert quickly into a warm (37 °C) drop of agar–agar.[a]

9. After the agar has cooled and hardened remove as much as possible of the agar around the filter.
10. Start dehydration using the graded series of ice-cold ethanol solutions: three times, 5 min for each step. It is important that a shorter time is not used.
11. Remove the absolute ethanol by adding propylene oxide (three times; 15 min), followed by propylene oxide/Epon (1:1) for 30 min.
12. Remove the mixture and leave the blocks overnight in pure Epon 812 on a rotating carousel in a desiccator with two dishes of crushed $CaCl_2$ to remove all moisture.
13. Place three drops of freshly prepared Epon into labelled BEEM capsules placed into a capsule holder.
14. Take out the filter piece with the PO and remove the old Epon on a blotting paper.
15. Insert the agar blocks vertically into the capsule with the filter side down.
16. Fill up the capsule to the rim with fresh Epon and polymerize the blocks at 60°C for two days in a $CaCl_2$ desiccated polymerization oven.

[a] Filters should not dry during cutting; they must be completely covered by agar–agar otherwise they will dissolve in the propylene oxide used for embedding.

Protocol 10. Histochemical staining for catalase with the alkaline DAB method

1. Follow *Protocol 9,* steps 1–3.
2. Rinse the filters three times in TS buffer.
3. Add the DAB solution and incubate for 1 h at 37°C in complete darkness with gentle shaking.
4. Rinse three times with 0.1 M cacodylate buffer, and proceed according to *Protocol 9,* step 4.

Protocol 11. Pre-embedding immuno-EM of PO fractions

1. Fix the PO during filtration in 0.25% GA in GB.
2. Rinse the filters by filtration with 800 μl of TBS buffer (TBS blocks free aldehyde groups).
3. Block non-specific protein binding sites by filtration of 800 μl of a 1% BSA solution in TBS.
4. Add 800 μl of the specific antibody solution (in our case an IgG frac-

Protocol 11. *Continued*

tion of a polyclonal rabbit anti-rat 70 kDa peroxisome membrane protein (PMP) antibody: 40 μg/μl, diluted 1:100 in TBS).

5. Filter half of the antibody solution, and leave the remaining half for overnight incubation at 4 °C in the filtration device.
6. Filter the remaining antibody solution, and remove all traces by filtration of 800 μl of TBSA buffer.
7. For visualization of immune complexes add:
   (a) Either 800 μl of a protein A–gold solution (size of the gold particles 10 nm) prepared according to refs 44 and 45. OD of a 1:25 dilution in TBSA = 0.5.
   (b) Alternatively, 800 μl of a polyclonal goat anti-rabbit antibody conjugated to 10 nm gold particles (Sigma Co.) diluted 1:50 to 1:100 in TBSA, supplemented with 1% goat serum.
8. Filter slowly for 60 min with reduced N_2 pressure.
9. Remove the unbound protein A or goat antibody by filtration of 800 μl TBSA.
10. Take out the filters, and fix the immunocomplexes for 10 min with 1.25% GA in TBS buffer.
11. Rinse the filters three times with TBS, and proceed according to *Protocol 9*, step 8.

Protocol 12. Pellet protocol for pre-embedding immuno-EM

If the filtration procedure is unsatisfactory, often the pores of filters become blocked, then centrifugation can be used for each of the filter steps described in *Protocol 11* as an alternative procedure.

1. After the immunological incubations resuspend the pellet very carefully using a glass rod; add a drop of agar–agar into the Eppendorf tube, and mix the PO into the agar with a spatula.
2. Immediately centrifuge for 2 min during the cooling of the agar to concentrate the PO on one side of the agar drop.
3. Cut off the bottom of the Eppendorf tube with a razor blade, and take out the pellet with forceps.
4. Cut the pellet into small pieces and post-fix with reduced OsO_4.
5. Proceed according to *Protocol 9*, step 10.

5.3 Results

In routine EM, 10–15 layers of PO are seen on the filter depending on the amount of protein loaded. PO are well preserved with a fine granular matrix surrounded by an intact single limiting unit membrane (*Figure 5A*). Nearly 15–20% of PO profiles contain crystalline cores in their matrix, and only a few free cores are seen. After DAB incubation, PO can be recognized as electron dense organelles with a granular matrix compartment (*Figure 5B*). Cores cannot be easily distinguished in these preparations. If pelleting is used instead of filtration usually the PO are clustered together and their morphology is impaired (*Figure 5C*).

In pre-embedding 70 kDa PMP preparations, PO are easily identified by the gold particles on the cytoplasmic face of their membranes (*Figure 6*). If preparations from control animals and animals treated with PO proliferators are compared, it can be clearly seen, that proliferated PO usually are more extracted than control PO (compare *Figure 6B* and *C*).

Post-embedding immuno-EM results of PO fractions are shown in ref. 46.

Acknowledgements

The original work in the laboratories of the authors has been supported by grants of the Deutsche Forschungsgemeinschaft, Bonn, FRG (Fa146/1–3, Vo317/3–1. Vo317/4–1, SFB352), and Landesforschungsschwerpunkt-Programm of the State of Baden-Wuerttemberg, FRG.

References

1. Böck, P., Kramar, R., and Pavelka, M. (1980). *Peroxisomes and related particles in animal tissues*. Cell Biol. Monographs, Vol. 7. Springer-Verlag, NY.
2. Fahimi, H. D. and Sies, H. (ed.) (1987). *Peroxisomes in biology and medicine*. Springer–Verlag, Heidelberg.
3. Gorgas, K. (1984). *Anat. Embryol.*, **169**, 261.
4. Gorgas, K. (1985). *Anat. Embryol.*, **172**, 21.
5. Yamamoto, K. and Fahimi, H. D. (1987). *J. Cell Biol.*, **105**, 713.
6. Zaar, K. (1992). *Eur. J. Cell Biol.*, **59**, 233.
7. Mannaerts, G. P. and Van Veldhoven, P. P. (1992). *Cell Biochem. Funct.*, **10**, 141.
8. Van den Bosch, H., Schutgens, R. B. H., Wanders, R. J. A., and Tager, J. (1992). *Annu. Rev. Biochem.*, **61**, 157.
9. Lock, E. A., Mitchel, A. M., and Elcombe, C. R. (1989). *Annu. Rev. Pharmacol. Toxicol.*, **29**, 145.
10. Fahimi, H. D., Baumgart, E., Beier, K., Pill, J., Hartig, F., and Völkl, A. (1993). In *Peroxisomes: biology and importance in toxicology and medicine* (ed. G. G. Gibson and B. Lake), pp. 395–424. Taylor and Francis Ltd., London.
11. Rickwood, D. (1984). *Centrifugation: a practical approach* (ed. D. Rickwood). IRL Press, Oxford.

12. De Duve, C., Pressman, B. C., Gianetto, R., Wattiaux, R., and Appelmans, F. (1955). *Biochem. J.*, **60**, 604.
13. Leighton, F., Poole, B., Beaufay, H., Baudhuin, P., Coffey, J. W., Fowler, S., *et al.* (1968). *J. Cell Biol.*, **37**, 482.
14. Verheyden, P. P., Fransen, M., Van Veldhoven, P. P., and Mannaerts, G. P. (1992). *Biochim. Biophys. Acta*, **1109**, 48.
15. Van Veldhoven, P. P., Debeer, L. I., and Mannaerts, G. P. (1983). *Biochem. J.*, **210**, 685.
16. Neat, C. E., Thomassen, M. S., and Osmundsen, H. (1980). *Biochem. J.*, **186**, 369.
17. Völkl, A. and Fahimi, H. D. (1985). *Eur. J. Biochem.*, **149**, 257.
18. Hajra, A. K. and Wu, D. (1985). *Annal. Biochem.*, **148**, 233.
19. Hartl, F., Just, W. W., Köster, A., and Schimassek, H. (1985). *Arch. Biochem. Biophys.*, **237**, 124.
20. Lüers, G., Hashimoto, T., Fahimi, H. D., and Völkl, A. (1993). *J. Cell Biol.*, **121**, 1271.
21. Cajaraville, M. P., Völkl, A., and Fahimi, H. D. (1992). *Eur. J. Cell Biol.*, **59**, 255.
22. Baudhuin, P., Beaufay, H., Rahman-Li, Y., Sellinger, O. Z., Wattiaux, R., Jaques, P., *et al.* (1964). *Biochem. J.*, **92**, 179.
23. Priest, D. G. and Pitts, O. M. (1972). *Anal. Biochem.*, **50**, 195.
24. Beaufay, H., Amar-Costesec, A., Feytmans, E., Thines-Sempoux, D., Wibo, M., Robbi, M., *et al.* (1974). *J. Cell Biol.*, **61**, 188.
25. Sottocasa, G. L., Kuylenstierna, B., Ernster, L., and Bergstrand, A. (1967). *J. Cell Biol.*, **32**, 415.
26. Cooperstein, S. J. and Lazarow, A. (1951). *J. Biol. Chem.*, **189**, 665.
27. Hogeboom, G. H. and Schneider, W. C. (1953). *J. Biol. Chem.*, **204**, 233.
28. Sigma: Techn. Bull. no 104.
29. Meisler, M. and Paigen, K. (1972). *Science*, **177**, 894.
30. Lowry, O. H., Rosebrough, N. J., Farr, A. L., and Randall, R. (1951). *J. Biol. Chem.*, **193**, 265.
31. Alexson, S., Fujiki, Y., Shio, H., and Lazarow, P. B. (1985). *J. Cell Biol.*, **101**, 294.
32. Leighton, F., Poole, B., Lazarow, P. B., and De Duve, C. (1969). *J. Cell Biol.*, **41**, 521.
33. Völkl, A., Baumgart, E., and Fahimi, H. D. (1988). *J. Histochem. Cytochem.*, **36**, 329.
34. Hajra, A. K. and Bishop, I. E. (1982). *Ann. N.Y. Acad. Sci.*, **386**, 170.
35. Fujiki, Y., Fowler, S., Shio, H., Hubbard, A. L., and Lazarow, P. B. (1982). *J. Cell Biol.*, **93**, 103.
36. Laemmli, U. K. (1970). *Nature*, **227**, 680.
37. Oakley, B. R., Kirsch, D. R., and Morris, N. R. (1980). *Anal. Biochem.*, **105**, 361.
38. Kyhse-Andersen, J. (1984). *J. Biochem. Biophys. Methods*, **10**, 203.
39. Beier, K., Völkl, A., Hashimoto, T., and Fahimi, H. D. (1988). *Eur. J. Cell Biol.*, **46**, 383.
40. Mayer, R. J. and Walker, J. H. (ed.) (1980). *Immunochemical methods in the biological sciences: enzymes and proteins*. Academic Press, London.
41. Fahimi, H. D. (1968). *J. Histochem. Cytochem.*, **16**, 547.
42. Fahimi, H. D. (1969). *J. Cell Biol.*, **43**, 275.
43. Baudhuin, P., Evrard, P., and Berthet, J. (1967). *J. Cell Biol.*, **32**, 181.

44. Slot, J. W. and Geuze, H. J. (1981). *J. Cell Biol.*, **90**, 533.
45. Slot, J. W. and Geuze, H. J. (1985). *Eur. J. Cell Biol.*, **38**, 87.
46. Fahimi, H. D. and Baumgart, E. (1993). In *Electron microscopic cytochemistry and immunocytochemistry in biomedicine* (ed. K. Ogawa and T. Barka), pp. 491–504. CRC Press, Boca Raton.

6

Lysosomes and endocytosis

TOR GJØEN, TROND OLAV BERG, and TROND BERG

1. Current models of endocytosis

Endocytosis is a multistep process whereby molecules are internalized into cells and subsequently transported to one of four destinations: lysosomes, the opposite side of the (epithelial) cell, storage granules, or back to the plasma membrane from which internalization took place. The various steps in the endocytic process are mediated by specialized organelles. Internalization may take place inside or outside coated pits and therefore involves two different areas of the plasma membrane (1). Clathrin-dependent internalization is a well-known process whereas the structure of the plasma membrane that mediates clathrin-independent internalization is not known. Invagination of coated pits leads to coated vesicles from which 'primary endosomes' are formed. Clathrin-independent endocytosis leads directly to formation of smooth-surfaced primary endosomes. Primary endosomes carry their cargo to early endosomes that consist of tubules and luminal parts. In some cells the early endosomes may gradually develop into multivesicular bodies. This 'maturation' involves formation of internal vesicles and loss of tubular extensions (2). The early endosomes serve as a sorting station. Some endocytic receptors are returned to the plasma membrane in 'receptosomes' whereas others follow a transcytotic route to the opposite side of the cells. Some receptors follow the ligand to the lysosomes. In other cells the transport from early to late endosomes may not primarily be due to a maturation process. Instead, specific carrier vesicles may bud off from early endosomes and these are subsequently moved along microtubules to late endosomes (3). The mechanism whereby receptor–ligand complexes are transported from endosomes to lysosomes is not known. Moreover, lysosomes may exist in different states of activity, and functions previously ascribed to lysosomes (e.g. proteolysis), may certainly also take place in endosomes (4). In some cells, such as macrophages, the dense lysosomes may primarily serve as storage granules while the main degradation takes place in a prelysosomal, late endosomal compartment. Transport to lysosomes may be mediated by vesicles budding off from the prelysosomal compartment. In other cells, late endosomes may fuse with lysosomes.

2. Analysis of endocytic uptake of ligands and pinocytosis markers

2.1 Receptor-mediated endocytosis

2.1.1 Analysis of binding

Physiological ligands for endocytic receptors usually occur in relatively low concentrations in extracellular fluids and the equilibrium dissociation constant for binding is often in the nanomolar range. To determine binding affinity and the functional number of endocytic receptors the most commonly used method is to incubate cells with increasing concentrations of ligand at low temperature (to avoid internalization), and then, when binding equilibrium is reached, measure cell-associated radioactivity. The binding data can be treated according to Scatchard (5) to obtain the equilibrium dissociation constant as well as the number of binding sites. Problems encountered include non-specific binding that must be corrected for. This can usually be done by measuring binding in presence of a surplus of unlabelled ligand. In cases in which the receptor is a lectin (the mannose receptor and the galactose receptor, for instance) unspecific binding can be measured by adding an excess of the oligosaccharide for which the receptor is specific, for instance lactose in the case of the galactose receptor. Several receptors (the LDL receptor, the mannose receptor, the galactose receptor) are calcium-dependent, and binding in absence of Ca^{2+} (or in presence of sufficient EGTA) is therefore non-specific. *Protocol 1* presents an example of how to determine the number of binding sites and equilibrium constant for binding of asialoglycoprotein (asialoorosomucoid) to the galactose receptor in rat liver parenchymal cells.

Protocol 1. Determination of the equilibrium constant (association constant) for binding of [^{125}I]asialoorosomucoid to the galactose receptor and number of galactose receptors in rat liver parenchymal cells

Equipment and reagents

- Incubation buffer: 137 mM NaCl, 5.4 mM KCl, 0.34 mM Na_2HPO_4, 0.35 mM KH_2PO_4,. 0.81 mM $MgSO_4$, 40 mM Hepes, 2 mM $CaCl_2$ pH 7.6
- Low-speed centrifuge and rotor
- Bovine serum albumin
- [^{125}I]asialoorosomucoid and unlabelled asialoorosomucoid

Method

1. Incubate isolated hepatocytes at 0–4°C as shaking suspensions in incubation buffer (137 mM NaCl, 5.4 mM KCl, 0.34 mM Na_2HPO_4, 0.35 mM KH_2PO_4, 0.81 mM $MgSO_4$, 40 mM Hepes, 2 mM $CaCl_2$ pH 7.6)

with 1% (w/v) bovine serum albumin (BSA). Use 15 ml centrifuge tubes (0.4 ml in each tube, 5×10^6 cells/ml).

2. To two series of tubes add [^{125}I]asialoorosomucoid to give the following concentrations: 1 nM, 5 nM, 10 nM, 20 nM, 40 nM (three tubes for each concentration). To one series of tubes add unlabelled asialoorosomucoid to a final concentration of 0.5 μM.
3. After incubating the cells for 90 min add 10 ml of ice-cold incubation medium and centrifuge at 300 *g* for 2 min. Wash the cells twice by adding new ice-cold medium and recentrifuge.
4. Resuspend the final cell pellet in, for example, 1 ml of medium and measure radioactivity in an aliquot of the suspension.
5. Calculate amount of [^{125}I]asialoorosomucoid bound per litre of cell suspension by subtracting binding in presence of excess unlabelled ligand from binding measured in absence of unlabelled ligand. Express binding as moles per litre.
6. Plot binding at equilibrium (as moles per litre) against the ratio of bound ligand divided by free ligand (at equilibrium). Binding may be expressed as moles per litre per 10^9 cells whereas bound/free ligand should be expressed as $1/10^9$ cells. Equilibrium constant and the number of binding sites may be obtained from the slope of the line and the intersection with the abscissa respectively. Computer programs are available for obtaining binding affinity (association constant) and number of binding sites.

Endocytic receptors are internalized and recycled back to the cell surface. For receptors mediating internalization of nutrients (LDL receptor, transferrin receptor) this process takes place constitutively. As a result, a considerable proportion of the receptor is at any given time located intracellularly and will under normal binding conditions not be measured, since internalization is blocked at low temperature. Analysis of the total number of receptors involved in binding may be performed in permeabilized cells. This has been done successfully by Weigel and co-workers for the galactose receptor in rat liver parenchymal cells (6). The cells were permeabilized by treatment with digitonin (0.055% (w/w) in medium added from a stock solution of 14 mg/ml in ethanol). This treatment evidently gave the ligand access to intracellular receptors without releasing the receptors themselves from the cell.

2.1.2 Analysis of recycling

Although the acidity of the endosomes usually reduces the binding affinity of the ligands for their specific receptors a considerable proportion of the internalized ligand may be recycled ('retroendocytosed') to the cell surface bound

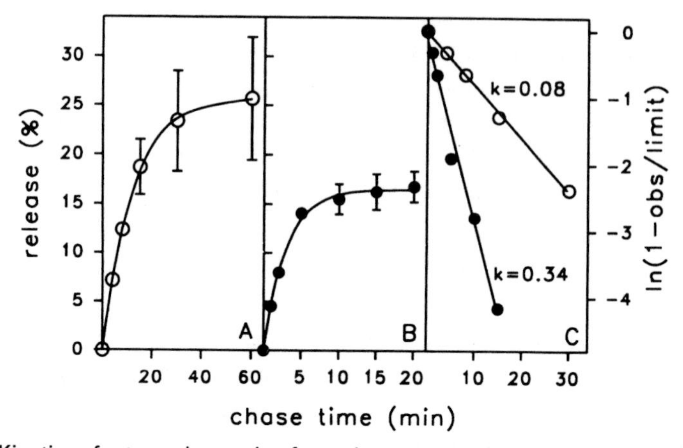

Figure 1. Kinetics of retroendocytosis of neoglycoproteins from rat liver cells. Cells were pulsed for 10 min at 37 °C with 1.5 nM [^{125}I]-labelled neoglycoproteins, washed with cold dissociating buffer, and incubated further at 37 °C in dissociating buffer (containing 5 mM EGTA). At the time points indicated, release of endocytosed neoglycoprotein was measured as described in *Protocol 2*. (A) Release of galactosylated, Gal_{28}–BSA from parenchymal cells. (B) Release of mannosylated, Man_{21}–BSA from endothelial cells. (C) Transformation of data in (A) (open symbol) and (B) (closed symbol) according to the equation $y = \ln(1 - \text{obs/limit})$; k = rate constants for the release. The figure shows mean values ± standard deviation for at least three identical experiments. From ref. 7.

to its receptor. To measure retroendocytosis it is necessary to develop a method for releasing surface-bound ligand. Three such methods may be envisaged. First, if the binding is calcium-dependent, the ligand may be dissociated by removing calcium (e.g. by adding EGTA). Secondly, if the binding affinity is reduced by lowering the pH, the ligand may be removed by an acid wash. Surface binding may conceivably also be removed by protease treatment. When a method for removing surface-bound ligand is available, retroendocytosis may be measured after first giving the cells a pulse of surface-bound ligand at low temperature. Following surface binding, the cells are allowed to internalize the ligand and then, following removal of surface-bound ligand the cells are incubated at optimal temperature (*c.* 37 °C) and release of retroendocytosed ligand may then be measured by adding the appropriate dissociating agent (7). The proportion of retroendocytosed ligand may be formidable, as much as 60% of internalized asialoglycoprotein may be released from isolated rat hepatocytes by adding EGTA. *Protocol 2* presents an example of how to quantitate retroendocytosis of labelled asialoorosomucoid in suspensions of rat liver parenchymal cells. See Section 5.2 for information on tyramine cellobiose (TC) labelled ligands. *Figure 1* shows recycling of glycoproteins in liver parenchymal cells and liver endothelial cells.

Protocol 2. Quantitation of retroendocytosis of [^{125}I]TC asialoorosomucoid in rat hepatocytes

Equipment and reagents

- Low-speed centrifuge and rotor
- Microcentrifuge
- [^{125}I]TC asialoorosomucoid
- EGTA
- Dibutyl phthalate

Method

1. Incubate cells with [^{125}I]TC labelled ligand at 37 °C (see *Protocol 8*).
2. After selected time intervals remove extracellular and surface-bound ligand by three washes (centrifugation at 300 *g* for 2 min) in ice-cold dissociating medium (medium containing 10 mM EGTA).
3. Incubate the cells further at 37 °C in the presence and absence of 10 mM EGTA.
4. At different time points, remove 200 μl aliquots of cells and layer on top of 300 μl dibutyl phthalate in a small (750 μl) centrifuge tube. Centrifuge for 30 sec at 7000 r.p.m. in a microcentrifuge (Beckman).
5. Radioactivities in supernatant and pellet are measured and amount of retroendocytosed ligand is expressed as per cent of total cell associated radioactivity at the start of the chase period.

If a method for measuring surface-bound ligand is available it is also possible to determine the rate of internalization of surface-bound ligand precisely. Again, the cells are allowed to bind ligand at low temperature and then following washing, the cells are incubated at an optimal temperature and after increasing time intervals, aliquots of cells are removed and cell surface-bound ligand is released/dissociated. In this way the amount of internalized (non-releasable) ligand may be measured and this allows determination of rate of internalization (8). This rate varies surprisingly much from cell to cell. For instance, the half-time for internalization of galactose receptors in rat hepatocytes is two to three minutes whereas the corresponding value for internalization via the mannose receptor in rat liver endothelial cells is 10–20 seconds. *Figure 2* shows the internalization kinetics for two different glycoproteins in rat liver parenchymal cells and endothelial cells.

2.1.3 Analysis of degradation

Most endocytosed ligands are transported to lysosomes and degraded by acid hydrolases. To measure degradation of protein ligands the most convenient method is to label tyrosine residues with radioactive iodine. Degradation may then be followed as formation of radioactivity soluble in trichloroacetic acid. It takes time to transport ligand from the cell surface to lysosomes and lyso-

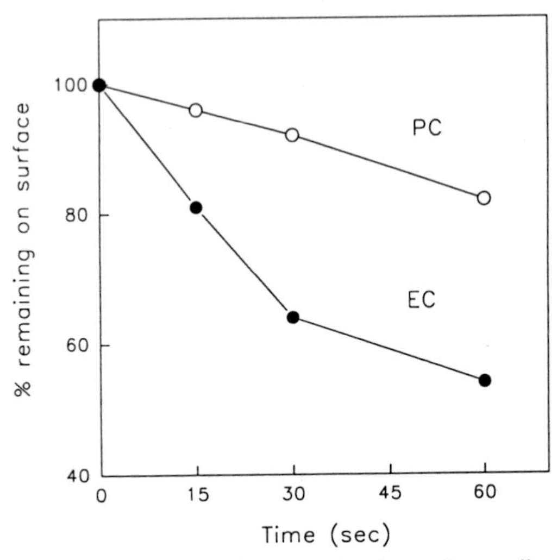

Figure 2. Kinetics of internalization of glycoproteins in rat liver cells. Rate of internalization in liver parenchymal cells (open symbols) and endothelial cells (closed symbols) of [^{125}I]-labelled asialoorosomucoid or ovalbumin respectively. 5 × 10^6 cells/ml were incubated with 5 nM of their respective ligands for 30 min at 4°C before removal by washing of unbound ligand and change to 37°C medium. At intervals, the internalization was determined by washing the cells with 5 mM EGTA in PBS, which removes the remaining surface-bound ligand.

somal degradation starts following a lag phase that may vary from cell to cell. In rat hepatocytes degradation of asialoorosomucoid starts after a lag phase of 15 minutes (4). To verify that degradation really is lysosomal, inhibitors against lysosomal degradation may be employed (9). Such inhibitors include protease inhibitors (leupeptin, pepstatin) and weak bases. However, degradation of endocytosed ligands may start in prelysosomal compartments (endosomes) and to check this possibility it may be necessary to combine degradation measurements with subcellular fractionation. The length of the lag phase before degradation starts may also be suggestive of prelysosomal degradation.

2.1.4 Analysis of clathrin-dependent and clathrin-independent endocytosis

Although all known receptor-mediated endocytosis takes place via coated pits there are several reports of endocytosis taking place outside coated pits and one may therefore differentiate between clathrin-dependent and clathrin-independent endocytosis. Examples of molecules taken up via clathrin-independent mechanisms include the plant toxin ricin and solutes taken up via fluid phase (10). Fluid phase endocytosis has been reported to be clathrin-independent in rat hepatocytes (11). Three methods (see *Protocol 3*) for distinguishing between clathrin-dependent and clathrin-independent endocytosis have been reported: first, a hypertonic medium may bring clathrin-dependent endocytosis to a halt whereas clathrin-independent endocytosis may continue unaffected by this treatment (12). Secondly, potassium depletion has been reported to stop clathrin-dependent endocytosis selectively, leaving clathrin-independent endocytosis unaffected (13). A third method is to acidify the cytosol. Acidification of the cytosol inhibits receptor-mediated endocytosis from coated pits by immobilizing the coated pits at the plasma membrane (12,14). Endocytosis via clathrin-independent pathways is only moderately reduced. The usefulness of all these methods have been supported by electron microscopy.

Protocol 3. Quantitation of clathrin-independent endocytosis in suspended rat hepatocytes

Reagents

- Incubation buffer (see *Protocol 1*)
- BSA
- NH_4Cl
- 0.14 M KCl, 2 mM $CaCl_2$, 1 mM $MgCl_2$, 1 mM amiloride, 20 mM Hepes pH 7.0
- K^+-free buffer: 140 mM NaCl, 20 mM Hepes, 1 mM $CaCl_2$, 1 mM $MgCl_2$, 1 mg/ml D-glucose pH 7.4
- KCl
- Sucrose

A. *Acidification of cytosol*

1. Suspend the cells in incubation medium (137 mM NaCl, 5.4 mM KCl, 0.34 mM Na_2HPO_4, 0.35 mM KH_2PO_4, 0.81 mM $MgSO_4$, 40 mM Hepes, 2 mM $CaCl_2$ pH 7.6) with 1% (w/w) BSA containing 15 mM NH_4Cl for 30 min at 37 °C.
2. Wash the cells in 0.14 M KCl containing 2 mM $CaCl_2$, 1 mM $MgCl_2$, 1 mM amiloride, and 20 mM Hepes pH 7.0.
3. Resuspend the cells in the same buffer.
4. Measure endocytic uptake of the labelled ligand in these cells and in untreated cells maintained in control medium.

B. *Potassium depletion*

1. Wash the cells three times with ice-cold K^+-free buffer (140 mM NaCl, 20 mM Hepes, 1 mM $CaCl_2$, 1 mM $MgCl_2$, 1 mg/ml D-glucose pH 7.4), and incubate the cells for 5 min in this buffer diluted 1:1 with water (hypotonic shock).
2. Wash the cells three times in K^+-free buffer and incubate for 15 min at 37 °C in K^+-free buffer.
3. Measure endocytic uptake of labelled ligand in the cells treated with potassium-free buffer at 37 °C and in control cells processed as above but with 10 mM KCl added to the buffer.

Protocol 3. *Continued*

C. *Hypertonic medium*

1. Prepare a 1.6 M sucrose solution in water and incubation medium with all solutes at double concentrations (purchased media may be concentrated twofold in a rotary evaporator).
2. Mix the sucrose solution with concentrated medium 1:1 to obtain a stock hypertonic medium with 0.8 M sucrose.
3. Add the hypertonic medium to the cells to give a final sucrose concentration of 0.2 M. (This concentration inhibits receptor-mediated endocytosis in rat hepatocytes by more than 90%, whereas fluid phase endocytosis is left unaffected.) No pre-incubation is required as the effect on the cells is instantaneous.

2.2 Fluid phase endocytosis

Fluid phase endocytosis is an inevitable consequence of absorbtive (receptor-mediated) endocytosis since some extracellular fluid will be trapped in the newly formed primary endosome. A considerable amount of this fluid will be released as a result of retroendocytosis. This release may in fact be one of the functions of retroendocytosis. The amount of fluid taken up by endocytosis varies from cell to cell. In the liver, for instance, the Kupffer cells have a higher pinocytic activity than the parenchymal cells and the endothelial cells. To measure fluid phase endocytosis (pinocytosis), fluid phase markers are needed that do not absorb or bind to the plasma membrane. Furthermore, the markers should not diffuse through biological membranes and should not be degradable by lysosomal enzymes. There are few molecules available that meet these requirements (see *Table 1*). In many cell types, horse-radish peroxidase (HRP) has been used successfully as a marker for fluid phase endocytosis. Usually cells are incubated in presence of > 1 mg HRP/ml and the marker is measured enzymatically using diaminobenzidine (DAB) and H_2O_2

Table 1. Markers for fluid phase endocytosis

Marker	Cell type	Reference
[^{14}C]raffinose	Hepatocytes	18
[^{14}C]sucrose	Fibroblasts	19
[^{125}I]BSA	Macrophages, endothelial cells, fibroblasts	20,21
Lucifer yellow	Hepatocytes, fibroblasts	12,19
Polyvinylpyrrolidone	Hepatocytes, macrophages, fibroblasts	16,17,19
HRP	Macrophages, BHK cells	15,22,23

as substrates (15). The method is extremely sensitive and the HRP behaves as a true fluid phase marker in many cell types. However, HRP (usually from yeast) is bound to lectins (mannose receptors) and is therefore useless as a pinocytosis marker in macrophages and other cells expressing the mannose receptor (e.g. liver endothelial cells). Alternative markers include radioactive sucrose or raffinose (e.g. [^{14}C]sucrose and [^{3}H]raffinose). These molecules do not easily penetrate biological membranes and are not degraded by lysosomal enzymes. It has been shown that both sucrose and raffinose may leak through the plasma membrane of freshly prepared hepatocytes and may be in cells incubated at low temperature. Care should therefore be taken to avoid incubation of cells in presence of the labelled markers at low temperature. Lucifer yellow has been shown to fulfil the requirements of a fluid phase marker in macrophages and hepatocytes (12). Lucifer yellow is measured spectrofluorimetrically and allows extremely sensitive determinations of fluid phase endocytosis. Since the biological membrane may become permeable to small molecules under certain conditions it is probably an advantage to use macromolecular markers of fluid phase endocytosis. We found that [^{125}I]polyvinylpyrrolidone, first introduced by Lloyd and co-workers (16), is a very useful marker in rat liver parenchymal cells (17). It is not commercially available at present and the labelling procedure is cumbersome. Native albumin is taken up by fluid phase endocytosis: it is, however, easily denatured and may then be taken up by absorbtive endocytosis to varying degrees. We have found that albumin labelled with [^{125}I]tyramine cellobiose may be a useful marker of fluid phase endocytosis.

3. Endocytosis of plasma membrane components

An important but somewhat neglected function of endocytosis is its participation in the catabolism and turnover of plasma membrane proteins that are not directly involved in endocytosis. To study endocytic uptake of plasma membrane proteins and lipids it is necessary to label these components. Hayalett and Thilo (24) have developed a method for labelling of galactose residues in the plasma membrane glycoproteins. This method first employs β-galactosidase to remove terminal galactose residues from the glycoproteins. Subsequently labelled galactose is introduced by means of galactosyl transferase. The intracellular transport of proteins thus labelled may subsequently be followed by means of subcellular fractionation.

Hoekstra and Kok (25) have developed fluorescent lipid probes to label phospholipids in the plasma membrane, whereas plasma membrane proteins may be labelled by iodination, biotinylation, or specific antibodies. Most of these labelling techniques have recently been described by Ellis *et al.* in another issue of this series (26), and will therefore not be discussed here.

4. Analysis of intracellular transport of ligands/markers and endocytic receptors by means of subcellular fractionation

Methods used for studying the intracellular transport of endocytosed ligands and/or receptors include electron microscopy (combined with immunocytochemistry or colloidal gold labelling), fluorescence microscopy (using fluorescent labelled ligand and/or fluorescent labelled antibodies), and subcellular fractionation. These techniques have been combined with the use of more or less specific inhibitors against various cell components/organelles suspected of being involved in endocytosis. It is of course necessary to combine techniques to obtain information about intracellular transport. The advantage of using subcellular fractionation is that this technique gives an overall view of the intracellular distribution of a given ligand and/or receptor. Most studies using subcellular fractionation to follow endocytosis have used rat liver as an experimental system. This chapter will mainly focus on the use of subcellular fractionation for studying hepatic endocytosis, although other experimental systems will be referred to. The cellular heterogeneity of the liver however may lead to methodological problems when hepatic endocytosis is studied by means of subcellular fractionation. First, the ligand in focus may be taken up in more than one cell type. Secondly, organelles may not have the same physical properties (size, buoyant density, charge) in different cells. These problems may be circumvented to some extent if the ligand in question is taken up only in one cell type (hepatocytes, endothelial cells, Kupffer cells, or stellate cells). An alternative to studying uptake in whole liver is to employ isolated cells as an experimental system. Methods are available for preparing isolated hepatocytes (27), endothelial cells (28), Kupffer cells (28), and stellate cells (29). Subcellular fractionation of isolated hepatocytes may be achieved by using methods developed for whole liver with relatively small modifications.

4.1 Differential centrifugation

Following homogenization of cells and tissues a preliminary differential centrifugation is usually needed. Differential centrifugation may in addition give useful information about the size distribution of components. Classical differential centrifugation of liver (see *Protocol 4*) gives five fractions: a nuclear fraction (N fraction), a heavy mitochondrial fraction (M fraction), a light mitochondrial fraction (L fraction), a particulate or microsomal fraction (P fraction), and a final supernatant or soluble fraction (S fraction). These fractions are obtained from the homogenate by sequentially increasing the centrifugal field (30,31).

If contamination of cytoplasmic components is a potential problem, the original homogenate may initially be layered on top of a layer of metrizamide or Nycodenz. Seglen and co-workers have developed a technique whereby isolated rat hepatocytes are first electropermeabilized/electrodisrupted (32). The resulting cell corpses may be layered on top of a layer of metrizamide, and following centrifugation at low speed the cell corpses, completely devoid of cytosol, are pelleted underneath the metrizamide layer. These cell corpses may have lost small, early endocytic vesicles, but other organelles including early and late endosomes and autophagic vacuoles remain associated with them, possibly bound to the cytoskeleton.

Analysis of the distribution of endocytosed ligand at various times after initiation of uptake in whole liver or in isolated hepatocytes has revealed that early (during the first minutes) after uptake, the ligand is mainly in the P fraction indicating that it is contained in a relatively small vesicle, probably a primary endosome (31,33). The ligand is subsequently found in heavier fractions, contained in early and late endosomes. Finally the distribution of ligand coincides with that of a lysosomal enzyme. The uptake in lysosomes may, however, be difficult to detect if the ligand is rapidly degraded, and if a 'trap label' is not used.

Protocol 4. Subcellular fractionation of rat liver parenchymal cells by differential centrifugation

Equipment and reagents

- Dounce homogenizer
- Low-speed, high-speed and ultra-centrifuge and rotors
- Sucrose solution: 0.25 M sucrose, 10 mM Hepes–NaOH, 1 mM EGTA pH 7.2

Method

1. Homogenize 5×10^7 rat liver parenchymal cells in 5 ml ice-cold 0.25 M sucrose, 10 mM Hepes–NaOH, 1 mM EGTA pH 7.2 with five strokes in a tight-fitting Dounce homogenizer.
2. Centrifuge the homogenate for 2 min at 2000 *g* in 10 ml centrifuge tubes. Keep the supernatant and rehomogenize the pellet in 5 ml buffer by four strokes in a tight-fitting Dounce homogenizer. Centrifuge again at 2000 *g* for 2 min and combine the supernatants.
3. Centrifuge the post-nuclear supernatants (9.5 ml) at 6750 *g* for 4 min. The pellet is resuspended using a vortex mixer in 4 ml ice-cold sucrose solution and recentrifuged at 6750 *g* for 4 min.
4. The combined supernatants following sedimentation of the M fraction are centrifuged at 22 000 *g* for 9 min to sediment the L fraction. The fluffy layer on top of the L fraction is removed by leaving 1 ml of the supernatant in the tube and careful agitation of the tube. The fluffy layer is then combined with the remainder of the supernatant.
5. The final centrifugation takes place at 48 000 *g* for 70 min and results in a P fraction and a final supernatant, the S fraction.

Protocol 4. *Continued*

6. Resuspend the pellets in equal volumes of ice-cold sucrose solution and analyse fractions for degraded and undegraded ligand, marker enzymes, and protein.

4.2 Isopycnic centrifugation in gradients of sucrose, Nycodenz, or Percoll

Differential centrifugation may be a necessary preliminary step in subcellular fractionation, to get rid of, for example cytosol and nuclei. Further separation and/or purification of endocytic organelles may be obtained using isopycnic centrifugation in density gradients. Density gradients (see *Protocol 5*) may be isotonic throughout the gradient (Nycodenz gradients, Percoll gradients) or osmolarity may increase in parallel with density (sucrose gradients). Because of differences in osmolarity it may be possible to obtain separation in an isotonic gradient if Nycodenz, Iodixanol, or Percoll gradients are used (see Chapter 7, Section 1.1 for more information) and not in a hypertonic gradient (of e.g. sucrose), and vice versa. This may be illustrated by an experiment aimed at following the intracellular transport of retinyl ester in association with chylomicron remnants in rat liver parenchymal cells. The intracellular

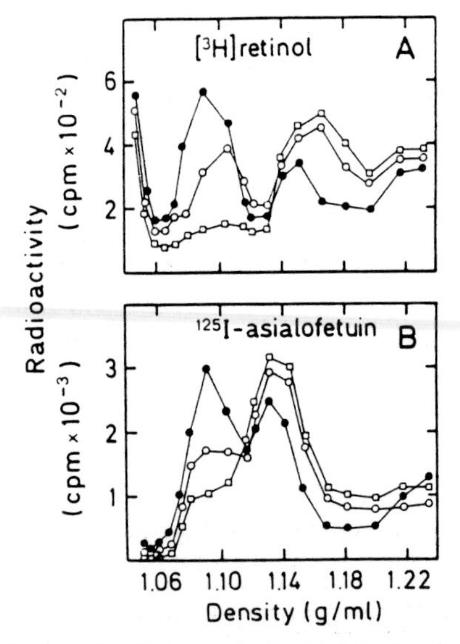

Figure 3. Comparison of the subcellular transport of endocytosed [^{125}I]TC–asialofetuin and [^{3}H]retinyl ester labelled chylomicrons. Parenchymal liver cells were loaded with (A)[^{3}H]retinyl ester labelled chylomicrons (loaded 12 min *in vivo*) or (B) [^{125}I]TC–asialofetuin (loaded 5 min *in vivo*). Following incubation for 0 min (closed circle), 15 min (open circle), and 45 min (open square), the cells were fractionated on Nycodenz gradients 85 000 *g* for 5 h). From ref. 34.

transport was studied by means of subcellular fractionation. It was found, using sucrose gradients, that labelled retinol became associated with an organelle banding at about 1.15 g/ml, at about the same density as that of endosomes. However, when Nycodenz gradients were employed it was found that the labelled retinol was well separated from endosomes (labelled with endocytosed [^{125}I]asialoorosomucoid) (*Figure 3*). The vitamin was instead associated with components that coincided in the gradient with markers for the endoplasmic reticulum (34).

Morphological studies using immunocytochemistry in combination with electron microscopy and fluorescence microscopy have indicated that compounds taken up by endocytosis are carried from the plasma membrane via primary endosomes to early endosomes, and further via endocytic carrier vesicles to late endosomes, and eventually to lysosomes. Ideally, if subcellular fractionation were to be used for analysis of endocytosis it should separate the various steps in the endocytic pathway because of differences in density and/or size and/or charge of the organelles involved. A complete analysis of the endocytic events has not yet been achieved, however a partial analysis is feasible.

Protocol 5. Preparation of gradients for isopycnic separation of organelles

There are several ways of preparing density gradients. Two consistent methods for preparing gradients of sucrose and Nycodenz are described below (they are designed for 38 ml centrifuge tubes in a SW28 Beckman rotor).

Equipment and reagents

- Biocomp Gradient Master™ (Nycomed)
- Syringe with a long cannula
- 21%, 32%, 42%, and 54% (w/w) sucrose in 10 mM Hepes–NaOH, 1 mM EDTA pH 7.2
- Nycodenz
- 1 M Hepes–NaOH, 1 mM EDTA pH 7.2
- Homogenization buffer: 0.25 M sucrose, 10 mM Hepes, 1 mM EDTA
- 15% and 54% (w/w) sucrose in homogenization buffer

Method

A. *Diffusion method*

1. Prepare four sucrose solutions (21%, 32%, 42%, and 54%, w/w) in 10 mM Hepes–NaOH, 1 mM EDTA pH 7.2.
2. Load the tubes with four successive 8 ml layers of sucrose solutions of increasing density by use of a syringe equipped with a long cannula. Plug the tubes and tilt them carefully to a horizontal position to allow diffusion. Return the tubes to an upright position after 1 h.
3. The sample to be fractionated can now be loaded on top of the gradient and the run started.

Protocol 5. *Continued*

B. *Biocomp Gradient Master*

1. **Nycodenz.** Overlay 16 ml of a dense solution of Nycodenz (dissolve 43 g Nycodenz in 100 ml of distilled water, then add 1 ml of 1 M Hepes–NaOH, 100 mM EDTA pH 7.2) with 16 ml homogenization buffer (0.25 M sucrose, 10 mM Hepes, 1 mM EDTA). Rotate at 25 r.p.m. for 2 min at 73°.
2. **Sucrose.** Overlay 16 ml 54% (w/w) sucrose with 16 ml 15% (w/w) sucrose in homogenization buffer. Rotate at 25 r.p.m. for 5 min at 55°, followed by 30 r.p.m. for 20 sec at 80°.

Subcellular fractionation both in isotonic gradients and hypertonic gradients has separated endosomes and lysosomes. In sucrose gradients endosomes from rat liver parenchymal cells band at 1.10–1.15 g/ml whereas lysosomes are denser (about 1.20 g/ml in liver). In some cases it has been possible to get further subfractionation of both endosomes and lysosomes. Using sucrose gradients and Nycodenz gradients early endosomes have been shown to be denser than late endosomes (31). The early primary endosomes from rat hepatocytes have a density of 1.16 g/ml whereas early endosomes band at about 1.11 g/ml in sucrose gradients. Subsequently the ligand is transferred to a denser endosome with the appearance of multivesicular body. Fusion with lysosomes leads to an organelle with an intermediate density. This organelle develops into a dense lysosome (*Figure 4*). Subcellular fractionation of liver following intravenous injection of [^{125}I]tyramine cellobiose labelled ovalbumin which is taken up selectively into liver endothelial cells has shown that this ligand is sequentially associated with three types of vesicles with increasing density in sucrose gradients. During the first minute the ligand bands at 1.14 g/ml. Then during the next five minutes it is associated with endosomes banding at 1.17 g/ml. Finally, starting at about six to nine minutes the ligand accumulates in lysosomes banding at a relatively high density (1.22 g/ml) (*Figure 5*) (35).

Analysis of endosomes by subcellular fractionation of MDCK cells has also revealed that early endosomes may be denser than late(r) endosomes (36). The reasons for this may be that the amount of membrane relative to the volume may be reduced in the late endosome. The density of the plasma membrane in liver cells is about 1.15 g/ml.

Subcellular fractionation has also revealed subgroups of lysosomes. Pertoft *et al.* (37) found buoyant and dense subgroups of lysosomes using Percoll density gradients. The buoyant lysosomes seemed to be an intermediate station between endosomes and dense lysosomes. In rat hepatocytes isopycnic centrifugation in Nycodenz and metrizamide gradients has separated active (buoyant) and resting (dense) lysosomes (38). A similar distribution may be seen in J774 macrophages.

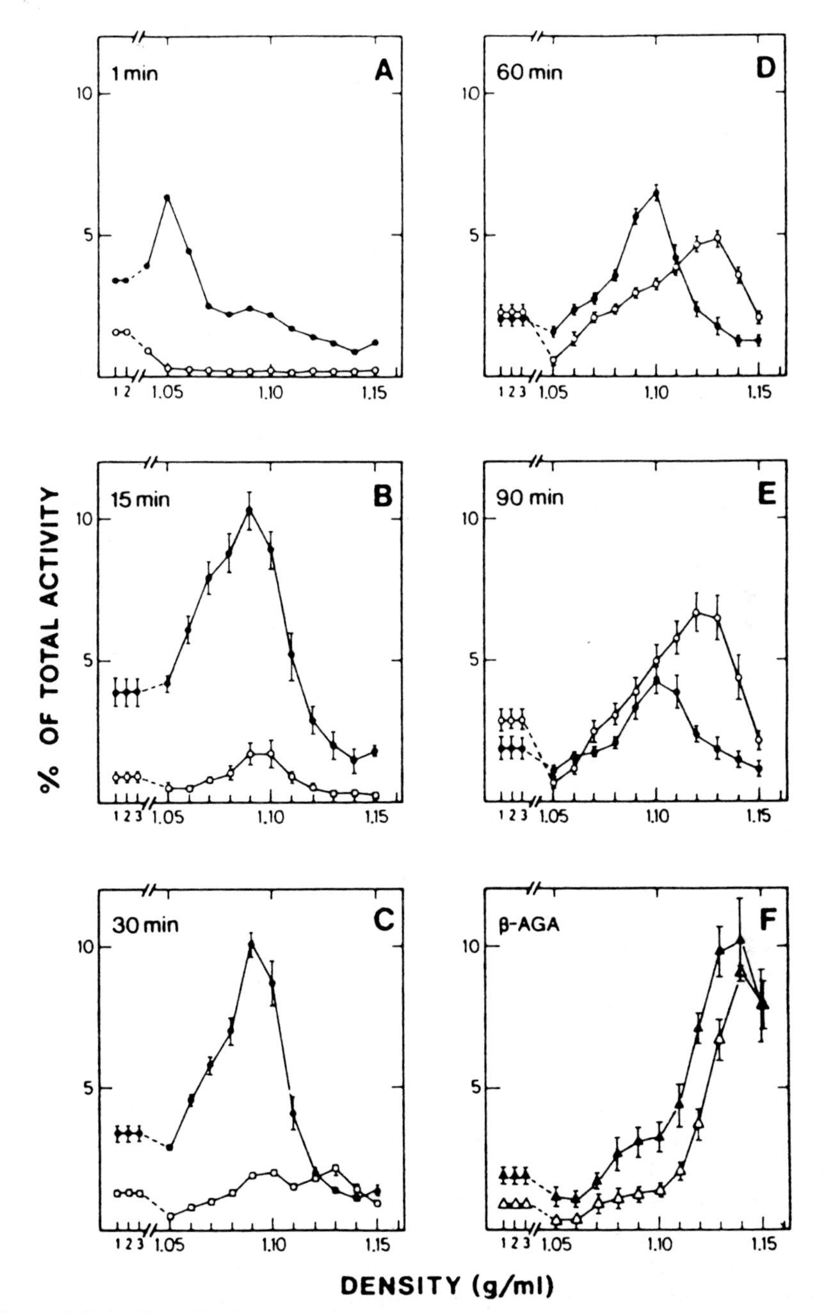

Figure 4. Subcellular fractionation of rat hepatocytes containing degraded (open circle) and undegraded (closed circle) [^{125}I]TC–AOM. The cells were first incubated at 4°C for 1 h in the presence of 50 nM [^{125}I]TC–AOM, and after removing extracellular ligand incubated at 37°C. At the indicated times cell aliquots were removed and fractionated in linear Nycodenz gradients. Panel F shows distribution of β-AGA in gradient A (open triangle) and gradient E (closed triangle). Activities are expressed as per cent of total in gradient. From ref. 4.

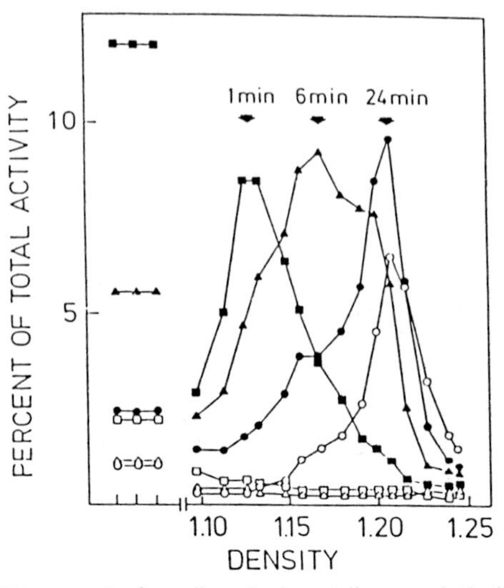

Figure 5. Intracellular transport of ovalbumin in rat liver endothelial cells. Post-nuclear liver fractions from rat injected with [^{125}I]TC–ovalbumin 1 min (squares), 6 min (triangles), and 24 min (circles) prior to sacrifice were loaded on to linear sucrose gradients and centrifuged for 4 h at 85 000 *g*. Undegraded (closed symbols) and degraded (open symbols) ligand were determined in each fraction and presented as per cent of total recovered radioactivity in the gradient as a function of density (except for the three upper fractions not entering the gradient itself).

Numerous studies have employed Percoll density gradients for separating endocytic organelles. Centrifugation in Percoll solution leads to a gradient in which the density increases in a sigmoid fashion down the gradient. There is a sharp increase in density at the top and towards the bottom of the gradient. As a result of this endocytic organelles tend to be separated into two groups, a light group consisting of endosomes and a dense group, mainly lysosomes. The fractionation in Percoll is, however, very simple. The homogenate may be mixed with Percoll with a given concentration and the solution is then centrifuged for about one hour in a fixed-angle rotor at about 50 000 *g* (26).

Isotonic gradients can also be used to separate endocytic vesicles of various sizes based on differences in their sedimentation rate. Kindberg *et al.* (39) separated the primary endosomes formed from the plasma membrane (by chasing the endocytic marker for 1 min) from the larger early endosomes (15 min chase) where dissociation of receptor and ligand takes place, by fractionation of hepatocyte homogenates on Nycodenz gradients centrifuged at 85 000 *g* for 45 min (*Figure 6*). Under these conditions, the primary endosomes banded around 1.06 g/ml, whereas the larger endosomes banded at 1.11 g/ml. When the same gradients were run for 3 h, large and small endosomes comigrated to 1.11 g/ml. Small vesicles therefore need several hours of

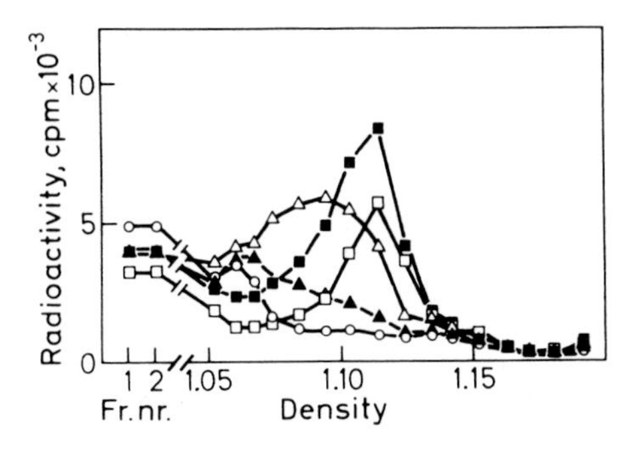

Figure 6. Distribution of ^{125}I-labelled AF containing endocytic vesicles in Nycodenz gradients after short (45 min) centrifugation time. Following incubations with ^{125}I-labelled AF at 4 °C for 60 min, cells were washed and then incubated at 37 °C for 0.5 min (○), 1 min (▲), 2.5 min (△), 15 min (■), and 30 min (□). Post-nuclear fractions were subsequently prepared and layered on top of linear Nycodenz gradients. The tubes were centrifuged at 45 min at 85 000 *g* (reproduced with permission from ref. 39).

centrifugation to reach their equilibrium density, a property that can be exploited in cell fractionation.

4.3 Continuous flow electrophoresis (CFE)

The use of CFE for organelle separation has recently been reviewed by Schmid (40) and the reader should refer to this for a thorough description of the method. Because of the cost of the equipment and its general lack of availability it is not widely used for the analysis of endocytic events, however Marsh *et al.* (41) have used it to separate endosomes and lysosomes from fibroblasts. A post-nuclear supernatant was centrifuged at 100 000 *g* for 30 min on to a cushion of 2 M sucrose and the banded material harvested. This was treated with *N*-tosyl-L-phenylalaninechloromethylketone (TPCK) trypsin for 5 min at 37 °C before inactivation with soya bean trypsin inhibitor. After repeated centrifugation at 1000 *g* for 5 min to remove any aggregates, the suspension was injected into the flowing electrophoresis separation buffer (0.25 M sucrose, 1 mM EDTA, 10 mM triethanolamine, 10 mM acetic acid pH 7.4) with an applied field of 1100–1750 V. Satisfactory resolution of the endosomes and lysosomes only occurred after trypsin treatment, otherwise contamination by other subcellular membranes is significant.

5. Labelling of ligands and markers

5.1 Use of [^{125}I]

To follow intracellular transport of endocytosed molecules it is necessary to identify these molecules. Horse-radish peroxidase (HRP) is a commonly used marker for fluid phase endocytosis (in cells not expressing the mannose receptor) and can be measured enzymatically (15). HRP can also conveniently be attached to protein ligands, and since receptor-mediated endocytosis is much more efficient than fluid phase endocytosis the uptake of the ligand may be measured indirectly by measuring uptake of HRP bound to the ligand. Lucifer yellow is fluorescent and may therefore be measured by spectrofluorimetry. The most commonly used method to label the endocytic marker/ ligand is by means of radioactivity, and the most widely used radioisotope for labelling endocytosed proteins is radioactive iodine [^{125}I] or [^{131}I]. These two isotopes may be measured separately and it is therefore possible to label two different proteins with [^{125}I] and [^{131}I], respectively, and follow the intracellular transport of the two proteins at the same time. *Protocol 6* describes the direct labelling of proteins with [^{125}I] (42), using sodium hypochlorite as an oxidizing agent.

Protocol 6. Labelling of proteins with [^{125}I]

Equipment and reagents

- PD10 column
- Phosphate-buffered saline (PBS)
- Na[^{125}I]
- 1 mM NaOCl
- 20 mg/ml $NaBH_4$ in 1 mM NaOH
- 50 mM KI, 0.5% (w/v) BSA, 50 mM PBS pH 7.4
- Incubation buffer (see *Protocol 1*)

Method

1. To 10 μg protein (1 μg/ml in PBS) add 40 MBq Na[^{125}I] solution and 5 μl of a 1 mM NaOCl solution and mix. Add two additional 5 μl aliquots of NaOCl at 30 sec intervals.
2. Add 20 μl $NaBH_4$ (20 mg/ml in 1 mM NaOH) and mix. Incubate for 5 min.
3. Stop reaction by the addition of 500 μl 50 mM KI, 0.5% (w/v) BSA in 50 mM PBS pH 7.4.
4. Remove non-bound iodine by dialysis against 10 litres of 20 mM PBS or by gel filtration on a PD10 column.
5. Dilute the labelled ligand in incubation medium and store at −20 °C.

5.2 Use of residualizing labels

Most proteins are rapidly degraded in lysosomes, and since radiolabelled amino acids (iodotyrosine) are transported through the lysosomal membrane extremely rapidly, it will not be possible to measure actual uptake of labelled molecules in lysosomes. A ligand to be used as a marker for the endocytic pathway should therefore either be relatively resistant to degradation, or, alternatively, the labelled degradation products formed should be trapped in the lysosomes/prelysosomes. HRP is relatively resistant to lysosomal hydrolases, and since it can be covalently attached to various ligands without interfering with ligand binding and HRP activity, it may be a useful marker to follow intracellular transport of endocytosed ligands. An alternative approach is to attach covalently a non-degradable radiolabelled group/adduct to the ligand which will be trapped in the degradative compartment following degradation of the ligand. A ligand that releases labelled degradation products that are trapped at the site of degradation offers the advantage that it may label the actual degradative compartment independently of other markers such as enzymes. It was shown using this principle that degradation of asialofetuin in the liver actually started in endosomes and not in the lysosome proper (4). Two useful methods for introducing non-degradable groups into protein ligands have been developed by Pittman *et al.* (43). Both methods take advantage of the fact that the chemical linkage between tracer and ligand is not hydrolysed by lysosomal hydrolases. The first probe to be developed was [^{14}C]sucrosyl labelled low density lipoprotein (LDL) (44). Following lysosomal degradation [^{14}C]sucrose is trapped in the lysosomes and can be used as a lysosomal marker in subcellular fractionation (see *Protocol 7*).

Protocol 7. Labelling with [^{14}C]sucrose

The method is suitable for labelling of a 50 kDa protein.

Equipment and reagents

- PD10 column
- [^{14}C]sucrose
- 0.125 N NaOH
- 62.5 mM cyanuric chloride in acetone
- 0.167 M acetic acid
- PBS

Method

1. Lyophilize 0.625 μmol [^{14}C]sucrose (250 μCi; 400 Ci/mole specific activity) and redissolve in 10 μl water.
2. Add 15 μl 0.125 N NaOH (three molar equivalents) and 20 μl 62.5 mM cyanuric chloride in acetone (two molar equivalents).
3. Incubate for 10–15 sec before addition of 15 μl 0.167 M acetic acid (four molar equivalents) and immediately transfer the mixture to 2 ml of the protein to be labelled (5 mg/ml dissolved in PBS pH 7.2).
4. Incubate for at least 3 h (or overnight) at room temperature and remove unbound sucrose by dialysis or gel filtration on a PD10 column.

Instead of sucrose, raffinose can be used and this trisaccharide can be coupled directly to the protein without cyanuric chloride activation (45). It is possible to obtain ligands with much higher specific radioactivities by attaching

^{125}I-labelled tyramine cellobiose to lysine groups in proteins (46). We have found such ligands to be very useful for following uptake and intracellular transport of endocytosed ligands (see *Protocol 8*). By measuring acid soluble radioactive degradation products, these labelled degradation products may serve as markers for the degradative compartments in which degradation actually takes place. Such measurements showed, for instance, that degradation of ligands may be initiated in prelysosomal endocytic compartments (4). Techniques have also been developed by Smedsrød and co-workers to attach [^{125}I]tyramine cellobiose to glycosaminoglycans (47).

Protocol 8. Labelling with [^{125}I]tyramine cellobiose (^{125}I-TC)

Equipment and reagents

- Glass vials
- PD10 column
- 1 mg/ml iodogen in methylene chloride
- 0.1 M Na_2CO_3
- Tyramine cellobiose
- [^{125}I]
- 0.1 M KI
- 0.1 M $NaHSO_3$
- 3 nM cyanuric chloride in acetone

Method

1. Coat small glass vials with 10 μg iodogen dissolved in methylene chloride (1 mg/ml). Let the solvent evaporate and wash the vials thoroughly with distilled water.
2. Dissolve the protein to be labelled in 0.1 M Na_2CO_3 (10 mg/ml). If the protein is already dissolved in, for example PBS, dilute 1:1 with 0.2 M Na_2CO_3.
3. For each nanomole protein to be labelled mix 10 nmol (1 μl) TC with 3 MBq [^{125}I] in the iodogen coated vial and incubate at room temperature for 30 min.
4. Transfer the labelled TC from the iodogen coated vial (take care not to transfer iodogen with the pipette) to another vial containing 5 μl 0.1 M KI and 10 μl 0.1 M $NaHSO_3$.
5. Activate the labelled adduct by addition of 2 μl 3 nM cyanuric chloride (dissolved in acetone) per microlitre TC and incubate for 3 min.
6. Add the protein to be labelled to the reaction mixture and incubate for 1 h. Separate non-bound iodine by exhaustive dialysis or gel filtration on a PD10 column.

5.3 Labelling of receptors

The fate of the endocytic receptors varies for different groups of receptors. Some receptors follow the ligand to the lysosome whereas other receptors are transcytosed, or, most commonly, recycled to the plasma membrane. The proportion of receptors that are present on the cell surface and intracellularly may change as a result of endocytosis or because the cells are exposed to signal molecules or drugs. To follow the intracellular transport of the receptors by subcellular fractionation it is necessary to label the receptor on the cell surface and subsequently determine the distribution of labelled receptor in subcellular fractions. The most direct method is of course to label the receptor with the ligand in which case it is necessary to determine the amount of ligand–receptor complex at various times after internalization. This can be done by means of gel filtration. Kindberg *et al.* (39) showed that asialofetuin was associated with its receptor up to about five minutes after internalization. Since the association is pH dependent, however, the ligand will dissociate from the receptor in the early endosomes. Transferrin is bound to its receptor continuously during internalization and recycling, and the ligand may therefore serve as a marker for the receptor in subcellular fractions. Cross-linking of ligand to receptor has in some cases been a very useful method for following the intracellular route of the receptor. Two examples serve to illustrate the usefulness of this method. Weigel and co-workers used *N*-hydroxysuccinimide ester to cross-link asialoorosomucoid to the galactose receptor in rat hepatocytes (48). Nielsen studied the intracellular pathway followed by the insulin receptor in isolated rabbit kidney proximal tubules by cross-linking with disuccinimidyl suberate (49).

To determine the distribution of receptors in subcellular fractions immunoprecipitation may be employed if antibodies against the receptor are available.

6. Markers for organelles involved in endocytosis

Two working hypotheses were central in early studies using subcellular fractionation. These were the postulate of biochemical homogeneity and the postulate of single location (30). If the shuttle model of endocytosis were valid one would expect that certain proteins were located exclusively in the main 'stations' along the endocytic pathway, the coated pits/vesicles, the early endosomes, the endocytic carrier vesicle, the late endosome, and the lysosome. It has turned out to be difficult to find reliable markers for the various endocytic compartments and it is therefore necessary to combine the use of several markers to identify the endocytic organelles.

6.1 The plasma membrane

It is important in subcellular fractionation experiments to have a reliable marker for the plasma membrane, since this is the first compartment with which a labelled ligand associates via specific receptors. The plasma membrane is of course heterogeneous with specialized regions such as coated pits. In addition, in polarized cells the apical and the basolateral plasma mem-

Table 2. Markers for various endocytic compartments

Marker	Reference
Plasma membranes	
Ligands at 4 °C	50
Iodination with lactoperoxidase	26
Biotinylation	51
5′-Nucleotidase	52
Early endosomes	
Ligands internalized < 10 min	53,54
Transferrin receptor	55
Rab 4, rab 5	56
Late endosomes	
Ligands internalized and chased for 10–30 min	53
M6P-R	57
Rab 7	56
Lysosomes	
Ligands internalized and chased > 2 h	4
Proteases	58
Acid hydrolases	59
LGPs	60

brane may differ with respect to sedimentation properties. Possible markers against the plasma membrane include plasma membrane enzymes, endocytic ligands, and other plasma membrane proteins that may be labelled selectively (see *Table 2*). Plasma membrane enzymes that have been used as markers in subcellular fractionation include 5′-nucleotidase, alkaline phosphodiesterase, and adenylate cyclase. 5′-Nucleotidase is mainly associated with the apical plasma membrane in hepatocytes and will therefore reflect the distribution of a part of the plasma membrane which is not involved in endocytosis of ligands/markers studied so far. Internalization of ligands and receptors does not take place at 0 °C in cells from homeothermic animals (mammals, birds). Therefore, the subcellular distribution of a ligand bound at 0 °C will reflect the distribution of the part of the plasma membrane with which the ligand is associated. Alternative means of labelling the plasma membrane are to introduce [^{14}C]galactose in the oligosaccharide moiety of plasma membrane glycoproteins and/or to allow binding of, for example ^{125}I-labelled ricin to terminal galactose residues of plasma membrane glycoproteins. Both methods allow general labelling of the plasma membrane. Very high specific activities may be reached since the number of glycoproteins that may be labelled is very high ($> 10^8$/cell). Methods for labelling of plasma membrane components have recently been reviewed by Ellis *et al.* (26) in this series.

6.2 Endosomes

It has been notoriously difficult to find reliable markers for particular groups of endosomes. Therefore, to get information about the sequential intracellular transport of a ligand, the early and late endosomes have been defined by the kinetics of appearance of ligand in organelles that may differ in size and/or density. As discussed in Section 4.2, there is a tendency for early endosomes to be more buoyant than late endosomes in density gradients. Data from subcellular fractionation can subsequently be combined with data obtained with electron microscopy of cells and fractions, using immune cytochemistry and/or colloidal gold labelling. Recently it has been shown that a number of small GTP-binding proteins are involved in discrete steps in the intracellular transport of vesicles involved in secretion and endocytosis. For instance, it has been demonstrated that rab 5 and rab 7 are associated with respectively early and late endosomes (61) whereas rab 9 is associated with the *trans*-Golgi (62). These rab proteins may be the sort of markers that have been sought for, and subcellular fractionation has, on the other hand, been instrumental in defining the subcellular location of the rab proteins. Gorvel and coworkers (61) showed by subcellular fractionation in sucrose density gradients that rab 5 and rab 7 labelled selectively early and late endosomes, respectively. In this study the late endosomes were more buoyant than the early ones. The transferrin receptor recycles via early endosomes and labelled transferrin may therefore serve as a marker for the early endocytic components in cells expressing the transferrin receptor. The cation-independent mannose 6-phosphate receptor (MPR) has been shown to be present mainly in the late endosomal compartment, presumably because this compartment exchanges the MPR with the *trans*-Golgi.

6.3 The lysosomes

The MPR does not seem to follow the lysosomal enzymes to the lysosomes (63). Therefore, if antibodies against the MPR are available it should be possible using immunoprecipitation to determine the distribution of the late endosomal compartment(s) in subcellular fractions. Recent experiments have shown the MPR to be present even in early endosomes (63). However, it is evidently transported rapidly further along the endocytic pathway. The transit time of the receptor in the late compartments is sufficiently long-lasting to accumulate the bulk of the receptor in this compartment. Most of the lysosomal hydrolases in cells are associated with the lysosomes proper and should therefore be useful markers for these organelles. However, lysosomal enzymes are to varying extent present also in endosomes and prelysosomal compartments and in some cases alternative markers are needed. Useful lysosomal markers are labelled degradation products that are trapped in lysosomes following uptake of a ligand to which the 'trap label' is covalently linked. A very useful trap label is [^{125}I]tyramine cellobiose. [^{125}I]tyramine cellobiose is trapped in the degradative compartment and may therefore serve as a marker for the organelle in which degradation actually takes place (46). If degradation takes place in a prelysosomal compartment the labelled degradation product ([^{125}I]tyramine cellobiose attached to a short peptide) would presumably be transported along with other components (undergraded ligand, lysosomal enzymes) to the terminal lysosome.

7. Induction of density shift of components of the endocytic pathway to determine a possible co-localization of ligand and other markers

It is a common experience that various groups of endocytic organelles (endosomes, lysosomes) co-localize more or less completely following subcellular fractionation by isopycnic centrifugation. It may therefore be difficult to decide whether two different ligands/receptors (or other components) co-localize or are in fact present in different organelles. This problem could conceivably be solved if the density of one or more of the organelles involved in endocytosis could be changed selectively. Two types of approaches have been tested more or less successfully. First, the endocytic organelles including lysosomes may be allowed to accumulate (via endocytosis) a compound that changes the density of the organelle in question. Secondly, a ligand may be exploited to induce a density shift of the organelle with which it is associated. The latter technique may allow one to decide 'who is with whom' in the endocytic pathway.

Several compounds have been found to induce density shift of lysosomes. The non-ionic detergent Triton WR-1339 was in the early 1960s found to accumulate in rat liver lysosomes following intravenous injection (64). The lysosomes gradually became more buoyant and reached a density of about 1.10 g/ml in sucrose density gradients. The density of the lysosomes reflects the density of the detergent. The mechanism whereby Triton WR-1339 is taken up in, for example liver cells has not been clearly shown, it may possibly be bound to lipoproteins (low density lipoprotein) and subsequently be endocytosed via the LDL receptor in the cells.

Other compounds that may induce formation of denser lysosomes include yeast invertase (65), the iron-containing complex Jectofer (66), the silver-containing complex Neosilvol (67), and colloidal gold (68). Yeast invertase, injected intravenously, has been shown to accumulate in lysosomes of liver endothelial cells. The enzyme is resistant to lysosomal degradation, and as a result these lysosomes become denser in sucrose density gradients.

It is possible to bind colloidal gold to various ligands and the complexes may be taken up by receptor-mediated endocytosis. Colloidal gold has been used successfully to induce density shift of several organelles involved in endocytosis. Hopkins and co-workers allowed colloidal gold–ricin to bind to the plasma membrane of fibroblasts (at low temperature) and were subsequently able to separate plasma membrane from other cell organelles in sucrose density gradients (68). Endosomes were made selectively denser following incubation at 16 °C with transferrin–gold. At this temperature transport from endosomes to lysosomes is blocked and the ligand–gold complex therefore accumulates in the endosomes, rendering these organelles selectively denser. The lysosomes were made selectively denser by first giving the cells a pulse of colloidal gold, and then incubating the cells for several hours at 37 °C in the absence of extracellular gold–ligand complex. In these experiments the density shift induced in the various organelles (plasma membrane, coated pits, endosomes, lysosomes) was sufficiently extensive to allow their sedimentation to the bottom of a sucrose gradient. This system may therefore be used both analytically, to determine whether components co-localize, and preparatively, as a step in a purification procedure.

7.1 Density shift induced in endosomes that contain ligand in complex with horse-radish peroxidase or colloidal gold

A very useful method for inducing density shift of endosomes is based on the fact that HRP in presence of H_2O_2 can catalyse polymerization of diaminobenzidine (DAB). Therefore, if HRP is taken up in endosomes/lysosomes by endocytosis it may be exploited to induce a density shift in the organelle in which HRP is present. Furthermore, the density shift may be induced in a cell-free system, and it is therefore not necessary to perturb the function of the organelles by accumulation of endocytosed material. To induce a density shift selectively in the endocytic pathway followed by ligands internalized via receptor-mediated endocytosis, HRP may be bound to the ligand in question. A density shift of the endocytic organelles in which the HRP–ligand complex is present may then be obtained at any given time (see *Protocols 9* and *10*). Density modification by DAB has been an extensively used method. The method is particularly useful for deciding whether two different components are present in the same endocytic compartment. It can be shown, for instance, that asialoorosomucoid departs from the transferrin pathway at an early point in the endocytic pathway (69). The method has also been used to determine the intracellular meeting-place of the biosynthetic pathways leading from the endoplasmic reticulum to the lysosomes and the endocytic pathway.

Another approach to obtain information about co-localization of ligands or markers is to allow one of the molecules in question to bind to colloidal gold (see *Protocol 11*). Following internalization the amount of colloidal gold accumulating in endosomes may be sufficiently high to cause a density shift of these organelles. The density distribution of another ligand will change if it is present in the same compartment as the one bound to colloidal gold (*Figure 7*).

Protocol 9. Induction of density shift by HRP and diaminobenzidine (DAB)

Preparation of a ligand–HRP conjugate.

Reagents

- HRP (Type VI, Sigma)
- 0.3 M $NaHCO_3$ pH 8.1
- 0.1% (w/w) 1-fluoro-2,4-dinitrobenzene in ethanol
- 60 mM $NaIO_4$
- 0.16 M ethylene glycol
- 10 mM Na_2CO_3 pH 9.5
- $NaBH_4$
- Phosphate-buffered saline

Method

1. Dissolve 5 mg HRP (Type VI from Sigma) in 1 ml freshly made 0.3 M $NaHCO_3$ buffer pH 8.1.
2. Add 100 μl 0.1% (w/w) 1-fluoro-2,4-dinitrobenzene (in ethanol) and mix with a magnetic stirrer for 1 h at room temperature.
3. Add 1 ml 60 mM $NaIO_4$ and mix for 30 min at room temperature.
4. Add 1 ml 0.16 M ethylene glycol and mix for 1 h at room temperature.
5. Dialyse against 3 × 1 litre 10 mM Na_2CO_3 buffer pH 9.5 at 4°C (the last time overnight).
6. Mix 3 ml of the activated HRP with 5 mg protein (salt-free lyophilized or in 1 ml Na_2CO_3 buffer pH 9.5). Mix for 2–3 h at room temperature.
7. Add 5 mg $NaBH_4$ and incubate at 4°C at least 3 h or overnight.
8. Dialyse against PBS at 4°C. If a precipitate forms, this can be removed by centrifugation.
9. Optional: remove uncomplexed HRP by gel filtration and read A_{280} (total protein) and A_{403} (activated HRP).

Protocol 10. Density shift using HRP–asialoorosomucoid conjugate

Equipment and reagents

- Ultra-centrifuge and swinging-bucket rotor
- 22 μm Millipore filter
- HRP–asialoorosomucoid
- [^{125}I]TC asialoorosomucoid
- 0.25 M sucrose, 10 mM Hepes–NaOH, 1 mM EGTA, 0.5% (w/w) BSA pH 7.2
- Imidazole
- 5 mg/ml 3,3′-diaminobenzidine
- Thimerosal
- H_2O_2
- Sucrose solutions

Method

1. Incubate the rat liver parenchymal cells with about 1 μM HRP–asialoorosomucoid and subsaturating amounts of [^{125}I]TC labelled

asialoorosomucoid (e.g. 10 nM) at 37 °C. The density of the endosomes is shifted rapidly following the addition of the HRP conjugate. In order to accumulate sufficient amounts of conjugate to shift the density of lysosomes an additional chase period of more than 1 h is necessary.

2. Remove external and plasma membrane-bound tracer by washing three times with buffer containing EGTA to remove calcium (0.25 M sucrose, 10 mM Hepes–NaOH, 1 mM EGTA, 0.5% (w/w) BSA pH 7.2). Homogenize in the same buffer supplemented with 3 mM imidazole, and prepare a post-nuclear fraction by differential centrifugation.
3. To 2 ml post-nuclear supernatant add 1.34 ml DAB solution (5 mg/ml 3,3′-diaminobenzidine, 12.5 mg/ml thimerosal, in the sucrose/ imidazole buffer, adjust pH to 7.4 with NaOH and filter through a 22 μm Millipore filter) and incubate in the dark at 25 °C before addition of 10.5 μl H_2O_2. Incubate for another 15 min at 25 °C.
4. Prepare control fractions incubating with DAB without H_2O_2.
5. Treated and control fractions are placed on top of suitable linear sucrose gradients and centrifuged at 85 000 *g* for 4 h.

Protocol 11. Induction of density shift in rat liver parenchymal cells by Au–asialoorosomucoid conjugates

Equipment and reagents

- High-speed centrifuge and rotors
- Colloidal gold (see below)
- 10% NaCl
- 0.25 M sucrose, 10 mM Hepes–NaOH, 1 mM EGTA pH 7.2
- Sucrose or Nycodenz

Method

1. Add 20 μl of a serially diluted protein solution to 180 μl gold solution, mix thoroughly. After 5 min 20 μl 10% NaCl is added. The lowest protein concentration that prevents a colour change from red to blue is taken as the stabilizing protein concentration. The amount of protein needed for stabilization is dependent on the size and number of colloidal gold particles.
2. Centrifuge the various sizes of protein coated gold particles as follows:
   - 5 nm: 45 000 *g* for 45 min
   - 10 nm: 45 000 *g* for 30 min
   - 15 nm: 12 000 *g* for 45 min
   - Supernatant is discarded and the 'loose' part of the pellet is resuspended in buffer to the desired final concentration.

Protocol 11. *Continued*

3. To load endocytic organelles, 50 μl gold conjugates are added per millitre cell suspension. The density of the endosomes is shifted rapidly following the addition of the Au conjugate. In order to accumulate sufficient amounts of conjugate to shift the density of lysosomes an additional chase period of more than 1 h is necessary.
4. Wash the cells twice in 0.25 M sucrose, 10 mM Hepes–NaOH, 1 mM EGTA pH 7.2 before homogenization. Organelles containing gold can be separated from the homogenate by centrifugation in sucrose or Nycodenz gradients as described in Protocol 5.

8. Use of specific perturbants to accumulate ligands and/or receptors at defined intracellular sites

The possible occurrence and importance of a given step in the intracellular transport of endocytosed ligand–receptor complexes could conceivably be evaluated if a specific inhibitor against the particular step existed (see *Table 3*). It has already been mentioned that potassium depletion or a hypertonic medium may inhibit or stop endocytosis via coated pits. Internalization via both clathrin-dependent and clathrin-independent pathways are blocked by reducing the incubation temperature to 0°C.

Intracellular transport of endosomes (or the maturation of endosomes) is dependent on intact microtubuli in many cells and may therefore be inhibited by microtubular drugs (colchicine, vinblastine, nocodazole). The specific step affected may be the transport of a vesicle mediating transfer of material from early to late endosomes (the endocytic carrier vesicle). Microtubular drugs may therefore lead to accumulation of material in early endosomes. A similar effect is obtained by incubating mammalian cells at 16°C. At this temperature transport from early to late endosomes is arrested and endocytosed ligand will accumulate in the early endosomes.

Although many compounds obstruct lysosomal degradation of endocytosed substrates there are few if any compounds that selectively act on the lysosomes. Weak bases accumulate in all acidic vesicles and will therefore interfere with ligand dissociation in addition to inhibiting lysosomal degradation. Moreover, lysosomal drugs may interfere with transport and/or fusion processes. Leupeptin, a thiol protease inhibitor, very efficiently inhibits lysosomal degradation of proteins (leading to denser organelles in gradients) (85). The same protease inhibitor will, however, also inhibit the transport of endocytosed material from endosomes to lysosomes, leading to accumulation of endocytosed material in late endosomes (86).

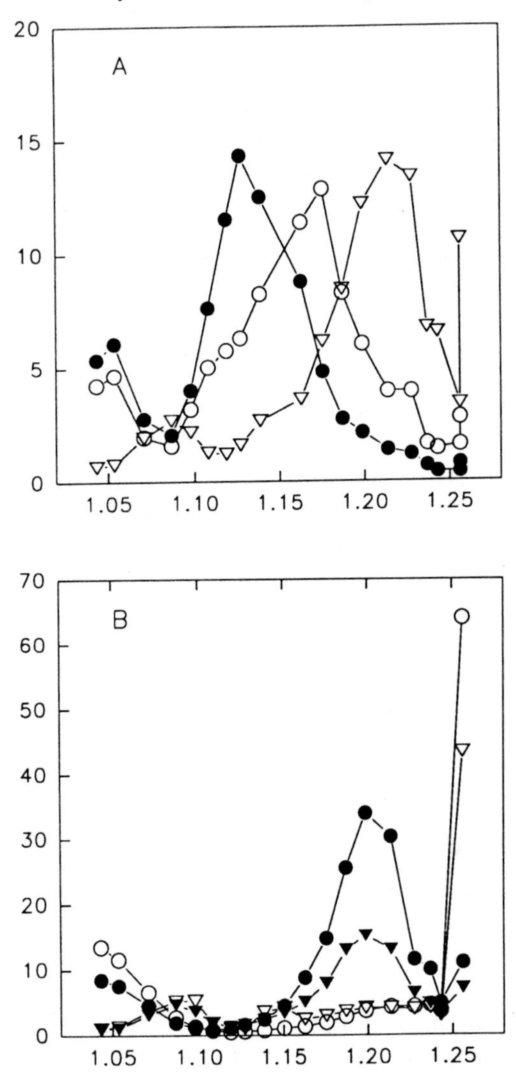

Figure 7. Induction of density shift of lysosomes in rat hepatocytes using colloidal Au coupled to AOM. Freshly prepared rat liver parenchymal cells (5×10^6 cells/ml) were incubated with 10 nM [^{125}I]TC–AOM (closed circle) or 19 nM [^{125}I]TC–AOM–Au (open circle). The cells were then washed twice in ice-cold sucrose and homogenized in a Dounce homogenizer. The post-nuclear supernatants were layered on top of linear sucrose gradients and centrifuged at 85 000 *g* for 6 h. After fractionation the gradients were analysed for acid-precipitable and acid-soluble activity in addition to the lysosomal enzyme β-acetylglucosaminidase (closed triangle in gradients with [^{125}I]TC–AOM and open triangle in gradients with [^{125}I]TC–AOM–Au). (A) shows distribution of acid-precipitable activity after 15 min incubation; (B) shows distribution of acid-soluble activity after 120 min incubation.

Table 3. Perturbants of intracellular transport and ligand processing

Inhibitor	**Site of action**	**Reference**
Colchicine	Microtubuli	70,71
Nocodazole	Microtubuli	72,73
Vinblastine	Microtubuli	74,75
Brefeldin A	Coat proteins	50,76
Monensin	Acidic organelles	77–79
Chloroquine	Acidic organelles	71
Leupeptin	Thiol proteinases	32,80
Vanadate	Phosphorylation (+/−)	81,82
Okadaic acid	Phosphorylation (+)	83,84

9. Use of lysosomal enzyme substrates to disrupt lysosomes selectively

One way to determine whether endocytosed ligands or other molecules are in lysosomes or in prelysosomal compartments is to rupture the lysosomes selectively by means of certain lysosomal enzyme substrates that diffuse across the lysosomal membrane in uncharged form and once inside the lyso-

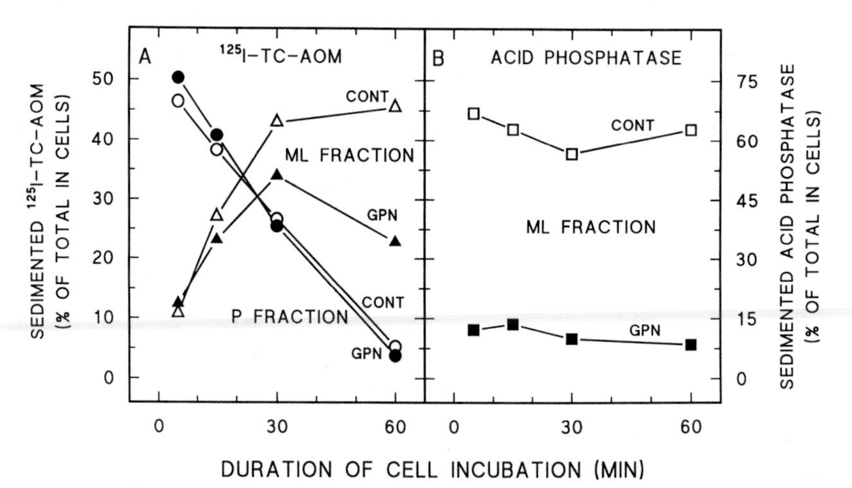

Figure 8. Selective disruption of lysosomes from rat liver hepatocytes using GPN. Hepatocytes (10^7 cells/ml) were incubated with 10 nM [^{125}I]TC–AOM at 37 °C. At the indicated times after initiation of ligand uptake aliquots of cells were removed and cell homogenates prepared and incubated in the presence (filled symbols) or absence (open symbols) of 0.5 mM GPN for 10 min at 37 °C. The homogenates were subsequently fractionated by differential centrifugation into a nuclear fraction, a ML fraction, a P fraction, and a soluble fraction as described in *Protocol 4*. (A) Radioactivity in the P fraction (circles) and the ML fraction (triangles) in control (open symbols) and GPN treated (filled symbols) disruptates. (B) Acid phosphatase activity in ML fractions prepared from control and GPN treated homogenates. All values are expressed as per cent of the total activities from which the fractions were prepared.

somes are degraded to hydrophilic products that cause lysis of the lysosomes (87). In low concentration such substrates can be used to measure lysosomal hydrolysis of substrates in cell-free systems. Numerous lysosomal enzyme substrates that may cause rupture of lysosomes have been tested. A particularly useful substrate is glycylphenylalanyl-2-naphthylamide. This peptide is a cathepsin C substrate. It has been shown by Wattiaux and co-workers to rupture rat liver lysosomes effectively (88). Methyl esters of amino acids can be used in the same way (89).

To determine the step(s) in endocytosis at which glycylphenylalanyl-2-naphthylamide (GPN) acts it was incubated with homogenates from hepatocytes that had internalized a pulse of surface-bound [^{125}I]tyramine cellobiose labelled asialoorosomucoid. It was found that GPN did not release labelled ligand from homogenates until the ligand had entered a degradative compartment (*Figure 8*). This compartment accumulated acid soluble degradation products formed from [^{125}I]TC–AOM and was shown by subcellular fractionation to be a genuine lysosome. Thus GPN may be employed to determine whether a given compound is present in prelysosomal or lysosomal compartments.

Protocol 12. Use of glycyl-L-phenylalanyl-2-naphthylamide (GPN) or methionine-*O*-methylester (MOM) to distinguish between endosomal and lysosomal localization of endocytosed [^{125}I]tyramine cellobiose labelled asialoorosomucoid ([^{125}I]TC–AOM)

Equipment and reagents

- High-speed centrifuge and rotor
- [^{125}I]asialoorosomucoid (TC labelled)
- Glycyl-L-phenylalanyl-2-naphthylamide
- Methionine-*O*-methylester
- DMSO
- Isotonic sucrose: 0.25 M sucrose, 10 mM Hepes, 1 mM EDTA pH 7.3

Method

1. Incubate rat liver hepatocytes with [^{125}I]TC-AOM at 37 °C.
2. At increasing time intervals, remove cells and prepare post-nuclear fractions or MLP fraction by differential centrifugation (to prepare MLP fraction, centrifuge the post-nuclear supernatant at 48 000 *g*, 70 min).
3. Dissolve GPN and MOM in DMSO at 75 mM and 2 M respectively and dilute both substrates 100-fold with isotonic sucrose (0.25 M sucrose, 10 mM Hepes, 1 mM EDTA pH 7.3).
4. Mix the GPN/sucrose solution with half of its volume with cell fractions to a final concentration of 0.5 mM GPN (and 0.67% DMSO). The GPN should be used immediately after being dissolved in DMSO, as it

Protocol 12. *Continued*

is unstable in solution according to the manufacturer (Sigma Chemicals). Incubate the fraction (post-nuclear fraction or MLP fraction) with GPN at 37 °C for 5 min. Mix MOM/sucrose solution with an equal volume of cell fraction to a final concentration of 10 mM MOM and 0.5% (w/w) DMSO. Incubate for 30 min at 0 °C.

5. After treatment with GPN or MOM the fractions are diluted with ice-cold isotonic sucrose solution and centrifuged at 48 000 *g* for 70 min. Acid soluble and acid precipitable radioactivities along with lysosomal enzymes are measured in supernatants and sediments. The effects of the substrates are expressed as the increase in unsedimentable radioactivity or enzyme activity in fractions incubated with substrates as compared to fractions incubated without substrates.

10. Concluding remarks

The mechanisms whereby cells carry out receptor-mediated endocytosis and other forms of endocytosis have been studied intensely during the last 20 years. The movement of ligand and/or receptors through the various cellular compartments has been traced by both morphological and by biochemical methods. Electron microscopy and fluorescence microscopy combined with immunocytochemistry have traced the various pathways followed by endocytic receptors and ligands and studies employing such methods have provided information that has been used to develop models for the endocytic process.

As a tool for cell biological research in general and membrane traffic in particular, the centrifuge still occupies a central position in many laboratories. The separation of organelles according to their size, density, or surface properties can rapidly be achieved by the methods described above. Still, the models of endocytosis are currently in a relatively fluid state. A major reason for this is the lack of reliable markers for the sequence of endosomes that are involved in transport of ligands and receptors. The newly discovered small GTP-binding proteins (rabs) and other proteins involved in selective fusion/transport between endocytic vesicles may be the sort of specific markers that will be helpful in further work with subcellular fractionation.

References

1. Van Deurs, B., Petersen, O. W., Olsnes, S., and Sandvig, K. (1989). *Int. Rev. Cytol.*, **117**, 131.
2. Murphy, R. F. (1991). *Trends Cell. Biol.*, **1**, 77.
3. Gruenberg, J., Griffiths, G., and Howell, K. E. (1989). *J. Cell Biol.*, **108**, 1301.

4. Berg, T., Kindberg, G. M., Ford, T., and Blomhoff, R. (1985). *Exp. Cell Res.*, **161**, 285.
5. Scatchard, G. (1949). *Ann. N.Y. Acad. Sci.*, **51**, 660.
6. Medh, J. D. and Weigel, P. H. (1991). *J. Biol. Chem.*, **266**, 8771.
7. Magnusson, S., Færevik, I., and Berg, T. (1992). *Biochem. J.*, **287**, 241.
8. Wiley, H. S. (1985). *Current topics in membrane transport*, pp. 369–412. Academic Press, New York.
9. Seglen, P. O. and Bohley, P. (1992). *Experientia*, **48**, 158.
10. Sandvig, K., Olsnes, S., Petersen, O. W., and Van Deurs, B. (1987). *J. Cell Biol.*, **105**, 679.
11. Oka, J. A. and Weigel, P. H. (1989). *Biochem. Biophys. Res. Commun.*, **159**, 488.
12. Oka, J. A., Christensen, M. D., and Weigel, P. H. (1989). *J. Biol. Chem.*, **264**, 12016.
13. Hansen, S. H., Sandvig, K., and Van Deurs, B. (1993). *J. Cell Biol.*, **121**, 61.
14. Van Deurs, B., Petersen, O. W., Olsnes, S., and Sandvig, K. (1989) *Int. Rev. Cytol.*, **117**, 131.
15. Steinman, R. M., Brodie, S. E., and Cohn, Z. A. (1976). *J. Cell Biol.*, **68**, 665.
16. Davies, D. E. and Lloyd, J. B. (1990). *J. Immunol. Methods*, **132**, 111.
17. England, I. G., Naess, L., Blomhoff, R., and Berg, T. (1986). *Biochem. Pharmacol.*, **35**, 201.
18. Gordon, P. B., Hoyvik, H., and Seglen, P. O. (1987). *Biochem. J.*, **243**, 655.
19. McKinley, D. N. and Wiley, H. S. (1988). *J. Cell Physiol.*, **136**, 389.
20. Lucas, K. L., Baynes, J. W., and Thorpe, S. R. (1990). *J. Cell Physiol.*, **142**, 581.
21. Holmes, J. M. and Morgan, E. H. (1989). *Am. J. Physiol.*, **256**, G1022.
22. Steinman, R. M. and Cohn, Z. A. (1972). *J. Cell Biol.*, **55**, 616.
23. Griffiths, G., Back, R., and Marsh, M. (1989). *J. Cell Biol.*, **109**, 2703.
24. Haylett, T. and Thilo, L. (1991). *J. Biol. Chem.*, **266**, 8322.
25. Hoekstra, D. and Kok, J. W. (1992). *Biochim. Biophys. Acta*, **1113**, 277.
26. Magee, A. I. and Wileman, T. (1992). In *Protein targeting: a practical approach* IRL Press, Oxford.
27. Seglen, P. O. (1976). *Methods Cell Biol.*, **13**, 29.
28. Smedsrod, B. and Pertoft, H. (1985). *J. Leukoc. Biol.*, **38**, 213.
29. Blomhoff, R. and Berg, T. (1990). In *Methods in enzymology*, Vol. 190, p. 58. Academic Press, New York and London.
30. De Duve, C. (1965). *Harvey Lect.*, **59**, 49.
31. Courtoy, P. J. (1993). In *Endocytic components: identification and characterization* (ed. J. J. M. Bergeron and J. R. Harris), pp. 29–68. Plenum, New York.
32. Seglen, P. O., Gordon, P. B., and Hoyvik, H. (1986). *Biomed. Biochim. Acta*, **45**, 1647.
33. Tolleshaug, H., Berg, T., Nilsson, M., and Norum, K. R. (1977). *Biochim. Biophys. Acta*, **499**, 73.
34. Blomhoff, R., Eskild, W., Kindberg, G. M., Prydz, K., and Berg, T. (1985). *J. Biol. Chem.*, **260**, 13566.
35. Eskild, W., Kindberg, G. M., Smedsrød, B., Blomhoff, R., Norum, K. R., and Berg, T. (1989). *Biochem. J.*, **258**, 511.
36. Bomsel, M., Prydz, K., Parton, R. G., Gruenberg, J., and Simons, K. (1989). *J. Cell Biol.*, **109**, 3243.
37. Pertoft, H., Warmegard, B., and Hook, M. (1978). *Biochem. J.*, **174**, 309.

38. Kindberg, G. M., Refsnes, M., Christoffersen, T., Norum, K. R., and Berg, T. (1987). *J. Biol. Chem.*, **262**, 7066.
39. Kindberg, G. M., Ford, T., Blomhoff, R., Rickwood, D., and Berg, T. (1984). *Anal. Biochem.*, **142**, 455.
40. Schmid, S. L. (1993). In *Subcellular biochemistry: endocytic components: identification and characterization* (ed. J. J. M. Bergeron and J. R. Harris), pp. 1–28. Plenum Press, New York.
41. Marsh, M., Schmid, S., Kern, H., Harms, E., Male, P., Mellman, I., *et al.* (1987). *Proc. Natl. Acad. Sci. USA*, **77**, 6139.
42. Redshaw, M. R. and Lynch, S. S. (1974). *J. Endocrinol.*, **60**, 527.
43. Pittman, R. C. and Taylor, C. A. (1986). In *Methods in enzymology* (ed. J. C. Albers and J. P. Segrest), Vol. 129, p. 612. Academic Press, Orlando.
44. Carew, T. E., Pittman, R. C., and Steinberg, D. (1982). *J. Biol. Chem.*, **257**, 8001.
45. Van Zile, J., Henderson, L. A., Baynes, J. W., and Thorpe, S. R. (1979). *J. Biol. Chem.*, **254**, 3547.
46. Pittman, R. C., Carew, T. E., Glass, C., Green, S. R., Taylor, C. A. J., and Attie, A. D. (1983). *Biochem. J.*, **212**, 791.
47. Dahl, L. B., Laurent, T. C., and Smedsrod, B. (1988). *Anal. Biochem.*, **175**, 397.
48. Herzig, M. C. and Weigel, P. H. (1989). *Biochemistry*, **28**, 600.
49. Nielsen, S. (1993). *Am. J. Physiol.*, **264**, C810.
50. Prydz, K., Hansen, S. H., Sandvig, K., and Van Deurs, B. (1992). *J. Cell Biol.*, **119**, 259.
51. Ellis, J. A., Jackman, M. R., and Luzio, J. P. (1992). *Biochem. J.*, **283**, 553.
52. Yu, F., Ando, S., and Hayashi, S. (1992). *Comp. Biochem. Physiol. [B]*, **103B**, 321.
53. Berg, T., Kindberg, G. M., Ford, T., and Blomhoff, R. (1985). *Exp. Cell Res.*, **161**, 285.
54. Schmid, S. L., Fuchs, R., Male, P., and Mellman, I. (1988). *Cell*, **52**, 73.
55. Hughson, E. J. and Hopkins, C. R. (1990). *J. Cell Biol.*, **110**, 337.
56. Zerial, M. and Stenmark, H. (1993). *Curr. Opin. Cell Biol.*, **5**, 613.
57. Griffiths, G., Hoflack, B., Simons, K., Mellman, I., and Kornfeld, S. (1988). *Cell*, **52**, 329.
58. Bohley, P. and Seglen, P. O. (1992). *Experientia*, **48**, 151.
59. Baufay, H. (1972). *Lysosomes a laboratory handbook*, pp. 1–25. North-Holland, New York.
60. Fukuda, M. (1991). *J. Biol. Chem.*, **266**, 21327.
61. Gorvel, J.-P., Chavrier, P., Zerial, M., and Gruenberg, J. (1991). *Cell*, **64**, 915.
62. Lombardi, D., Soldati, T., Riederer, M. A., Goda, Y., Zerial, M., and Pfeffer, S. R. (1993). *EMBO J.*, **12**, 677.
63. Ludwig, T., Griffiths, G., and Hoflack, B. (1991). *J. Cell Biol.*, **115**, 1561.
64. Leighton, F., Poole, B., Beaufay, H., Baudhuin, P., Coffey, J. W., Fowler, S., *et al.* (1968). *J. Cell Biol.*, **37**, 482.
65. Kindberg, G. M., Stang, E., Andersen, K. J., Roos, N., and Berg, T. (1990). *Biochem. J.*, **270**, 205.
66. Arborgh, B., Berg, T., and Ericsson, J. L. (1974). *Acta Pathol. Microbiol. Scand. [A]*, **82**, 747.
67. Arborgh, B., Glaumann, H., Berg, T., and Ericsson, J. L. (1974). *Exp. Cell Res.*, **88**, 279.

68. Beaumelle, B. D., Gibson, A., and Hopkins, C. R. (1990). *J. Cell Biol.*, **111**, 1811.
69. Stoorvogel, W., Geuze, H. J., Griffith, J. M., Schwartz, A. L., and Strous, G. J. (1989). *J. Cell Biol.*, **108**, 2137.
70. Kolset, S. O., Tolleshaug, H., and Berg, T. (1979). *Exp. Cell Res.*, **122**, 159.
71. McAbee, D. D., Clarke, B. L., Oka, J. A., and Weigel, P. H. (1990). *J. Biol. Chem.*, **265**, 629.
72. Hunziker, W., Male, P., and Mellman, I. (1990). *EMBO J.*, **9**, 3515.
73. Thyberg, J. and Moskalewski, S. (1989). *J. Submicrosc. Cytol. Pathol.*, **21**, 259.
74. Gordon, P. B., Hoyvik, H., and Seglen, P. O. (1992). *Biochem. J.*, **283**, 361.
75. Oka, J. A. and Weigel, P. H. (1983). *Biochim. Biophys. Acta*, **763**, 368.
76. Damke, H., Klumperman, J., von Figura, K., and Braulke, T. (1991). *J. Biol. Chem.*, **266**, 24829.
77. Berg, T., Blomhoff, R., Naess, L., Tolleshaug, T., and Drevon, C. A. (1983). *Exp. Cell Res.*, **148**, 319.
78. McAbee, D. D., Oka, J. A., and Weigel, P. H. (1989). *Biochem. Biophys. Res. Commun.*, **161**, 261.
79. Melby, E. L., Prydz, K., Olsnes, S., and Sandvig, K. (1991). *J. Cell Biochem.*, **47**, 251.
80. Clarke, B. L. and Weigel, P. H. (1989). *Biochem. J.*, **262**, 277.
81. Kindberg, G. M., Gudmundsen, O., and Berg, T. (1990). *J. Biol. Chem.*, **265**, 8999.
82. Oka, J. A. and Weigel, P. H. (1991). *Arch. Biochem. Biophys.*, **289**, 362.
83. Holen, I., Gordon, P. B., and Seglen, P. O. (1993). *Eur. J. Biochem.*, **215**, 113.
84. Lucocq, J., Warren, G., and Pryde, J. (1991). *J. Cell Sci.*, **100**, 753.
85. Tolleshaug, H. and Berg, T. (1981). *Exp. Cell Res.*, **134**, 207.
86. Zachgo, S., Dobberstein, B., and Griffiths, G. (1992). *J. Cell Sci.*, **103**, 811.
87. Reeves, J. P. (1979). *J. Biol. Chem.*, **254**, 8914.
88. Jadot, M., Bielande, V., Beauloye, V., Wattiaux-de Coninck, S., and Wattiaux, R. (1990). *Biochim. Biophys. Acta*, **1027**, 205.
89. Goldman, R. and Kaplan, A. (1973). *Biochim. Biophys. Acta*, **318**, 205.

7

The membranes of the secretory and exocytic pathways

JOHN M. GRAHAM

1. Introduction

This chapter is concerned primarily with isolation of plasma membrane, Golgi membranes, and endoplasmic reticulum from rat liver. The first sections concentrate on the preparative isolation of these three major membrane systems using routine centrifugation techniques, and by immunoaffinity and density perturbation techniques. The later sections are concerned with the use of centrifugation to analyse the events associated with exocytic or secretory pathways in other animal cells and in yeast, which in addition to the endoplasmic reticulum, Golgi membranes, and plasma membrane, involve a number of compartments including the *trans*-Golgi network and a variety of vesicle populations which shuttle between membranes and membrane domains.

1.1 Gradient media

Most of the protocols for the isolation and analysis of the membrane compartments of the secretory pathway use sucrose as a gradient medium. Its cheapness and the huge literature on the use of sucrose gradients makes this a first choice for many workers. Significant disadvantages to the use of sucrose are the osmolarity and viscosity (see *Table 1*) of the aqueous solutions used for gradient formation. The high osmolarity means that vesicles will lose water and their density will approach that of the membrane itself; any opportunity to resolve vesicles on the basis of their native density (membrane plus water compartment) is thus compromised. The high viscosity means that biological particles tend to move rather slowly through sucrose gradients and hence the need for long run times. Sucrose/D_2O gradients reduce but not alleviate this problem (see Section 3.2).

Iodinated density gradient media such as metrizamide, Nycodenz (see Sections 2.4.2 and 3.1) can provide gradients which are iso-osmotic up to densities of 1.19 and 1.16 g/ml respectively, while the latest addition to this group of compounds, iodixanol (essentially a dimer of Nycodenz) provides

Table 1. Physical properties of sucrose solutions

% (w/w)	% (w/v)	Molarity (M)	Density (g/ml)	RI (η)[a]	mOsm[b]	mPas[c]
5.0	5.1	0.149	1.018	1.3403	160	1.73
10.0	10.4	0.303	1.038	1.3479	330	2.06
15.0	15.8	0.464	1.059	1.3557	470	2.52
20.0	21.6	0.632	1.081	1.3639	660	3.14
25.0	27.5	0.806	1.104	1.3723	860	4.06
30.0	33.8	0.988	1.127	1.3811		5.43
35.0	40.3	1.177	1.151	1.3902		7.16
40.0	47.0	1.375	1.176	1.3997		11.44
45.0	54.1	1.581	1.203	1.4096		
50.0	61.4	1.796	1.230	1.4200		33.18
55.0	69.1	2.020	1.258	1.4307		
60.0	77.1	2.255	1.287	1.4418		159.1

[a] RI = refractive index at 20°C.
[b] Osmolarity (mOsm) measured by freezing point depression.
[c] Viscosity (mPas) measured at 5°C.

iso-osmotic solutions at all densities, as does the colloidal silica medium Percoll (see *Table 2–5*). These media not only provide the low osmolarity necessary for the fine resolution of organelles and vesicles, they also have a much lower viscosity than sucrose solutions, so centrifugation times can be reduced (1–5). A useful attribute of both iodixanol and Percoll is their ability to form self-generated gradients (1,4,5): all the other media are used as preformed discontinuous or continuous gradients. The employment of self-generated gradients not only simplifies the formation of continuous gradients, it improves the reproducibility of their density profiles. Methods which employ metrizamide or Nycodenz can generally be carried out with iodixanol, although significant reductions in banding density may be observed of particles whose density in Nycodenz is above 1.16 g/ml (metrizamide, above 1.19 g/ml) because of the reduced osmolarity of iodixanol solutions at higher densities (5).

There are also some pertinent considerations regarding choice of density medium when it is necessary to harvest material from gradients prior to analysis. All of the media except Percoll are true solutes, so membranes can be harvested from solutions of sucrose and iodinated density gradient media simply by dilution of the medium with two to three volumes of buffer and pelleting the membrane at the appropriate centrifugation force. The lower viscosity of the iodinated density gradient media compared to sucrose makes this procedure easier; the one caveat with iodixanol is that the centrifugation force × time should be less than 8×10^6 g.min to prevent self-generation of a gradient. For most spectrophotometric assays for marker enzymes or for SDS–PAGE, however, it is not necessary to remove sucrose or any of the iodinated density gradient media (4). Percoll, however, needs to be removed before spectrophotometric analysis because of its light scattering properties. Because the

Table 2. Physical properties of metrizamide (2)

% (w/v)	Density (g/ml)	Refr. Ind	Osmolarity (mOsm)	Viscosity (mPas)
0	0.998	1.3330	0	1.0
10	1.051	1.3483	107	1.3
20	1.106	1.3646	180	1.6
30	1.161	1.3809	247	2.3
40	1.216	1.3971	320	3.6
50	1.271	1.4133	385	6.0
60	1.326	1.4295	440	11.0

Table 3. Physical properties of Nycodenz (2)

% (w/v)	Density (g/ml)	Refr. Ind	Osmolarity (mOsm)	Viscosity (mPas)
10	1.052	1.3494	112	1.3
20	1.105	1.3659	211	1.4
30	1.159	1.3824	299	1.8
40	1.212	1.3988	388	3.2
50	1.265	1.4153	485	5.3
60	1.319	1.4318	595	9.5

Table 4. Physical properties of iodixanol (4)

% (w/v)	Density (g/ml)	Refr. Ind	Osmolarity (mOsm)	Viscosity (mPas)
10	1.052	1.3490	38	1.5
20	1.105	1.3649	80	1.9
30	1.159	1.3809	115	2.5
40	1.213	1.3968	150	5.0
50	1.266	1.4128	200	7.2
60	1.320	1.4287	260	17.0

Table 5. Physical properties of iso-osmotic Percoll[a] (diluted with 0.25 M sucrose)

% (v/v)	Density (g/ml)	% (v/v)	Density (g/ml)
100	1.148	50	1.089
90	1.136	40	1.078
80	1.125	30	1.066
70	1.113	20	1.054
60	1.101	10	1.043

[a] Percoll is a sol of polyvinylpyrrolidone coated silica particles. Density = 1.130 g/ml, osmolarity = 25 mOsm, viscosity = 10 mPas. For membrane fractionation 9 vol. of Percoll is routinely diluted with 1 vol. 2.5 M sucrose to provide a working stock. In the table 100% (v/v) refers to this working stock which was diluted with 0.25 M sucrose to provide a series of dilutions. Data from ref. 2.

speeds used to pellet the Percoll particles may also cause the sedimentation of some of the membrane particles, loss of material may occur at this stage.

2. Isolation of plasma membrane, endoplasmic reticulum, and Golgi from rat liver

Rat liver has proved one of the most widely-used and reliable sources of all three membranes. The ability to homogenize the tissue gently under both iso-osmotic and hypo-osmotic conditions promotes the recovery of large sheets of plasma membrane (contiguous and bile canalicular domains) which are rapidly sedimenting and relatively easy to separate from other material; purification of the sinusoidal domain is rather less easy by straightforward density gradient centrifugation, although immunoaffinity and density perturbation methods are available (see Sections 2.5 and 2.6). Also the well-organized and large Golgi apparatus in hepatocytes facilitates its isolation by standard centrifugation methods and the extensive nature of the rough endoplasmic reticulum membranes in liver makes their isolation relatively easy. In contrast the plasma membrane from cultured cells, even from polarized monolayer cells, tends to fragment more easily and the endomembranes, by and large, are less extensive.

2.1 Plasma membrane

The protocols are based on one of the earliest plasma membrane methods which was worked out by Neville (6) who used 1 mM $NaHCO_3$ pH 8.0 as the homogenization medium, which causes the cells to swell and render them very susceptible to homogenization. While Neville's original method involved the preferential harvesting of the middle region of a tripartite low speed pellet, most modern methods rely on discontinuous sucrose barrier separation of this pellet. Generally, in techniques using a bicarbonate medium, the entire nuclear pellet is processed, while other techniques using an iso-osmotic sucrose homogenization medium (see *Protocol 2*) often include a step to remove most of the nuclei prior to gradient purification (7,8). *Protocol 1* is adapted from Meier *et al.* (9).

The plasma membrane sheets band at and just above the 42–53% interface. Meier *et al.* (9) used a Sorvall TX-28 zonal rotor for the isopycnic banding of the membrane sheets (total volume 1350 ml), and for scaling up the method to livers from ten or more animals, the use of a zonal rotor simplifies the task considerably. To achieve good separation of these membranes it is important to use relatively large volumes of medium to minimize aggregation. The TZ-28 rotor is only capable of 20000 r.p.m. and the centrifugation time using this rotor must be increased to 2 h, and together with the loading and unloading times, the total time for using this rotor is probably nearer 3 h. The livers from four to five animals could be quite easily handled using two fully-loaded runs with an SW28 type rotor. It is not advisable to reduce the total volumes by increasing the concentration of material as this severely curtails yield and purity. Wisher and Evans (10) also used hypotonic 1 mM $NaHCO_3$ for the homogenization of rat liver but included 0.5 mM $CaCl_2$ since this increases the amount of plasma membrane recovered in the low speed pellet. The addition of divalent cations stabilizes the plasma membrane so that it resists fragmentation during homogenization. The use of a low speed A-type zonal rotor and a continuous (but not linear) gradient of sucrose (6–60%, w/v), at 3900 r.p.m., to separate the plasma membrane sheets on the basis of sedimentation rate (10) is probably the best method, but the equipment is no longer commercially available. The efficacy of the separation (as with all sedimentation rate procedures) depends on the provision of a very narrow sample zone on top of the gradient; an A-type zonal rotor makes this very easy to achieve with a large volume of sample; while even the largest swinging-bucket rotor would necessitate the use of many tubes. The method could however be adapted to the TZ-28 rotor.

Protocol 1. Isolation of plasma membrane using a hypo-osmotic medium

The protocol is given for 10 g of rat liver. All operations are carried out at 0–4 °C.

Equipment and reagents

- Loose-fitting Dounce homogenizer (Wheaton type B, 0.1–0.3 mm clearance) of about 40 ml capacity
- Low speed centrifuge with swinging-bucket rotor
- 1 mM $NaHCO_3$ pH 8.0
- Ultracentrifuge with 30–40 ml swinging-bucket rotor (e.g. Beckman Type 28 or equivalent)
- Sucrose solutions (%, w/v): 71%, 53%, and 42%, in 10 mM Tris–HCl pH 7.6 and 0.25 M sucrose

Method

1. Mince the liver finely with scissors or a razor blade and wash the pieces in 1 mM $NaHCO_3$ three times (use about 30 ml each time).
2. Homogenize the liver in 60 ml $NaHCO_3$ using six or seven strokes of the pestle.
3. Dilute to 180 ml with $NaHCO_3$ and filter twice through two layers of cheesecloth or nylon mesh.
4. Centrifuge at 1500 g_{av} for 15 min, then resuspend the pellet in 40 ml 71% (w/v) sucrose, and stir for 15 min.
5. Transfer 4 × 10 ml of the suspension to tubes for the Beckman SW28 swinging-bucket rotor, overlay with 12 ml each of 53% and 42% (w/v) sucrose, and top up the tube with 0.25 M sucrose.
6. Centrifuge at 100 000 g_{av} for 1 h.

Protocol 2. Using an iso-osmotic homogenization medium (adapted from ref. 7)

Carry out all operations at 4°C. The procedure is given for one liver.

Equipment and reagents

- Loose-fitting Dounce homogenizer (Wheaton type B, clearance 0.1–0.3 mm)
- Low speed refrigerated centrifuge with swinging-bucket rotor
- Ultracentrifuge with 30–40 ml swinging-bucket rotor (e.g. Beckman Type 28, or equivalent)
- Syringe (20 ml) with metal filling cannula
- Medium A: 0.25 M sucrose in 1 mM $MgCl_2$, 10 mM Tris–HCl pH 7.4
- Medium B: 2.0 M sucrose in 1 mM $MgCl_2$, 10 mM Tris–HCl pH 7.4
- Nylon mesh (750 μm)

Method

1. Blanch the liver of an anaesthetized rat by perfusing about 20 ml of medium A through the portal vein. Anaesthetization of experimental animals must be carried out by licenced and trained operators.
2. Excise and weigh the liver, then chop finely with scissors, and suspend the mince in the same medium (containing any suitable mixture of protease inhibitors if required) at about 4 ml/g of liver.
3. Homogenize in the Dounce homogenizer using ten strokes of the pestle. Try to avoid frothing.
4. Remove connective tissue and unbroken cells by filtering through nylon mesh (750 μm). Do not force the homogenate through the filter, facilitate filtration by gentle stirring with a glass rod.
5. Dilute the filtrate with medium A to 5 ml/g liver and divide between two 50 ml tubes for centrifugation at 280 g_{av} for 5 min.
6. Aspirate the supernatant using a syringe and metal cannula; do not decant it as the pellet is very loosely packed. If required the two pellets can be resuspended in half the original volume of medium, rehomogenized (three strokes of the pestle), the centrifugation repeated, and the supernatant combined with the first.
7. Centrifuge the 280 *g* supernatant at 1500 g_{av} for 10 min and decant the supernatant.
8. Resuspend the pellet in 25 ml medium A using the Dounce homogenizer (five strokes of the pestle) and mix well with 2 vol. of medium B. Adjust the density if necessary to 1.18 g/ml (η = 1.4106).
9. Divide the suspension between two tubes for a Beckman SW28 swinging-bucket rotor, fill by overlayering with about 2 ml of medium A, and centrifuge at 113 000 g_{av} for 1 h.
10. Collect the plasma membrane sheets which band at the interface;

dilute to 40 ml with homogenization medium, and harvest them by centrifugation at 3000 g_{av} for 10 min.

11. Resuspend the pellets in the required volume (approx. 1 mg protein/ml) of medium.

The large volumes of medium used in the preparation should not be reduced as this limits the recovery and purity of the membranes.

2.2 Separation of plasma membrane domains

This section is devoted exclusively to the fractionation of the plasma membrane domains of rat liver on a preparative scale. The fractionation of the plasma membrane domains from cultured cells such as Caco-2 and MDCK, and in particular the use of these cells in studies relating to the sorting of proteins destined for domain-specific insertion is covered in Section 3.5.

2.2.1 Separation of bile canalicular and contiguous domains

The sheets of liver plasma membrane (see previous section) can be used to prepare the basolateral (contiguous) and apical (bile canalicular) membrane domains: either sonication or vigorous homogenization have been used to disrupt the membrane. The membranes should be resuspended in buffered 0.25 M sucrose without any divalent cations. Generally the plasma membrane sheets (from one rat liver) are suspended in 2–5 ml. Hubbard *et al.* (7) preferred using sonication in a bath sonicator (output 80 W) containing an ice slurry; they found that between 3 and 20 × 10 sec 'bursts' of sonication with 1 min 'rests' were required to effect vesiculation, which can be monitored by phase-contrast microscopy. Wisher and Evans (10) favoured 20 strokes of a tight-fitting Dounce homogenizer (0.076 mm clearance). This method has also been used for hamster liver (11), although 30 strokes of the pestle was found to be more efficacious, while Meier *et al.* (9) used as many as 50 strokes. A Polytron homogenizer, 12 mm tip at setting 8 (8) can also be used.

Whatever method is used to fragment the membranes, the two domains are subsequently separated on a sucrose gradient. Hubbard *et al.* (7) used a continuous linear sucrose gradient (0.46–1.42 M) loading the sample on top (4 ml sample per 32 ml gradient) and centrifuged at 72 000 g_{av} for 20 h. Discontinuous gradients on the other hand tend to cover a much narrower range and may be more useful for routine practice: 37%, 43%, and 49% (w/v) sucrose at 95 000 g_{av} for 3–16 h (10), or 35%, 39%, and 44% (w/v) sucrose at 196 000 g_{av} for 3 h (9). Meier *et al.* (9) also scaled down the volume of the gradients used, these workers layered 3.5 ml sample on top of a total gradient volume of 9 ml. The resolution of the two membrane domains was not compromised by using smaller discontinuous gradients and in high performance small volume swinging-bucket rotors buoyant density separation can be achieved in a few hours. The canalicular membrane bands predominantly at the sample/gradient interface and the basolateral (contiguous) domain bands at the 39–44% interface (9).

The identification of these domains is relatively simple: their enzymic characteristics have been well documented from histochemical localization at the electron microscope level. It is widely accepted that the apical membrane, not only from liver but also from other organized tissues such as kidney and intestine, is enriched in enzymes such as leucine aminopeptidase and alkaline phosphodiesterase. Basolateral (contiguous) membranes can be identified by the Na^+/K^+-ATPase.

2.2.2 Sinusoidal domain

Upon homogenization of rat liver the sinusoidal membrane vesiculates and becomes a component within a mixed population of smooth vesicles which are part of the total microsomal fraction. These vesicles are difficult to resolve because although they may be derived from a variety of sources, smooth endoplasmic reticulum, Golgi membranes, other domains of the plasma membrane, endosomes, etc., they all have rather similar densities—at least in sucrose gradients. In a continuous sucrose gradient, Golgi and sinusoidal plasma membrane overlapped very significantly (10). The isolation of the sinusoidal membrane by immunoaffinity techniques is considered later in Section 2.5.

2.3 Endoplasmic reticulum (ER)

Both the rough and the smooth ER form vesicles of the correct orientation (i.e. cytoplasmic face outermost) upon homogenization of a tissue. They sediment predominantly at about 100 000 *g* for 30–40 min although some of the largest and densest vesicles will sediment at about 30 000 *g*. Routinely, after the light mitochondrial fraction has been pelleted, the microsomes are purified either from the light mitochondrial supernatant directly or from a subsequent 100 000 *g* pellet.

Protocol 3. Isolation of endoplasmic reticulum in a discontinuous sucrose gradient

The method is adapted from ref. 12. Perform all operations at 0–4 °C.

Equipment and reagents

- Polytron 20ST homogenizer (or equivalent)
- High speed centrifuge with 8 × 50 ml fixed-angle rotor
- Ultracentrifuge with swinging-bucket rotor (Beckman SW28 or equivalent)
- Medium A: 0.5 M sucrose, 5 mM $MgCl_2$, 1% (w/v) dextran (M_r = 225 000), 37 mM Tris–maleate buffer pH 6.4
- Sucrose gradient solutions: 1.3 M, 1.5 M, and 2.0 M

Method

1. Mince rat liver using scissors or a razor blade and add 2 vol. of medium A.

2. Homogenize using the Polytron 20ST homogenizer at 6000 r.p.m. for 45 sec (alternatively an Ultra-Turrax type will suffice). If a Potter–Elvehjem homogenizer is the only type available then the volume of buffer will need to be at least doubled.
3. Centrifuge the homogenate at 6000 g_{av} for 15 min to remove nuclei, plasma membrane sheets, Golgi, heavy mitochondria, and cell debris.
4. Dilute the supernatant with medium 1:5 and centrifuge at 10 000 g_{av} to remove the remaining mitochondria, lysosomes, and peroxisomes.
5. In a swinging-bucket rotor layer the supernatant (10 ml) on to a discontinuous sucrose gradient of equal volumes (9 ml) of 2.0, 1.5, and 1.3 M sucrose, and centrifuge at 85 000 g_{av} for 90 min.
6. Harvest the material at the sample–1.3 M sucrose interface.

This interfacial material consists of both smooth and rough elements of the endoplasmic reticulum. The rough ER can be purified from the smooth ER in a continuous sucrose gradient (*Protocol 4*). One of the problems of standard preparations of the rough ER is contamination from peroxisome cores, which co-band with rough ER; although this can be eliminated if the rough ER is density-perturbed by stripping off most of the ribosomes using pyrophosphate (13). A widely used alternative to a continuous sucrose gradient for the simultaneous preparation of smooth and rough membranes from a post-mitochondrial supernatant is to use a caesium chloride-containing sucrose barrier (*Protocol 5*).

Protocol 4. Purification of smooth and rough endoplasmic reticulum (adapted from ref. 13)

The method is for about 10 g of rat liver. Carry out all operations at 0–4°C.

Equipment and reagents

- Ultracentrifuge with swinging-bucket rotors (Beckman SW28 and SW40Ti or equivalent)
- Two-chamber gradient maker
- Post-mitochondrial supernatant or resuspended microsomal fraction in 0.25 M sucrose (25–30 ml)
- Sucrose solutions: 13.6%, 17.5%, 50%, 51.5% (w/v)—these can be made up containing any required organic buffer (10 mM) at pH 6.5–7.5
- Stripping solution: 68% (w/v) sucrose, 40 mM pyrophosphate, 3 mM imidazole–HCl pH 8.2

Method

1. Layer the microsomal suspension over a linear 17.5–51.51% (w/v) sucrose gradient (ρ = 1.07–1.20 g/ml) with a small cushion of 51.5% sucrose at the bottom. Use the ratio of sample:gradient:cushion of about 3:2:1.

Protocol 4. *Continued*

2. Centrifuge at 100 000 g_{av} for 16 h; at higher speeds e.g. 300 000 g the time can be reduced to 3 h.
3. The smooth ER bands broadly within the gradient. Aspirate a volume equal to the sample plus gradient, and resuspend the pellet of rough microsomes in the residual medium.
4. Adjust the density to 1.25 g/ml in a volume of 14 ml and add 5 ml of stripping solution.
5. Make a 13.6–50% (w/v) linear sucrose gradient for the Beckman SW40Ti swinging-bucket rotor and underlay it with the sample (gradient: sample volume ratio of about 3:1) and centrifuge at 200 000 g_{av} for about 2 h.

The stripped vesicles band between densities 1.10 and 1.19 g/ml, while peroxisomal and mitochondrial contaminants band below this region.

Protocol 5. Separation of rough and smooth endoplasmic reticulum (adapted from ref. 14)

Equipment and reagents

- Ultracentrifuge and suitable fixed-angle rotor (10–14 ml tubes), e.g. Beckman 50Ti
- Post-mitochondrial supernatant in 0.25 M sucrose
- Sucrose solutions: 0.6 M and 1.3 M in 15 mM CsCl, 5 mM Tris–HCl pH 7.4

Method

1. In tubes for the fixed-angle rotor layer 3 ml and 1.5 ml respectively of the 1.3 M and 0.6 M sucrose solutions.
2. Fill the tube with the post-mitochondrial supernatant and centrifuge at 100 000 g_{av} for 90 min.
3. Harvest the smooth membranes from the 0.6–1.3 M interface and the rough microsomes from the pellet

2.4 Golgi membranes

2.4.1 Using sucrose gradients

Golgi membranes from actively secreting tissues such as rat liver form a major defined organelle and are relatively easy to isolate so long as the method of tissue disruption is maintained as mild as possible so that the Golgi stacks are retained. In other cells, particularly tissue culture cells, the Golgi is much less well organized, less prominent, and correspondingly more difficult to isolate. The density of Golgi membranes, at least those from liver, and particularly of those membranes of the *trans*-Golgi, tends to be lower than other membranes which helps their isolation considerably.

If rat liver is homogenized in buffered iso-osmotic sucrose using no more than six strokes of the pestle of a Potter–Elvehjem homogenizer then the majority of the Golgi membranes will sediment at approximately 15 000 *g* for 20 min (i.e. in the light mitochondrial pellet). One of the most widely-used methods for recovering the Golgi membranes from the light mitochondrial pellet, involves suspending the latter in dense sucrose and overlayering it with a discontinuous sucrose gradient and allowing the Golgi membranes to float upwards during centrifugation. This method, described in *Protocol 6*, is of particular use if other subcellular fractions need to be investigated as the sucrose gradient achieves partial resolution of mitochondria, lysosomes, and any plasma membrane in the sample. Simplified sucrose barrier methods are given in *Protocols 7* and *8*.

Protocol 6. Isolation of Golgi membranes from a rat liver light mitochondrial fraction in discontinuous sucrose gradients (adapted from refs 15 and 16)

Equipment and reagents

- Low speed centrifuge with suitable swinging-bucket rotor
- High speed centrifuge with 8 × 50 ml fixed-angle rotor
- Ultracentrifuge with suitable swinging-bucket rotor (e.g. Beckman SW41Ti or equivalent)
- Potter–Elvehjem (P/E) homogenizer, clearance 0.09 mm (about 40 ml)
- Refractometer
- Loose-fitting Dounce homogenizer (about 40 ml)
- Rat liver homogenate prepared in 0.25 M sucrose, 10 mM Hepes–NaOH pH 7.6, 1 mM EDTA (about 40 ml for a 12 g liver), using approx. six strokes of the pestle (500 r.p.m.) of the P/E homogenizer
- Sucrose solutions: 2.0 M, 1.33 M, 1.2 M, 1.1 M, 0.77 M, 0.25 M, all in 10 mM Hepes–NaOH pH 7.4

Method

1. Centrifuge the homogenate at 1000 g_{av} for 10 min to remove the nuclei.
2. Keep the supernatant on ice, resuspend the nuclear pellet in the homogenization medium (about 20 ml), and resuspend in the loose-fitting Dounce homogenizer, using about four gentle strokes of the pestle.
3. Recentrifuge at 1000 g_{av} for 10 min.
4. Combine the supernatants and centrifuge at 3000 g_{av} for 10 min to pellet the heavy mitochondria. Remove as much of the supernatant as possible and centrifuge it at 15 000 g_{av} for 20 min. Pour off the supernatant.
5. Resuspend the light mitochondrial pellet in 1–2 ml of 0.25 M sucrose by vortex mixing; then add 8 ml of 2.0 M sucrose and homogenize in

Protocol 6. *Continued*

the loose-fitting Dounce homogenizer, and adjust the concentration of sucrose to 1.55 M. Check that the refractive index is η = 1.4080 and re-adjust if necessary.

6. Transfer 4–5 ml of the light mitochondrial suspension to a 13–14 ml tube for a swinging-bucket rotor and overlay it with the following sucrose solutions: 2.5 ml of 1.33 M, 2 ml of 1.2 M and 1.1 M, and 1 ml of 0.77 M and 0.25 M sucrose. Centrifuge at 160 000 g_{av} for 1 h.
7. Harvest the material which bands at the top two interfaces and any intervening material.
8. Dilute it with 1 vol. of buffer and ,sediment the Golgi membranes at 180 000 g_{av} for 1 h.

Centrifugation of homogenates or post-nuclear supernatants on discontinuous sucrose gradients is widely used not only for the isolation of a particular membrane type but also to study the synthesis and translocation of proteins between the various compartments of the endoplasmic reticulum, Golgi, plasma membrane, etc. These studies will be dealt with in more detail in Section 3. Generally discontinuous sucrose gradients tend to suffer from a lack of resolution, especially for the separation of multiple membrane types, where they can be used more effectively in an analytical mode rather than a preparative one. Nevertheless some useful information can be gained from using this technique.

Multiple discontinuous sucrose gradients (19%, 23%, 27%, 31%, 35%, and 43%) have for example, identified three subcellular fractions containing the $G_{S\alpha}$ subunit of the guanine nucleotide-binding regulatory protein in lymphoma cells (17). The major fraction, which was the densest, not unexpectedly co-sedimented with a plasma membrane marker but two minor lighter fractions were also identified: one contained an endoplasmic reticulum marker and the lightest may have been the Golgi. Although the Golgi membranes from these cells apparently split into two subfractions of widely different densities, the $G_{S\alpha}$ subunit was associated only with the minor lighter fraction. Although discontinuous sucrose gradients often allow analysis of membrane processing the resolution rarely permits a preparative approach. Low density membranes from adipocytes on 0.1–1.5 M discontinuous sucrose gradients (60 min at 90 000 g) demonstrate clearly that the three vesicle populations bearing the alpha subunits of G_s and G_i and the glucose transporter have distinct banding density profiles, but the overlaps of these profiles would not allow the isolation of a specific vesicle type (18). Sucrose gradients (ρ = 1.06–1.25 g/ml) have also been used for resolving Golgi and plasma membrane vesicles (19). By and large, the newer gradient media such as metrizamide, Nycodenz, and iodixanol provide greater resolving power (see

Section 1.1) and the use of these materials in a general membrane fractionation mode is given in Section 2.4.2

Protocol 7. Rapid density barrier isolation of Golgi membranes from rat liver (version A)

This method is adapted from ref. 20. Carry out all operations at 0–4 °C.

Equipment and reagents

- Low speed or high speed centrifuge (5000 g_{av} required)
- Ultracentrifuge with swinging-bucket rotor, Beckman SW28 or equivalent
- 1.2 M sucrose
- Polytron homogenizer (or Ultra-Turrax type)
- Buffer: 0.5 M sucrose, 1% (w/v) dextran (M_r = 225 000), in 0.05 M Tris–maleate buffer pH 6.4

Method

1. Mince the liver finely using a razor blade and suspend in buffer (about 10 g liver per 20 ml buffer).
2. Homogenize in Polytron homogenizer at 10 000 r.p.m. for 30–50 sec.
3. Centrifuge for 15 min at 5000 g_{av} and very carefully aspirate the supernatant.
4. The Golgi membranes are contained in the upper 1/2–2/3 (yellow-brown portion of the bipartite pellet); resuspend them in some of the supernatant carefully avoiding resuspension of the lower part of the pellet which contains nuclei and whole cells.
5. Adjust the concentration of the resuspended material with medium to 6 ml per 10 g liver, and layer over 2 vol. of 1.2 M sucrose.
6. Centrifuge in swinging-bucket rotor at 120 000 g_{av} for 30 min; the Golgi membranes collect at the interface.

If it is an aim of the study to attempt to resolve the various domains of the Golgi membranes it may be necessary to remove the dextran that is used in this method. The dextran is added to the isolation medium to maintain the stacking and thus facilitate separation (20), but to isolate the various domains the Golgi must be unstacked by incubation in a mixture of 3 mg each of crude α-amylase Type X-A from *Aspergillus oryzae* and α-amylase Type III-A from barley with 4–5 ml of Golgi suspension at 4 °C for 45 min (21). This is followed by gentle liquid shear using a Pasteur pipette (tip i.d. = 1 mm) but a syringe attached to a metal cannula of the same internal diameter is generally easier to use. Whether the enzyme treatment has any effect on Golgi apparatus prepared in the absence of dextran is not entirely clear.

A modified version (*Protocol 8*) of this method was used by Balch *et al.* (22). The homogenate was produced by a 'cell cracker' which may not be available to the reader. It comprises a ball bearing within a socket (clearance 0.025 mm) through which the cell suspension is forced repeatedly (10–12 times)—an alternative is 20 strokes of a very tight-fitting Dounce homogenizer (22). The method avoids any preparatory differential centrifugation steps.

Protocol 8. Rapid density barrier isolation of Golgi membranes from rat liver (version B)

Equipment and reagents

- 'Cell cracker' or tight-fitting (clearance 0.025–0.075 mm) Dounce homogenizer (see above)
- Ultracentrifuge and suitable swinging-bucket rotor (e.g. Beckman SW28)
- Homogenization medium: 0.25 M sucrose, 10 mM Tris–HCl pH 7.4
- Sucrose solutions: 0.8 M, 1.2 M, 1.6 M, and 2.3 M sucrose, all in 10 mM Tris–HCl pH 7.4
- 100 mM EDTA

Method

1. Make the homogenate (in 0.25 M sucrose) 1.4 M with respect to sucrose by adding an equal volume of a 2.3 M sucrose, 10 mM Tris–HCl pH 7.4.
2. Then add 100 mM EDTA to a final concentration of 1 mM if it is not already present in the homogenate. Use a vortex mixer to ensure complete mixing.
3. Transfer 12 ml to tubes for the swinging-bucket rotor; overlay with 14 ml of 1.2 M sucrose and 8 ml of 0.8 M sucrose, underlay with 3 ml 1.6 M sucrose (all solutions containing 10 mM Tris–HCl pH 7.4), and centrifuge at 110 000 g_{av} for 2 h. For other rotors maintain the same ratio of sample and sucrose layers. The centrifugation speed and time can be adjusted as appropriate so long as the $g \times$ time is the same.
4. Remove the Golgi band at the 0.8–1.2 M sucrose interface.

This band contains virtually all of the mannosidase I and galactosyl transferase activities while the endoplasmic reticulum (glucosidase I) bands predominantly within the 1.4 M sucrose layer. It can therefore be used to isolate these two types of membranes simultaneously. However for the simultaneous isolation of Golgi together with separation of both rough and smooth ER, an interesting method which combines flotation of the Golgi away from a light mitochondrial supernatant together with the separation of rough and smooth ER on a sucrose/CsCl barrier has been described by Macintyre (23), for application to tissue culture cells (*Protocol 9*). A rather more simple procedure using a self-generated gradient of Percoll (24) has been used to obtain a partial separation of Golgi, ER, and plasma membrane from cultured cells (*Protocol 10*).

Protocol 9. Simultaneous isolation of Golgi and ER from cultured cells (adapted from ref. 23)

Equipment and reagents

- Ultracentrifuge with Beckman SW28 swinging-bucket rotor (or equivalent)
- Cell homogenate in 0.25 M sucrose
- 0.15 M CsCl
- Sucrose solutions: 8.4%, 32.5%, 45.2%, 77% (w/v) in 10 mM Tris–HCl pH 7.4; 1.3 M containing 15 mM CsCl, 10 mM Tris–HCl pH 7.4

Method

1. Centrifuge the cell homogenate at 11 000 g_{av} for 10 min and then mix the supernatant with 77% (w/v) sucrose (1:2).
2. In tubes for the swinging-bucket rotor layer 12 ml of the sample with 6.5 ml each of 45.2%, 32.5%, and 8.4% (w/v) sucrose and centrifuge at 90 000 g_{av} for 1 h.
3. Collect the Golgi membranes at the top two interfaces, dilute with buffer to 0.25 M sucrose, and harvest by centrifugation at 120 000 g_{av} for 1 h.
4. Aspirate from the top of the material remaining in the SW28 tube to leave 10 ml. Add 4.8 ml water and 1.6 ml 0.15 M CsCl.
5. After mixing well, transfer an aliquot to a tube for a fixed-angle rotor; overlay with an equal volume of 1.3 M sucrose/15 mM CsCl, and centrifuge at 140 000 g_{av} for 2 h.
6. Remove the material at the interface (smooth microsomes) and dilute to 0.25 M sucrose prior to harvesting by centrifugation, and resuspend the pellet (rough microsomes) in the same medium.

Protocol 10. Simultaneous isolation of Golgi, ER, and plasma membrane from cultured cells (adapted from ref. 24)

Equipment and reagents

- Ultracentrifuge with Beckman 65 fixed-angle rotor (8 × 13.5 ml) or equivalent
- DNase I in 0.25 M sucrose, 10 mM $MgCl_2$, 50% (w/v) glycerol, 10 mM Tris–HCl pH 7.4 (2000 U/mg), stored at −20°C
- Homogenate of cells in 0.25 M sucrose, 10 mM Tris–HCl pH 7.4 (buffer A)
- 90% (v/v) Percoll stock: 9 vol. Percoll plus 1 vol. 2.5 M sucrose

Method

1. Incubate the homogenate with DNase I (250 μg/mg cell protein) for 30 min at room temperature.

Protocol 10. *Continued*

2. Dilute the Percoll stock with buffer A to produce a 12.5% or 15% (v/v) Percoll suspension. The correct Percoll concentration required will vary with the material and must be deduced by experimentation.
3. Mix 2.5 ml of homogenate with 10 ml of the diluted Percoll and centrifuge at 28 000 g_{av} for 1 h in the Beckman 65 rotor.
4. Allow the rotor to decelerate from 2000 r.p.m. to rest without the brake.
5. Collect the gradient by tube puncture.

The ER, Golgi, and plasma membrane band broadly through the gradient, with median densities of approx. 1.05, 1.035, and 1.03 g/ml. Although the overlap of bands is considerable, the gradient may be useful to determine the location of a particular marker.

2.4.2 Using iodinated density gradients

The advantage of iodinated density gradient media over sucrose is the reduced osmotic activity of the former; so that particles are isolated more or less in their native hydrated state. Both metrizamide and Nycodenz have been used successfully, either as continuous or discontinuous preformed gradients, for the isolation of Golgi membranes and other subcellular organelles (25). In 5–15% (w/v) Nycodenz gradients (containing 10 mM Tris–HCl, 1 mM Ca^{2+}, 1 mM Mg^{2+}, and 75 mM sucrose) however, the density of the rough ER and Golgi is apparently reversed (26): the order of banding (increasing density) was rough ER, Golgi, plasma membrane. Peter *et al.* (26) also made the interesting observation that the subcellular fractionation was more reliable if isolated hepatocytes were used rather than the intact tissue. Graham *et al.* (25) however reported the more usual densities of the various membrane organelles, the Golgi having a lower density compared to other organelles in EDTA-containing Nycodenz gradients (25). Whether the particular gradient conditions used by Peter *et al.* (26) contributed to this significant difference is not clear.

The latest addition to the iodinated density gradient media is iodixanol which is capable of forming its own gradient (self-generated) in a centrifugal field (4). Graham *et al.* (5) used iodixanol to simplify the bulk preparation of Golgi membranes. The method minimizes the number of steps and the use of vertical rotors makes the procedure very efficient.

Protocol 11. Isolation of Golgi membranes in a self-generating gradient

Equipment and reagents

- Low speed and high speed centrifuges
- Medium A: 0.25 M sucrose, 1 mM EDTA, 10 mM Tris–HCl pH 7.4
- OptiPrep™ (60% w/v iodixanol)

- Ultracentrifuge (at least 50 000 r.p.m., preferably 65 000 r.p.m.) with vertical rotor (e.g. Beckman VTi65.1), near-vertical (e.g. Beckman NVT65), or shallow angle fixed-angle rotor (e.g. Beckman Type 90Ti)
- Medium B: 5 vol. OptiPrep plus 1 vol. 0.25 M sucrose, 6 mM EDTA, 60 mM Tris–HCl pH 7.4

Method

1. Homogenize the tissue or cells in medium A using standard methods; for liver use six or seven strokes of a Potter–Elvehjem homogenizer (500 r.p.m.).
2. Sediment the nuclei and heavy mitochondria at 3000 g_{av} for 10 min, and produce a light mitochondrial pellet from the supernatant at 15 000 g_{av} for 15 min.
3. Resuspend the pellet in medium A (about 20 ml for the material from one rat liver).
4. Add medium B to the resuspended light mitochondrial pellet to make the final concentration of iodixanol 15% (w/v).
5. Tansfer 10–11 ml to tubes for the Beckman NVT65, near-vertical rotor: top up with homogenization medium, and centrifuge at 350 000 g_{av} for 1.5 h.

It is possible to use the 3000 *g* supernatant rather than the light mitochondrial fraction in this separation. The Golgi bands near the top of the gradient while other membranes and organelles band towards the bottom of the gradient or form a pellet. With a vertical rotor it is necessary to include a small (1 ml) cushion of 25% iodixanol (medium B diluted with medium A) to prevent material from reaching the wall of the tube. In fixed-angle rotors, which have longer sedimentation path lengths, it is necessary to increase the run times to about 3 h. The density of all osmotically active organelles and membrane vesicles is much lower in iodinated density gradient media because of their much lower osmotic pressure (compared to that of sucrose). The density of Golgi membranes in iodixanol is about 1.06–1.09 g/ml (5). This method may also be useful for resolving other elements of the secretory system in the shallow gradient below the Golgi membranes.

2.5 Isolation of membranes by immunoaffinity

As molecules synthesized and modified within the cisternae of the cytoplasmic membrane systems are expressed at the external surface of the sinusoidal membrane (or other plasma membrane domain) and as vesicles formed during homogenization of a tissue or cells are predominantly right side out, antibodies to an externally expressed marker, linked to some dense bead can be used to isolate plasma membrane domains bearing these markers. The process is called immunoaffinity isolation. Luzio and Richardson (27) used this

method very successfully for isolation, for example, of cholinergic nerve terminals and the method has been reviewed by a number of workers (28,29). The method requires in addition, either a secondary antibody or protein A to act as a bridge between the bead and the probing primary antibody; alternatively (27) a derivatized cellulose (diazocellulose) can be used as the bead, others have used commercially available protein A–Sepharose (30).

Hubbard *et al.* (31) preferred the use of intact (protein A bearing) *Staphylococcus aureus* for the isolation of vesicles using antibodies to a variety of cell surface receptors or domains of these receptors (including the asialoglycoprotein receptor). The antibody binding capacities of the intact cells were found to be considerably higher than that of protein A bound to a polyacrylamide support. Using an antibody to the cytoplasmic tail of the receptor the authors were able to isolate intracellular pools of these receptors and to distinguish them from Golgi fractions. Better results are obtained if the antibody is reacted first with the membranes rather than with the *S. aureus* (31), moreover larger amounts of antibody are required in the second approach. The authors also found that although affinity purified antibody provides higher specificity, whole antiserum could be used, especially if the serum is first incubated with the *S. aureus*.

Although membrane fractions bound to beads can be harvested by centrifugation, the availability of magnetic beads (32) allows the separation of membranes very rapidly and effectively in a magnetic field: their disadvantage is their cost compared to that of the non-magnetic variety. The secondary antibody can be linked simply by hydrophobic interaction to a magnetic bead, but it is generally considered preferable to link it covalently: less linking antibody is required and apparently the antigen–antibody reaction is more efficient (32). The beads are transferred from an aqueous suspension through a series of acetone/water mixtures to pure acetone, prior to activation by a process called tosylation in which the surface hydroxyl groups are reacted with toluene-4-sulfonyl chloride in the presence of acetone and pyridine. After activation the beads are transferred back to water (through the acetone/water mixtures). The tosylated beads can be stored almost indefinitely in HCl at 4°C. During handling the beads must of course be protected from any ferromagnetic material and because they tend to stick to glass surfaces, the latter must be siliconized (32).

To bind the secondary antibody, the beads are suspended in borate buffer pH 9.5 and mixed with the antibody (5–15 μg IgG/mg beads, using 2 mg/ml of beads). To quench any remaining active sites the beads are then washed with bovine serum albumin-containing buffers (32). It is reported that about 700 000 IgG molecules are bound per bead. It is recommended that at least a twofold excess of primary antibody should be used for binding to the IgG–bead. This coupling is routinely carried out overnight just prior to use. Alternatively the primary antibody can be reacted with the unpurified membrane suspension rather than with the IgG–bead. In this case the unreacted

antibody must be removed by centrifugation and washing or by centrifuging the membrane-bound antibody away from the soluble form through a density barrier prior to addition of the IgG–bead.

Protocol 12. Isolation of the hepatocyte sinusoidal domain using beads, by centrifugation

Isolation of the sinusoidal membrane using the polymeric IgA receptor (called the secretory component [SC]) as the marker has been used by Sztul *et al.* (30).

Equipment and reagents

- Microcentrifuge
- High speed and ultracentrifuge and suitable fixed-angle rotors (e.g. 8 × 40–50 ml)
- Medium A: 0.25 M sucrose, 50 mM Tris–HCl pH 7.4
- Medium B: 2% (w/v) Triton X-100, 150 mM NaCl, 2 mM EDTA, 30 mM Tris–HCl pH 7.4
- Medium C: 1% (w/v) Triton X-100, 0.2% (w/v) SDS, 150 mM NaCl, 5 mM EDTA, 10 mM Tris–HCl pH 8.0
- Medium D: as medium C minus the detergents
- Anti-SC serum (and non-immune serum as control)
- Protein A–Sepharose beads

Method

1. Make a 30% (w/v) homogenate of rat liver in medium A.
2. Remove the nuclei, mitochondria, etc. by centrifugation at 10 000 g_{av} for 10 min, and centrifuge the supernatant at 120 000 g_{av} for 90 min to pellet the total microsomes.
3. Resuspend the microsomes in the buffered medium A (2 mg protein/ml).
4. Incubate 200 μl of a 60% (v/v) suspension of protein A–Sepharose beads in medium B at 24°C with 50 μl of anti-SC serum (or non-immune serum for 60 min).
5. Pellet the beads in a microcentrifuge and wash them four times in medium C, then wash them a further two times in medium D.
6. Incubate the beads in 500 μl of the microsome fraction for 2 h at room temperature.
7. Pellet the beads in a microcentrifuge, wash four times in the buffered sucrose medium, and then finally resuspend in this buffer for further analysis.

The binding of the surface receptors to a ligand (wheat germ agglutinin–Sepharose beads) also permits the resolution of inside out (non-binding) from right side out (binding) vesicles, e.g. the vesicles from brush border membrane (33).

2.6 Density perturbation methods

2.6.1 Lectin–gold

Gupta and Tartakoff (34) recognized that although standard density gradient protocols were effective at resolving most of the Golgi from rough microsomes, resolution from some plasma membrane remained a problem. They developed a method which utilized the presence on the surface of the plasma membrane of carbohydrate groups which bind wheat germ agglutinin and the ability of the agglutinin to bind to dense colloidal gold particles. The binding of wheat germ agglutinin (34) is not competed by sucrose and therefore fractionation in sucrose gradients is not a problem, but to improve its cross-linking to the gold colloid, Gupta and Tartakoff (34) increased its molecular size by linking covalently to bovine serum albumin. The authors bound the lectin–BSA–gold to intact rat myeloma cells for 30–60 min at 4 °C, then washed the cells to remove excess reagent. After breaking open the cells in a hypotonic buffer by Dounce homogenization, the homogenate was adjusted to 0.25 M with respect to sucrose and fractionated in a 20–40% sucrose gradient. Compared to untreated cells plasma membrane markers were shifted to the bottom of the gradient while Golgi markers were unaffected.

2.6.2 Horse-radish peroxidase methods

Density perturbation methods for removing endosomes from Golgi fractions depend on the ability to modulate the density of the endosomes once they have taken up a horse-radish peroxidase-linked ligand (35). In the presence of hydrogen peroxide, the enzyme oxidizes 3,3′-diaminobenzidine (DAB) and it is this reaction which is considered responsible for the resulting raised density of the endosomes.

Any suitable ligand can be chosen and linked to horse-radish peroxidase (HRP) and presented to either isolated cells or an intact tissue. Courtoy *et al.* (36) used neogalactosylalbumin as the ligand coupled to HRP, which was injected into rats intravenously (1 μg/g body weight) for 10 min prior to liver perfusion, excision, and homogenization. The crude Golgi/endosome fraction is then isolated and treated according to the following protocol (35,36). The membranes can be reacted while suspended in the gradient medium or they can be harvested and resuspended in 1.5 ml of a buffered 0.25 M sucrose medium.

Protocol 13. HRP–DAB–H_2O_2 perturbation

1. Prepare a solution of 7.2 mM DAB in an appropriate medium and remove any particles by passing it through a 0.45 μm filter. The A_{280} of the filtrate, diluted 50 times, should be about 1.63.
2. Incubate the membrane fraction with 5 ml of the DAB solution (final concentration 5 mM) and 11 mM H_2O_2 at 25°C for 15–30 min, in the

dark, with gentle agitation. The H_2O_2 should be prepared freshly from stock 30%, addition of 40 μl of a 6% solution should be sufficient. Controls can be incubated if required without the peroxide.

3. Recentrifuge the membrane suspension through a suitable gradient. In a linear sucrose gradient (ρ = 1.1–1.3 g/ml), centrifuged at 240 000 *g* for 100 min, the density of endosomes was shifted from about ρ = 1.13 g/ml to about 1.21 g/ml (36).

2.6.3 Digitonin

The ability of digitonin to form a stable covalent complex with cholesterol, has also been used to density perturb those membranes which are rich in this lipid (37). Membranes are suspended in 0.25 M sucrose to a volume equivalent to 1 ml per gram of liver, and 2.7 volumes of 0.235% (w/v) digitonin added dropwise with continuous stirring. The suspension is then allowed to stand at 0°C for 15 min, then excess soluble digitonin removed by dilution with buffer and pelleting, before recentrifuging the membranes through a suitable gradient. In sucrose gradients the mean density of membranes containing relatively low concentrations of cholesterol (endoplasmic reticulum and mitochondria) is unaffected, while the plasma membrane is shifted by about 0.03 g/ml. Golgi membranes are shifted by a modest 0.015 g/ml (37).

3. The secretory pathway

The variety of source, system, and gradient design are so varied that a comprehensive survey of the techniques is impossible, thus only some of the more widely-used approaches can be discussed in any detail. Although the methods have been formulated primarily for analytical purposes some can be used in a preparative mode for the study of the composition of specific compartments where the separation from other compartments is clear cut.

3.1 Resolution of Golgi domains

Continuous (free) flow electrophoresis has been used to fractionate Golgi domains (38) using fairly standard operational conditions: a chamber buffer containing 0.25 M sucrose, 0.5 mM $MgCl_2$, 10 mM triethanolamine, and 10 mM acetic acid buffered to pH 6.5, and an electrode buffer containing 100 mM of triethanolamine and acetic acid. A buffer flow rate of 2.75 ml/fraction/h is used at approx. 130 V/cm (167 mA) and at 6°C. The *cis*-elements are less electronegative than the *trans*-elements. But while the *cis*-marker, nucleoside (UDP) diphosphatase, shows only one major peak of activity, the *trans*-marker distribution (galactosyl transferase), although demonstrating a maximum close to the anode it trailed significantly into the less electronegative regions. Although continuous flow electrophoresis has the merits of maintaining the membranes in an iso-osmotic medium; tends not to cause aggregation, and whose efficacy is unaffected by the volume of sample, the apparatus is expensive and, compared to an ultracentrifuge, not widely available.

Partial separation of the *cis*-, medial-, and *trans*-domains of the Golgi can be obtained in a variety of sucrose gradients. Dunphy and Rothman (39) emphasized the need for relatively long periods of centrifugation (20 h) for the membranes to reach 'apparent equilibrium'. After 4 h they found that the various enzyme markers were diffusely distributed throughout the gradient. Goldberg and Kornfeld (40) used a different gradient system but their treatment of the cells prior to homogenization was similar (see below) to that of Fries and Rothman (41) and Dunphy *et al.* (42). Generally speaking it has been found that the buoyant density in sucrose of the *cis*-, medial-, and *trans*-domains decreases in that order. Analyses of the asparagine-linked oligosaccharide processing and of the enzymes involved in this processing show good separation of the endoplasmic reticulum, and of the *cis*- and *trans*-Golgi, but resolution of the *cis*- and medial-Golgi is less clear.

Markers for the *cis*-domain include *N*-acetylglucosamine-1-phosphodiester α-*N*-acetylglucosaminidase (an enzyme used in construction of mannose-6-P residues on lysosomal enzyme oligosaccharide side chains) and *N*-acetylglucosaminylphosphotransferase. The median-domain is characterized by *N*-acetylglucosaminyl transferase I and the *trans*-domain by galactosyl ovalbumin transferase, while glucosidase I and II are components of the ER.

In the methods of Goldberg and Kornfeld (40) and Dunphy and Rothman (39) cells were swollen in 10 ml of 15 mM KCl, 1.5 mM magnesium acetate (MgOAc), 1 mM dithiothreitol (DTT), 10 mM Hepes–KOH pH 7.5, and then after centrifugation, sufficient supernatant was removed to leave a volume equivalent to 3.5 times that of the pellet, which was resuspended in that volume. Presumably without swelling, the efficiency of cell breakage was not sufficient; the reduction in volume of the supernatant, once the cells have been swollen however will serve to protect the released organelles from the hypo-osmotic medium after homogenization. Goldberg and Kornfeld (40) used nitrogen cavitation (750 psi for 15 min) to disrupt macrophages, while Dunphy and Rothman (39) used 20 strokes of a tight-fitting Dounce homogenizer for CHO cells.

Protocol 14. Resolution of Golgi domains from culture cells (adapted from ref. 40)

Equipment and reagents

- Low speed centrifuge
- Ultracentrifuge with fixed-angle (e.g. Beckman 50Ti or equivalent) and swinging-bucket rotors (e.g. Beckman SW28 or equivalent)
- Tight-fitting Dounce homogenizer
- Gradient maker
- Hepes–KOH pH 7.5, dithiothreitol (DTT), KCl, MgOAc (concentrations will depend on homogenization protocol)

- Medium A: 10 mM NaCl, 1.5 mM $MgCl_2$, 10 mM Tris–HCl pH 7.4
- Medium B: 1 mM EDTA, 1 mM Tris–HCl pH 8.0
- Cell homogenate
- Sucrose solutions: 15%, 45%, 60% (w/w) sucrose in medium B

Method

1. After homogenization, adjust the suspension to 45 mM Hepes, 75 mM KCl, 5 mM MgOAc, 1 mM DTT pH 7.5.
2. Remove the nuclei by centrifugation at 600 g_{av} for 5 min and dilute the supernatant to 15 ml with medium A.
3. Centrifuge the supernatant at 150 000 g_{av} for 1 h.
4. Resuspend the pellet in 5 ml medium B with eight strokes of the pestle of a tight-fitting Dounce homogenizer and repeat the centrifugation.
5. Resuspend the pellet in 1.5 ml of 45% (w/w) sucrose using a Dounce homogenizer and mix with 4.5 ml of 60% (w/w) sucrose.
6. In 38.5 ml tubes for a Beckman SW28 swinging-bucket rotor (or equivalent) form a linear 15–45% (w/w) sucrose gradient (34 ml) and underlay it with the sample.
7. Centrifuge at 100 000 g_{av} for 16–20 h.

The ER peaks in the bottom third of the gradient while the Golgi domains are resolved within the second third of the gradient. Goldberg and Kornfeld (40) included a 1 ml 60% sucrose cushion to prevent any material from pelleting: if the tubes are harvested by upward displacement this is not necessary but with tube puncture, a cushion may be advantageous. The gradient used by Dunphy and Rothman (39) was a discontinuous one rather than a continuous one and comprised 4.5 ml of 55%, 1.5 ml of 40%, 2.5 ml of 35%, 30%, and 25%, and 1 ml of 20% (w/w) sucrose. These gradients can be scaled up or down in volume according to the rotor available and with the higher centrifugal forces of smaller volume rotors, the time can be proportionately decreased. Mundy and Warren (43) top loaded a post-nuclear supernatant on to a similar sucrose gradient. It also appears that these gradients may also serve to identify light vesicles which are responsible for transporting the processed molecules away from the Golgi—although they were not well resolved from these membranes (40).

Sarasate *et al.* (44) used a similar flotation through sucrose gradients for resolving the Golgi domains from a microsomal fraction of rat pancreas and using a combination of enzyme markers and temperature modulation of secretory protein processing, obtained three fractions which, in decreasing order of density, the authors identified as *cis*-, medial-, and *trans*-Golgi.

Sztul *et al.* (30) also reported that a discontinuous system could resolve a light and a heavy Golgi fraction, and supplemented this with a second gradient to isolate the rough endoplasmic reticulum (*Protocol 15*) which may be a useful approach for studying the synthetic and translocation events at the ER/Golgi level.

Protocol 15. Resolution of Golgi domains and ER from liver (adapted from ref. 30)

Equipment and reagents

- High speed centrifuge with 8 × 50 ml fixed-angle rotor
- Ultracentrifuge with 8 × 40 ml fixed-angle rotor and swinging-bucket rotor (Beckman SW28 or equivalent)
- Tight-fitting Dounce homogenizer (clearance 0.025–0.075 mm)
- Homogenate in buffered 0.25 M sucrose
- Sucrose solutions: 0.25 M, 0.86 M, 1.15 M, 1.22 M, 2 M in a suitable buffer

Method

1. Use a 25% (w/v) homogenate of liver and centrifuge it at 7000 g_{av} for 10 min to remove large particles, and then harvest the total microsomes from the supernatant by centrifugation at 100 000 g_{av} for 90 min.
2. Resuspend the 100 000 *g* pellet in 10 ml 1.22 M sucrose and in the swinging-bucket rotor layer on top 8 ml each of 1.15, 0.86, and 0.25 M sucrose.
3. Centrifuge at 100 000 g_{av} for 2.5 h and harvest the material at the 0.25–0.86 M and 0.86–1.15 M interfaces to give the light and heavy Golgi fractions.
4. Collect the material remaining at the bottom of the tube and after ensuring that the material is dispersed in the sucrose solution by homogenization in a Dounce homogenizer, adjust its density, using 2 M sucrose, to 1.14 M sucrose (η = 1.3886).
5. Layer 7 ml of this over 2 ml 1.9 M sucrose and overlay with 1 ml 0.86 M sucrose.
6. Centrifuge at 195 000 g_{av} for 16 h in a swinging-bucket rotor and harvest the rough microsomes from the 1.4–1.9 M interface.

The light Golgi fraction (30) contained Golgi vesicles loaded with lipoprotein particles, while the heavy fraction contained mainly Golgi cisternae, although a number of vesicles from a variety of sources contaminate this fraction. The rough ER fraction on the other hand is very pure.

Discontinuous (often six or seven step) sucrose gradients covering the range 20–55% (w/v), equivalent to a density range of 1.07–1.20 g/ml have been used by a number of workers to localize glycosylation functions to specific domains of the Golgi. Sugumaran *et al.* (45) resolved a number of enzymes involved in glycosylation and chondroitin sulfate synthesis, while

Trinchera *et al.* (46) were able to discriminate two glycosphingolipid sialyltransferases.

Iodinated density gradient media seem to offer a significant improvement in overall resolution of Golgi domains. Partial resolution of the domains of the Golgi involved in the sulfation of heparan sulfate and perhaps the transport vesicles responsible for its transport to the plasma membrane have been carried out by flotation through discontinuous metrizamide gradients (16). The method achieves a significant separation of the galactosyl transferase and *O*-sulfotransferase activities and of a lighter compartment which may be responsible for the transport of the completed molecule away from the Golgi.

Protocol 16. Subfractionation of Golgi domains in an iodinated density gradient medium

Equipment and reagents

- Any of the current iodinated density gradient media can be used, metrizamide, Nycodenz, or iodixanol: 10%, 15%, 20%, 25%, and 30% (w/v) solutions in 10 mM Tris–HCl pH 7.4 are required
- Ultracentrifuge with swinging-bucket rotor, Beckman SW41Ti or equivalent
- Tight-fitting Dounce homogenizer

Method

1. Prepare a Golgi fraction by one of the protocols above (*Protocols 6, 7, 8,* or *10*).
2. Dilute the fraction with 2 vol. of 10 mM Tris–HCl pH 7.4 and centrifuge it at 100 000 g_{av} for 30 min.
3. Resuspend the pellet in 3–4 ml of 30% gradient medium using about 15 strokes of tight-fitting Dounce homogenizer and transfer it to an appropriate centrifuge tube for a Beckman SW41Ti swinging-bucket rotor.
4. Overlayer the sample with 2.5 ml each of the other gradient solutions and centrifuge at 250 000 g for 4 h.

The low osmolarity of iodinated density gradient media seem to offer improved opportunities for resolving membrane vesicles which in sucrose are difficult to separate. In order of increasing density, the gradient resolves a putative *trans*-Golgi to surface transport vesicle, and galactosyl transferase and sulfotransferase-rich fractions. Evans and Enrich (47) used Nycodenz gradients very successfully for resolving endosomal vesicles containing different receptors and ligands.

3.2 Identification of vesicles involved in transport to the Golgi

An important part of the secretory pathway is the transfer of proteins from the rough endoplasmic reticulum to the *cis*-Golgi. Using sucrose–D_2O

gradients (and relatively low centrifugal forces) Lodish *et al.* (48) showed that this involves a low density vesicle quite distinct from its source and destination membranes. D_2O is a means of increasing the density of sucrose gradients without increasing their viscosity and so decreasing centrifugation times and centrifugal forces (*Protocol 17*). Resolution of such gradients also tends to be higher. An alternative procedure developed by Schweizer *et al.* (49) using the p53 protein as a marker used a self-generated Percoll gradient to distinguish an ER–Golgi intermediate from *cis*-Golgi and ER elements (*Protocol 18*).

Protocol 17. Use of sucrose–D_2O gradients to analyse vesicles involved in secretion (48)

The cells should be homogenized in a suitable buffer; Lodish *et al.* (48) allowed the cells to swell in 0.25 M sucrose, 0.01 M Hepes–NaOH pH 7.4 for 10 min and then disrupted them using 15 strokes of the pestle of a tight-fitting Dounce homogenizer.

Equipment and reagents

- Sucrose solutions in D_2O: 10%, 12.5%, 15%, 17.5%, 20%, 22.5%, 25%, 27.5%, 30%, and 50% (w/v), containing 10 mM Hepes–NaOH pH 7.4
- Ultracentrifuge with swinging-bucket rotor (Beckman SW41Ti or equivalent)

Method

1. Pellet the nuclei by centrifugation at 1000 g_{av} for 10 min.
2. Layer the post-nuclear supernatant over a discontinuous gradient of sucrose in D_2O, and centrifuge at 160 000 g_{av} for 3 h. In a Beckman SW41 swinging-bucket rotor use 2 ml of the sample and 1 ml each of the sucrose solutions. For other rotors scale up or down as appropriate.

By taking small fraction volumes (12 ml in 32 fractions) the high resolution of the gradient can be maintained. The low density RER–Golgi intermediate was found in the region 12.5–17.5% sucrose, while the ER itself banded around 22% sucrose. The gradient is also capable of a high degree of resolution of ER–Golgi–*trans*-Golgi network (TGN) events (48,50).

Sucrose–D_2O gradients have also been used in the analysis of *N*-linked oligosaccharide processing. The Man_8 GlcNAc–glycoprotein intermediate was detected in two peaks of Golgi (around 11% and 15% sucrose), along with the precursor, and not at all in the ER which banded around 40%; suggesting that the endomannosidase deglycosylation is a Golgi event, not an ER event (51).

Protocol 18. Percoll gradient to resolve ER–Golgi intermediates (adapted from ref. 51)

Equipment and reagents

- High speed centrifuge with 8 × 50 ml fixed-angle rotor (Sorvall SS34 or equivalent)
- Percoll stock: 9 vol. Percoll plus 1 vol. 2.5 M sucrose
- Cell homogenate in buffer A: 0.25 M sucrose, 1 mM EDTA, 10 mM tri-ethanolamine–acetic acid pH 7.4

Method

1. Pellet the nuclei at 500 g_{av} for 10 min.
2. Adjust the 500 *g* supernatant to 30 ml with buffer A and add 3.96 ml Percoll stock (ρ = 1.046 g/ml).
3. Centrifuge at 37 000 g_{av} for 41 min in the SS34 rotor.
4. Allow the rotor to decelerate from 4000 r.p.m. to rest without the brake.
5. Collect the gradient fractions by tube puncture in approx. 1 ml fractions.

In the Percoll gradient described in *Protocol 18* the major p53-containing compartment banded broadly in the middle, between the Golgi (lower density) and the endoplasmic reticulum (higher density). There was some evidence of fine heterogeneity within this p53 band but Schweizer *et al.* (49) did not investigate this aspect of the fractionation. Instead, to purify this compartment further, they harvested the median region of the Percoll gradient, adjusted it to 30% (w/w) metrizamide, and formed a discontinuous gradient with layers of 27% and 18.5% (w/w) metrizamide on top. After 18–20 h at 85 000 *g*, the p53 compartment banded at the interface between the two lighter layers. On a continuous 5–30% metrizamide gradient the p53 compartment appeared to be resolved further into two subfractions (49).

3.3 Identification of vesicles involved in transport from the Golgi

Tooze and Huttner (52) investigated late Golgi and post-Golgi events by studying the formation of secretory granules in the neuroendocrine PC12 cell line, using a combination of sedimentation velocity and equilibrium density gradients to resolve *trans*-Golgi network (TGN) and post-TGN secretory vesicles. The cells were variously treated for studying the incorporation of [^{35}S]sulfate and the protocol given below only details the fractionation procedure. To homogenize the cells, the authors used a 'cell cracker' (22) but a tight-fitting Dounce will suffice (see *Protocol 8*).

Protocol 19. Fractionation of secretory vesicles from cultured cells (adapted from ref. 52)

Equipment and reagents

- Low speed centrifuge
- Ultracentrifuge with swinging-bucket rotor (Beckman SW41Ti or equivalent)
- Tight-fitting Dounce homogenizer
- Gradient maker
- Medium A: 0.25 M sucrose, 1 mM EDTA, 1 mM MgOAc, 10 mM Hepes–KOH pH 7.2
- Sucrose solutions: 0.3 M, 0.5 M, 1.2 M, 2.0 M in 10 mM Hepes–KOH pH 7.2

Method

1. After the cells have been suitably washed (the authors used Tris-buffered saline containing protease inhibitors), pellet them at 700 *g* for 5 min and wash once in medium A.
2. Centrifuge the cells at 1700 g_{av} for 5 min, then resuspend the pellet in five to ten times its volume of medium A. It is very important to keep the volume of this to a minimum since the first stage of the purification (steps 4 and 5) is a sedimentation velocity one.
3. Homogenize the cells using 20 strokes of a tight-fitting Dounce homogenizer and pellet the nuclei and unbroken cells by centrifugation at 1000 g_{av} for 10 min.
4. Load approx. 1.3 ml of supernatant on to a 12.0 ml linear sucrose gradient, 0.3–1.2 M.
5. Centrifuge in a swinging-bucket rotor at 78 000 g_{av} for 15 min (at speed); then collect 1 ml fractions by upward displacement. As this is a sedimentation velocity separation, if a rotor other than the SW41Ti is used, it is important that the rotor dimensions should be close to the following: r_{min} = 67 mm; r_{av} = 110 mm; r_{max} = 153 mm. The post-TGN vesicles, which comprise immature granules and vesicles, band in the top 4 ml and the TGN in the 8–10 ml region.
6. Further fractionation of these two compartments can be attempted in equilibrium density gradients. Layer the pooled fractions on to separate 10 ml linear sucrose gradients (0.5–2.0 M) and centrifuge at 100 000 g_{av} for at least 4 h.
7. Collect the fractions, again by upward displacement.

Tooze and Huttner (52) studied the secretory pathways of two sulfated macromolecules, secretogranin II and heparan sulfate: equilibrium density gradients were unable to resolve more than one population of TGN vesicles, which contained both molecules, on the other hand two quite distinct populations of post-TGN vesicles were resolved, one containing the heparan sulfate, one containing the secretogranin, suggesting that post-TGN events effect the segregation of the two macromolecules.

For the resolution of TGN vesicles it may be beneficial to use a more shallow sucrose gradient for the equilibrium density step, e.g. 0.8–1.6 M sucrose, or to try one of the iodinated density gradient media such as Nycodenz or iodixanol to provide an iso-osmotic gradient in which the vesicles will be fully hydrated and not collapsed by the osmolarity of the sucrose. With iodixanol advantage can be taken of the ability of this medium to form self-generated gradients (4,5). The vesicle suspension might be adjusted to 15% or 20% iodixanol and centrifuged in a vertical rotor (e.g. Beckman VTi65.1 or equivalent) for 1–2 h. Useful gradients can be obtained under these conditions that the author has used successfully for endosome fractionation.

Sometimes both sucrose and Nycodenz gradients have been used in the analysis of secretory events. Linear sucrose gradients (10–45%) were found to produce a considerable overlap of the lysosomal β-*N*-acetylglucosaminidase and the *trans*-Golgi UDP-galactosyl-glycoprotein-galactosyl transferase which were resolved on a subsequent discontinuous 10–40% (w/v) Nycodenz gradient, centrifuged at 100 000 *g* for 10 h (53).

The IgA receptor can be used for the immunoaffinity isolation of specific cytoplasmic membranes if the antibody used is raised against the cytoplasmic domain of the receptor molecule. However, the immunoaffinity selection will only be effective if the contaminating membranes do not have the marker. During synthesis many of the synthetic compartments will display the cytoplasmic tail of the marker protein on their external surface and they will all be selected by the antibody. An antibody raised to the cytoplasmic tail of VSV G_2 protein has discriminated TGN vesicles carrying this protein from other TGN vesicles (54) but because the same tail is present in the *trans*- and medial-Golgi, the antibody reacted with vesicles from these compartments as well.

3.4 *In vitro* systems to study transfer of molecules between membranes

A common approach to investigating membrane transfer between compartments is to radiolabel a 'donor' membrane and to incubate it with an 'acceptor' membrane. This system has been used to study the transfer from endoplasmic reticulum transitional elements to the Golgi, which was immobilized on nitrocellulose strips (12,55,56).

Protocol 20. Immobilization of acceptor membranes (adapted from refs 12, 55, and 56)

Equipment and reagents

- Buffer: 2.5 mM MgOAc, 33 mM KCl, 33 mM Hepes–NaOH pH 7.0
- 5% (w/v) bovine serum albumin
- Nitrocellulose strips

Protocol 20. *Continued*

Method

1. Suspend the membranes in buffer (approx. 1–2 mg protein/ml) and incubate with 1 cm^2 nitrocellulose strips (10 strips/ml) for 1 h at 4°C with continuous shaking.
2. Transfer the strips to 5% (w/v) bovine serum albumin in the above buffer and continue the incubation for a further 1 h.
3. Rinse the strips with four changes of buffer.
4. Use the strips in a suitable incubation medium with the radiolabelled donor membranes under the chosen conditions and at the end, rinse again in the buffer four times. Air dry on filter paper and count as required.

In the experiments described by Nowack *et al.* (56), the donor membranes were labelled with [^{3}H]leucine and the transfer buffer contained, in addition to donor membranes, an ATP-regenerating system, a cytosol fraction, and any supplements as required, in a total volume of 1 ml; for more information the reader is referred to ref. 56.

3.5 Sorting of proteins destined for specific plasma membrane domains

Protein sorting to the different domains of the plasma membrane can occur either within the *trans*-Golgi network prior to directed transfer to the domains or if there is no sorting at this intracellular level, there could be a re-allocation of specific proteins from one domain of the plasma membrane to another by transcytosis (57). The predominant mechanism seems to depend on the cell type; in MDCK cells the former seems to operate, in hepatocytes the latter, and in the Caco-2 human adenocarcinoma cell line, both seem to be operable. The use of the latter cells has become increasingly popular as they can be grown as monolayers in special chambers which allow defined access to both the basolateral and apical surfaces. Ellis *et al.* (57) have devised a method for isolating the two plasma membrane domains from Caco-2 cells grown on permeable filter supports.

Protocol 21. Isolation of plasma membrane domains from filter grown Caco-2 cells (57)

Equipment and reagents

- Nitrogen pressure vessel for cell disruption (Artisan Industries, Waltham, Mass.)
- Low speed centrifuge
- Ultracentrifuge with swinging-bucket rotor (Beckman SW50.1 or equivalent)
- 1 M $MgCl_2$
- Buffer A: 0.25 M sucrose, 10 mM Tris–HCl pH 7.4
- Buffer B: as buffer A plus 5 mM EDTA
- Sucrose solutions: 34%, 40%, 54% (w/v) in 10 mM Tris–HCl pH 7.4

Method

1. Rinse the filters once in isotonic saline and once in buffer A.
2. Scrape the cells off the filter with a rubber policeman into 5 ml of buffer A.
3. Homogenize the cells by nitrogen cavitation using nitrogen at 3795 kPa (550 psi) for 10 min.
4. After allowing the foam to subside by gentle stirring, centrifuge the homogenate at 270 *g* for 10 min, and then recentrifuge the supernatant at 920 *g* for 10 min.
5. Add 1 M $MgCl_2$ to the supernatant (final concentration 10 mM) and stir for 15 min before centrifuging at 2300 g_{av} for 15 min.
6. Centrifuge this supernatant at 170 000 g_{av} for 45 min and resuspend the pellet in 0.5 ml of buffer B.
7. Layer the resuspended pellet over the following discontinuous gradient: 54%, 40%, and 34% (w/v) sucrose, 10 mM Tris–HCl pH 7.4 (1.4, 1.7, and 1.4 ml respectively), and centrifuge in a swinging-bucket rotor at 68 000 g_{av} for 4.5 h.

Bands are obtained at the three interfaces, the top is enriched in Golgi membranes, the middle in basolateral membranes, and the bottom in apical membranes. After dilution to approx. 0.25 M sucrose, the banded material can be harvested by centrifugation at about 40 000 g_{av} for 1 h. Ellis *et al.* (57) recommended making the middle band suspension 10 mM with respect to $MgCl_2$ and centrifuging at 2300 *g* for 15 min prior to harvesting the basolateral membranes.

Studies on the sorting of proteins to different plasma membrane domains in MDCK cells is facilitated by virus infection, since the influenza haemagglutinin protein is delivered to the apical domain (58), while the vesicular stomatitis virus (VSV) G protein is delivered to the basolateral domain. Wandinger-Ness *et al.* (58) took advantage of the ability of the virus-infected cells to release transport vesicles into the medium upon perforation of the cells, in the presence of an ATP-regenerating system which is required for vesicle formation. It is outside the scope of this chapter to describe in detail the process of growing cells on filter supports or the permeabilization process. Briefly, the cells are seeded on to 0.4 μm pore size filters which are held within special holders in a Petri dish. After virus infection and any required radiolabelling of macromolecules, the cells are permeabilized mechanically by contact with a nitrocellulose acetate filter (0.45 μm pore size) in the presence of Hepes-buffered potassium acetate. Permeabilization allows the recovery of the secretory vesicles from the incubation medium.

Protocol 22. Analysis of secretory vesicles from permeabilized MDCK cells (adapted from ref. 58)

Equipment and reagents

- Low speed centrifuge
- Ultracentrifuge with swinging-bucket rotor (Beckman SW40 or equivalent)
- Gradient maker
- Sucrose solutions: 0.25 M, 0.3 M, and 1.5 M sucrose in 1 mM EGTA, 1 mM DTT, 10 mM Hepes–NaOH pH 7.4
- ATP-regenerating solution: 38 mM potassium gluconate, 38 mM potassium aspartate, 1 mM DTT, 2 mM EDTA, 2.5 mM $MgCl_2$, 25 mM Hepes–NaOH pH 7.4, 1 mM ATP, 8 mM creatine phosphate, and 50 μg/ml creatine kinase

Method

1. Maintain vesicle formation by the permeabilized cells by incubation of the filter culture in the ATP-regenerating solution (10 ml in a 10 cm culture dish).
2. Incubate for 1 h at 37 °C, then remove the incubation medium and clarify it at 1000 g_{av} for 10 min.
3. In a suitable tube for the Beckman SW40 rotor overlay 2 ml of 0.25 M sucrose, with the 1000 *g* supernatant and centrifuge the membrane vesicles at 200 000 g_{av} for 3 h.
4. After aspiration of the supernatant, dry the walls of the tube and resuspend the vesicle pellet in 0.3 ml 1.5 M sucrose, by drawing up into and expelling the liquid from an automatic pipette.
5. Overlay this suspension with a 0.3–1.5 M linear sucrose gradient and centrifuge at 125 000 g_{av} for 12 h in the same rotor.

The linear sucrose gradient achieves a partial, but quite clear separation of the two types of vesicle (the peak density of the HA2 vesicles was 1.099 g/ml, while that of the VSV G vesicles was 1.113 g/ml). A better way of producing a preparative isolation is to purify a total vesicle fraction by flotation to the interface of 1.2–0.8 M discontinuous sucrose gradient (instead of the linear sucrose gradient) and then to use immunoaffinity isolation with antibodies to the cytoplasmic tails of the two proteins. Sztul *et al.* (30) have isolated transcytotic vesicles by immunoadsorption to protein A–Sepharose beads using an antibody to the cytoplasmic tail of polymeric IgA receptors. Both the binding of the beads to the antibody and the immunoisolation step (incubation with a crude vesicle fraction) were carried out at 4 °C overnight (30).

3.6 Secretory pathways in yeast

A number of proteins have been identified which appear to control specific events in the translocation of vesicles in the secretory pathway. The facility of modulation of the secretory pathway by genetic manipulation in yeast has made this organism very popular in these studies.

Bowser and Novick (59) have devised fractionation schemes for the investigation of vesicular traffic from the Golgi to the plasma membrane in *Saccharomyces cerevisiae*. Spheroplast formation was carried out by incubation in 1.4 M sorbitol, 10 mM NaN_3, 40 mM β-mercaptoethanol, 50 mM Tris–HCl pH 7.5 containing 0.125 mg/ml Zymolyase-100T at 37 °C for 45 min. Spheroplasts were harvested and homogenized in 0.8 M sorbitol, 1 mM EDTA, 20 mM triethanolamine–HCl pH 7.2 containing a cocktail of proteases.

Protocol 23. Analysis of secretory pathway in yeast (adapted from ref. 59)

Equipment and reagents

- Low speed and high speed centrifuges
- Ultracentrifuge with fixed-angle (8 × 40 ml) and swinging-bucket rotor (Beckman SW41Ti or equivalent)
- Wheaton type A Dounce homogenizer
- 1 M Mes–NaOH pH 6.5
- Lysis buffer: 0.8 M sorbitol, 1 mM EDTA, 20 mM triethanolamine–HCl pH 7.2
- Sucrose solutions: 34%, 43.5%, 50%, 54%, 57.5%, 61.5%, 69% (w/v) in 10 mM Mes–NaOH pH 6.5

Method

1. Homogenize the spheroplasts in the lysis buffer in a 40 ml Dounce homogenizer (Wheaton type A pestle) using 20 strokes.
2. Centrifuge at 450 *g* for 3 min, then rehomogenize the pellet in the same volume of lysis buffer, recentrifuge the suspension, and pool the two supernatants.
3. Add 1 M Mes–NaOH pH 6.5 so that the final concentration is 50 mM and centrifuge at 10 000 g_{av} for 10 min, and carefully decant the supernatant, keeping the pellet.
4. Centrifuge the supernatant from step 3 at 100 000 g_{av} for 1 h.
3. Resuspend the two pellets (steps 3 and 4) in 3 ml 69% (w/v) sucrose, by a few strokes of the pestle of a small Dounce homogenizer. In a suitable tube overlay the sample with 1.5 ml 61.5%, 1.5 ml 57.5%, 2.0 ml 54%, 2.0 ml 50%, 1.5 ml 43.5%, and 1.5 ml 34% sucrose solution.
6. Centrifuge at 170 000 g_{av} for 16 h and then collect fractions by tube puncture or upward displacement.

The gradient was used to localize Sec15p (in the 10 000 *g* pellet) to the plasma membrane (in wild-type cells) which banded broadly in the bottom third of the gradient: vanadate-sensitive Mg^{2+}-ATPase was used as a plasma membrane marker (59). Mitochondria banded towards the middle of the gradient, while the endoplasmic reticulum was distributed almost throughout the entire gradient, again, like the mitochondria, peaking towards the middle. The Golgi banded quite sharply near the top of the gradient and the Kex2 protease (a late Golgi marker) was very clearly less dense than the GDPase. This separation was even more marked with the 100 000 *g* pellet, although interestingly, the GDPase was associated with a lighter fraction in this case.

In some yeast membrane fractionations, a low speed pellet has been used as the source of endoplasmic reticulum (60). The homogenate is first centrifuged at 3000 g_{av} for 6 min and it is critical that this centrifugation be carried out as soon as possible after homogenization. The low speed pellet is then suspended in 77% (w/v) sucrose and overlaid with a discontinuous sucrose gradient of 69%, 62%, 54%, 47%, 40%, and 34%. The gradient is centrifuged at 170 000 g_{av} for 12 h. Over 70% of the endoplasmic reticulum sediments in the low speed pellet, along with about 40% of the plasma membrane and most of the mitochondria. The gradient is quite efficient at resolving the endoplasmic reticulum from the other membranes which partially overlap in the lighter regions of the gradient. The Sec53p protein was clearly located within the ER region, although a slight shift in this protein to a higher density than the NADPH–cytochrome *c* reductase was not commented upon by the authors and may have been better resolved in a more shallow gradient or within a gradient of lower osmolarity.

ER–Golgi transport in yeast has also been studied in discontinuous 34–61% (w/v) sucrose gradients, with the sample underlaid in 69% (61). While the mitochondria and plasma membrane band at a slightly lower density than the ER, the vacuole and Golgi markers are well separated, although both show a strongly biphasic distribution (peaks at low and high densities) not unlike that observed by Svoboda *et al.* (17). By using a very subtle adjustment to the gradient, i.e. the sample in 77% beneath a 34–69% (w/v) gradient, the ER subfractionated (on the basis of the NADPH–cytochrome *c* reductase) into two major components. This type of sucrose gradient has been shown to be of considerable use in the dissection of the secretory pathways of yeast: subfractionation of the late Golgi compartment, secretory vesicles, invertase-containing vesicles, and vacuolar membrane precursors (62–64) are possible.

3.7 Gel filtration

Although not a widely-used technique for membrane fractionation, Sephacryl S-1000 can be used in certain circumstances to fractionate membrane vesicles; it has been used to discriminate specific secretory vesicles in yeast (59,65).

Protocol 24. Fractionation of vesicles from yeast by gel filtration (adapted from refs 59 and 65)

Equipment and reagents

- Wheaton type A Dounce homogenizer
- Sephacryl S-1000 column (1.5 × 90 cm)
- Ultracentrifuge with fixed-angle rotor (approx. 12 × 14 ml)
- 1 M Mes pH 6.5
- Homogenization buffer: 0.8 M sorbitol, 1 mM EDTA, 20 mM triethanolamine pH 7.2
- Elution buffer: 0.8 M sorbitol, 1 mM EDTA, 10 mM triethanolamine, 10 mM Mes pH 6.5

Method

1. Suspend the cells in homogenization buffer containing any cocktail of proteases. Bowser and Novick (54) used an amount of yeast equivalent to an A_{599} of 500–800 units in 20 ml of buffer.
2. Homogenize the cells using 20 strokes of the pestle, then centrifuge at 750 *g* for 3 min. Resuspend the pellet in 4 ml buffer and repeat the centrifugation.
3. Combine the supernatants and add 1 M Mes pH 6.5 to a concentration of 10 mM, then from a post-mitochondrial supernatant (10 000 g_{av} for 10 min) sediment the microsomes at 100 000 g_{av} for 1 h. Resuspend the pellet in 1 ml buffer.
4. Layer this suspension on to a 1.5 × 90 cm S-1000 Sephacryl gel filtration column, elute with elution buffer at a flow rate of about 9 ml/h, and collect 4 ml fractions.

Bowser and Novick (59) were able to resolve quite clearly three major fractions containing:

- plasma membrane and Golgi vesicles
- vesicles bearing invertase and Kex2 protease
- Sec15p

which were eluted in that order from the column (total eluate volume was about 130 ml). The column also showed some resolution of the invertase and Kex2 protease-containing vesicles. The Sec15p eluted close to the major protein peak; the Sec4p protein on the other hand (63) elutes entirely with the invertase-containing secretory vesicles from the sec6-4 mutant.

References

1. Dobrota, M. and Hinton, R. (1992). In *Preparative centrifugation: a practical approach* (ed. D. Rickwood), p. 77. Oxford University Press, Oxford, UK.
2. Rickwood, D. (1983). In *Iodinated density gradient media: a practical approach* (ed. D. Rickwood), p. 1. Oxford University Press, Oxford, UK.

3. Graham, J., Bailey, D., Wall, J., Patel, K., and Wagner, S. (1983). In *Iodinated density gradient media: a practical approach* (ed. D. Rickwood), p. 91. Oxford University Press, Oxford, UK.
4. Ford, T. C., Graham, J. M., and Rickwood, D. (1994). *Anal. Biochem.*, **220**, 360.
5. Graham, J. M., Ford, T. C., and Rickwood, D. (1994). *Anal. Biochem.*, **220**, 367.
6. Neville, D. M. (1968). *Biochim. Biophys. Acta*, **154**, 540.
7. Hubbard, A. L., Wall, D. A., and Ma, A. (1983). *J. Cell Biol.*, **96**, 217.
8. Scott, L., Schell, M. J., and Hubbard, A. L. (1993). In *Methods in molecular biology* (ed. J. M. Graham and J. A. Higgins), Vol. 19, p. 59. Humana Press, Totowa, NJ, USA.
9. Meier, P. J., Sztul, E. S., Reuben, A., and Boyer, J. L. (1984). *J. Cell Biol.*, **98**, 991.
10. Wisher, M. H. and Evans, W. H. (1975). *Biochem. J.*, **146**, 375.
11. Graham, J. M. and Northfield, T. C. (1987). *Biochem. J.*, **242**, 825.
12. Paulik, M., Nowack, D. D., and Morré, D. J. (1988). *J. Biol. Chem.*, **263**, 17738.
13. Amar-Costesec, A., Godelaine, D., and Hortsch, M. (1988). In *Cell free analysis of membrane traffic* (ed. D. J. Morré, K. E. Howell, G. M. C. Cook, and W. H. Evans), p. 211. Alan R Liss Inc, NY.
14. Bergstrand, A. and Dallner, G. (1969). *Anal. Biochem.*, **29**, 351.
15. Fleischer, B. and Fleischer, S. (1970). *Biochim. Biophys. Acta*, **219**, 301.
16. Graham, J. M. and Winterbourne, D. J. (1988). *Biochem. J.*, **252**, 437.
17. Svoboda, P., Kvapil, P., Insel, P. A., and Ransnas, L. A. (1992). *Eur. J. Biochem.*, **208**, 693.
18. Schurmann, A., Monden, I., Joost, H. G., and Keller, K. (1992). *Biochim. Biophys. Acta*, **1131**, 245.
19. Nakagawa, Y., Purushotham, K. R., Wang, P. L., Fisher, J. E., Dunn, W. A., Schneyer, C. A. *et al.* (1992). *Biochem. Biophys. Res. Commun.*, **187**, 1172.
20. Morré, D. J., Cheetham, R. D., and Nyquist, S. E. (1972). *Prep. Biochem.*, **2**, 61.
21. Hartel-Schenk, S., Minnifield, N., Reutter, W., Hanski, C., Bauer, C., and Morré, D. J. (1991). *Biochim. Biophys. Acta*, **1151**, 108.
22. Balch, W. E., Dunphy, W. G., Braell, W. A., and Rothman, J. E. (1984). *Cell*, **39**, 405.
23. Macintyre, S. S. (1992). *J. Cell Biol.*, **118**, 253.
24. Morand, J. C. and Kent, C. (1986). *Anal. Biochem.*, **159**, 157.
25. Graham, J. M., Ford, T. C., and Rickwood, D. (1990). *Anal. Biochem.*, **187**, 318.
26. Peter, F., Nguyen, Van P., and Soling, H. D. (1992). *J. Biol. Chem.*, **267**, 10631.
27. Luzio, J. P. and Richardson, P. J. (1993). In *Methods in molecular biology* (ed. J. M. Graham and J. A. Higgins), Vol. 19, p. 141. Humana Press, Totowa, NJ, USA.
28. Howell, K. E., Gruenberg, J., Ito, K., and Palade, G. E. (1988). In *Cell free analysis of membrane traffic* (ed. D. J. Morré, K. E. Howell, G. M. C. Cook, and W. H. Evans), p. 77. Alan R Liss Inc, NY.
29. Luzio, J. P., Mullock, B. M., Branch, W. J., and Richardson, P. J. (1988). In *Cell free analysis of membrane traffic* (ed. D. J. Morré, K. E. Howell, G. M. C. Cook, and W. H. Evans), p. 91. Alan R Liss Inc, NY.
30. Sztul, E. S., Howell, K. E., and Palade, G. E. (1985). *J. Cell Biol.*, **100**, 1255.
31. Hubbard, A. L., Dunn W. A., Mueller, S. C., and Bartles, J. R. (1988). In *Cell free analysis of membrane traffic* (ed. D. J. Morré, K. E. Howell, G. M. C. Cook, and W. H. Evans), p. 115. Alan R Liss Inc, NY.

32. Howell, K. E., Schmid, R., Ugelstad, J., and Gruenberg, J. (1989). *Methods Cell Biol.*, **31**, 265.
33. Naim, H. Y. (1992). *Biochem. J.*, **286**, 451.
34. Gupta, D. and Tartakoff, A. M. (1989). *Methods Cell Biol.*, **31**, 247.
35. Courtoy, P. J., Quintart, J., and Baudhuin, P. (1984). *J. Cell Biol.*, **98**, 870.
36. Courtoy, P. J., Quintart, J., Draye, J.-P., and Baudhuin, P. (1988). In *Cell free analysis of membrane traffic* (ed. D. J. Morré, K. E. Howell, G. M. C. Cook, and W. H. Evans), p. 169. Alan R Liss Inc, NY.
37. Amar-Costesec, A., Wibo, M., Thinès-Sempoux, D., Beaufay, H., and Berthet, J. (1974). *J. Cell Biol.*, **62**, 717.
38. Morré, D. J., Morré, D. M., and Heidrich, H.-G. (1983). *Eur. J. Cell Biol.*, **31**, 263.
39. Dunphy, W. G. and Rothman, J. E. (1983). *J. Cell Biol.*, **97**, 270.
40. Goldberg, D. E. and Kornfeld, S. (1983). *J. Biol. Chem.*, **258**, 3159.
41. Fries, E. and Rothman, J. E. (1981). *J. Cell Biol.*, **90**, 697.
42. Dunphy, W. G., Fries, E., Urbani, L. J., and Rothman, J. E. (1981). *Proc. Natl. Acad. Sci. USA*, **78**, 7453.
43. Munday, D. I. and Warren, G. (1992). *J. Cell Biol.*, **116**, 135.
44. Sarasate, J., Bronson, M., Palade, G. E., and Farquhar, M. G. (1988). In *Cell free analysis of membrane traffic* (ed. D. J. Morré, K. E. Howell, G. M. C. Cook, and W. H. Evans), p. 129. Alan R Liss Inc, NY.
45. Sugumaran, G., Katsman, M., and Silbert, J. E. (1992). *J. Biol. Chem.*, **267**, 8802.
46. Trinchera, M., Pirovani, B., and Ghidoni, R. (1990). *J. Biol. Chem.*, **265**, 18242.
47. Evans, W. H. and Enrich, C. (1988). In *Cell free analysis of membrane traffic* (ed. D. J. Morré, K. E. Howell, G. M. C. Cook, and W. H. Evans), p. 155. Alan R Liss Inc, NY.
48. Lodish, H. F., Kong, N., Hirani, S., and Rasmussen, J. (1987). *J. Cell Biol.*, **104**, 221.
49. Schweizer, A., Matter, K., Ketcham, C. M., and Hauri, H.-P. (1991). *J. Cell Biol.*, **113**, 45.
50. Miller, S. G., Carnell, L., and Moore, H. H. (1992). *J. Cell Biol.*, **118**, 267.
51. Moore, S. E. and Spiro, R. G. (1992). *J. Biol. Chem.*, **267**, 8443.
52. Tooze, S. A. and Huttner, W. B. (1990). *Cell*, **60**, 837.
53. Maheshwari, R. K., Sidhu, G. S., Bhartiya, D., and Friedman, R. M. (1991). *J. Gen. Virol.*, **72**, 2143.
54. de Curtis, I. and Simons, K. (1988). In *Cell free analysis of membrane traffic* (ed. D. J. Morré, K. E. Howell, G. M. C. Cook, and W. H. Evans), p. 101. Alan R Liss Inc, NY.
55. Moreau, P., Rodriguez, M., Cassagne, C., Morré, D. M., and Morré, D. J. (1991). *J. Biol. Chem.*, **266**, 4322.
56. Nowack, D. D., Paulik, M., Morré, D. J., and Morré, D. M. (1990). *Biochim. Biophys. Acta*, **1051**, 250.
57. Ellis, J. A., Jackman, M. R., and Luzio, J. P. (1992). *Biochem. J.*, **283**, 553.
58. Wandinger-Ness, A., Bennett, M. K., Antony, C., and Simon, K. (1990). *J. Cell Biol.*, **111**, 987.
59. Bowser, R. and Novick, P. (1991). *J. Cell Biol.*, **112**, 1117.
60. Ruohola, H. and Ferro-Novick, S. (1987). *Proc. Natl. Acad. Sci. USA*, **84**, 8468.
61. Newman, A. P., Groesch, M. E., and Ferro-Novick, S. (1992). *EMBO J.*, **11**, 3609.

62. McCaffrey, M., Johnson, J. S., Goud, B., Myers, A. M., Rossier, J., Popoff, M. R., *et al.* (1991). *J. Cell Biol.*, **115**, 309.
63. Gould, B., Salminen, A., Walworth, N. C., and Novick, P. J. (1988). *Cell*, **53**, 753.
64. Yaver, D. S., Nelson, H., Nelson, N., and Klionsky, D. J. (1993). *J. Biol. Chem.*, **268**, 10564.
65. Walworth, N. C. and Novick, P. J. (1987). *J. Cell Biol.*, **105**, 163.

8

Isolation and purification of functionally intact mitochondria and chloroplasts from plant cells

ANTHONY L. MOORE and DAVID G. WHITEHOUSE

1. Introduction

The photosynthetic higher plant cell is compartmentalized into organelles such as chloroplasts, mitochondria, peroxisomes, and vacuoles and is unique amongst higher plant cells in containing two bioenergetic organelles. The success and the survival of the plant cell depends upon the co-ordination and integration of the differing metabolic activities of the different organelles. Clearly the metabolic activities and roles of the organelles will vary not only with the developmental state of that cell but also with the type of cell, e.g. mesophyll, epidermal, or bundle sheath. Photosynthetic higher plant cell systems offer the opportunity to isolate both mitochondria and chloroplasts from the same cell at a similar stage of development, a factor of some importance when attempting to present a complete picture of the bioenergetics of that cell (1, 2). There is a large diversity of plant cells each with differing characteristics and developmental patterns and a complete description of the life cycle of a plant cell will only emerge when the integrated pattern of both metabolism and development is known. An answer to this question necessitates the isolation and purification of functionally intact organelles.

2. Mitochondria

Plant mitochondria, like their mammalian counterparts, play central roles in the supply of ATP and carbon skeletons for biosynthetic purposes (1). The degree to which plant mitochondria are involved in these processes does appear to be developmentally regulated and depend not only upon the age of the plant cell but also upon type of tissue (i.e. floral, meristematic, etc.) and location (i.e. stem, leaf root, etc.) (3). Thus mitochondria isolated from green leaf tissue display considerable glycine decarboxylase activity, due to their involvement in the photorespiratory pathway, but possess low levels of cyanide resistance (which may increase with age of the tissue, however) (4). Alternatively, mitochondria isolated from the spadices of thermogenic tissues, such as *Arum maculatum*, are highly cyanide-resistant and often display rates of oxygen consumption in excess of 1–2 μmol/min/mg protein whilst oxidizing exogenous NADH as substrate (5). Thus it is of importance to have a clear idea of the particular mitochondrial function that is to be studied in order that the correct tissue be selected.

Plant mitochondria are admirably suited to be both a supplier of ATP and carbon skeletons since they possess a branched respiratory chain which allows them to function when cytosolic demands for ATP may be met by alternative sources such as chloroplasts (1,4,5). In addition to the four major respiratory chain complexes, namely the NADH: ubiquinone oxidoreductase (complex I), succinate dehydrogenase (complex II), ubiquinol: cytochrome *c* reductase (complex III or the bc_1 complex), and cytochrome *c* oxidase (complex IV), the majority of plant mitochondria possess an external NAD(P)H dehydrogenase located on the outer surface of the inner membrane, an internal rotenone-insensitive NADH dehydrogenase (that may be part of complex I) and a cyanide- and antimycin-insensitive alternative oxidase, an integral membrane protein that has its active site on the inner surface of the inner membrane (*Figure 1*). The internal rotenone-insensitive NADH dehydrogenase, the external NADH dehydrogenase, and the alternative oxidase are non-phosphorylating (i.e. non-protonmotive) routes of electron transport. Thus

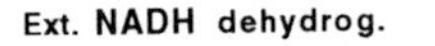

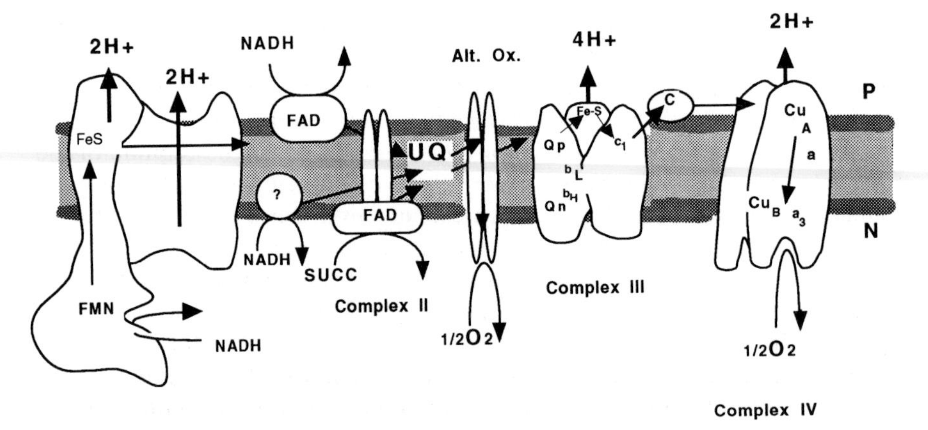

Figure 1. Diagrammatic representation of the plant mitochondrial respiratory chain. Abbreviations used: NADH, nicotinamide adenine dinucleotide (reduced); FMN, flavin mononucleotide; b_L and b_H, cytochromes b_{566} and b_{560}; c_1 and c, cytochromes c_1 and c; a and a_3, cytochromes a and a_3; UQ, ubiquinone pool; Fe-S, iron–sulfur proteins; Qp, site of ubiquinol oxidation; Qn, site of ubiquinone reduction; P and N refer to the positive and negative faces of the inner membrane from which protons are pumped.

substrates can be oxidized protonmotively either via complexes I, III, and IV, or non-protonmotively via engagement of the non-phosphorylating routes. The degree to which of these routes are engaged depends upon a variety parameters including the mitochondrial NADH/NAD ratio, the redox poise of the ubiquinone pool, and the cytosolic calcium concentration (1,2). As indicated above green leaf mitochondria also have the ability to oxidize glycine as a respiratory substrate due to the presence within the mitochondrial matrix of glycine decarboxylase (4). Activity of this enzyme and serine hydroxymethyltransferase results in the production of NADH, NH_3, and CO_2. In isolated mitochondria, the NADH is reoxidized by complex I but the degree to which this occurs *in vivo* is uncertain. In mitochondria isolated from mature plant cells glycine decarboxylase can account for as much as 50% of total mitochondrial protein (4).

2.1 General isolation principles

The isolation of mitochondria from plant cells that display similar biochemical and morphological characteristics to those observed *in vivo* requires considerable expertise. Although it is relatively easy to prepare a crude mitochondrial fraction by differential centrifugation, marked changes in functional capabilities and morphological appearance are often observed in such fractions suggesting that some degree of structural damage has occurred during the isolation procedure. Most of the problems involved in the isolation of intact and fully functional mitochondria occur during the homogenization phase because of the high shearing forces required to rupture the plant cell wall. Such forces tend to have a deleterious effect on other subcellular organelles, such as the vacuole, resulting in the release of degradative enzymes and secondary products, such as flavonoids and phenolic compounds, which may severely impair mitochondrial integrity or result in functional inactivation. Hence it is important to ensure that isolation and wash media are alkaline and contain, in addition to an osmoticum, such as sucrose, mannitol, or sorbitol, either PVP and/or cysteine (or 2-mercaptoethanol). The effects of homogenization on mitochondrial structure and function thus range from the undesirable to the totally destructive. Obviously techniques that reduce all or any of these problems are desirable and for the isolation of functionally intact mitochondria the most crucial aspect of the whole procedure is therefore the grinding of the tissue. Consequently the grinding procedure must be kept to an absolute minimum in order to maintain structural and functional integrity. Although longer blending improves the final yield, it drastically reduces the percentage of intact mitochondria. Hence the key point is to sacrifice quantity for quality. Numerous techniques can be used to break the cell wall, but the methods outlined in the protocols described below tend to be the mildest.

In this section specific techniques for the isolation and purification of functionally active and intact mitochondria from potato tuber and green leaf tissue are described. As indicated above careful choice of the plant material is required in order to ensure that the isolated mitochondria possess the requisite functional characteristics. Since isolation techniques tend to vary with the type of tissue used as the source of mitochondria, the reader is referred to other articles (6–11) for a detailed analysis of techniques employed for the isolation and purification of mitochondria from specific tissues. All the values of g are g_{av}.

Protocol 1. Isolation of mitochondria from potato tubers

Equipment and reagents

- Plants: use fresh potatoes (*Solanum tuberosum*) that have been stored at 4 °C for at least 24 h—cold storage results in the conversion of starch to sucrose which tends to aid separation of mitochondria from other organelles
- Isolation medium (1.5 litre/1.5 kg tissue): 0.3 M mannitol, 7 mM L-cysteine, 2 mM EDTA, 0.1% (w/v) defatted bovine serum albumin (BSA), 0.6% (w/v) polyvinylpyrrolidone (PVP-40, M_r 40 000), 40 mM MOPS–KOH pH 7.6[a]
- Cheesecloth (or muslin)
- Wash medium (50 ml): 0.3 M mannitol, 2 mM EDTA, 0.1% (w/v) BSA, 40 mM MOPS–KOH pH 7.6
- Purification medium (50 ml): 0.3 M sucrose, 0.1% (w/v) BSA, 5 mM MOPS–KOH pH 7.6, and 22% (v/v) Percoll (Pharmacia, Uppsala, Sweden or Sigma, St. Louis, MO)
- A paint brush for resuspending the mitochondrial pellets
- A domestic liquidizer, e.g. Moulinex juice extractor (type 140) or hand-held Moulinex or Braun mixer

Method

1. Peel and wash fresh potatoes (1.5 kg) in cold water and homogenize using a juice extractor (Moulinex type 140) or other mild method of homogenization (hand-held Moulinex or Braun mixer). In the case of the juice extractor, collect the juice directly into 1.50 litres of isolation medium maintained at 4 °C. During this operation, maintain the pH at pH 7.6 by dropwise addition of 10 M KOH. If a hand-held homogenizer is used, dice the potatoes directly into the isolation medium and homogenize by macerating for short periods (10–15 sec) at 4 °C.[b,c]
2. Rapidly squeeze the juice/homogenate through four layers of cheesecloth and centrifuge the filtered suspension in 500 ml centrifuge bottles at 400 g (e.g. Beckman JA10 rotor) for 5 min at 4 °C. Discard the pellets.
3. Centrifuge the supernatant at 3000 g for 10 min and discard the pellet that contains nuclei, unbroken cells, and cell fragments.
4. Centrifuge the supernatant at 17 700 g for 10 min.
5. Following centrifugation, aspirate the supernatant and gently resuspend each pellet, which is highly enriched in mitochondria, in approx. 4 ml of wash medium using a paint brush. Pool the resuspended pellets, dilute to approx. 80–90 ml with wash medium, and

transfer the suspension to two 50 ml centrifuge tubes.

6. Harvest the washed mitochondria by centrifugation at 12 000 *g* for 10 min and gently resuspend the pellet in a small volume of wash medium. Store at approx. 50 mg protein/ml.
7. To purify mitochondria, layer 2–3 ml of the washed mitochondria on 36 ml of the purification medium in two 50 ml tubes and centrifuge at 22 300 *g* for 30 min.
8. Recover the purified mitochondria as a broad band in the bottom third of the gradient below a yellowish band containing numerous small vesicles and above the peroxisomal pellet.
9. Collect the mitochondrial fraction, after careful aspiration of the upper part of the gradient, using a long-tipped Pasteur pipette. Dilute 10–15 times with wash medium to dilute the Percoll concentration.
10. Harvest purified mitochondria by centrifugation at 12 000 *g* for 15 min, and resuspend the pellet in minimum volume (0.5–1.0 ml) of wash medium to give a protein concentration of approx. 20–40 mg/ml based on Lowry *et al.* (12) protein assay. It may be necessary to repeat this step if the pellet is loose, indicative that Percoll is still present.

[a] All media are made up fresh just prior to the isolation procedure. BSA may be added as the last ingredient and left to dissolve without stirring.
[b] This procedure may also be used to isolate mitochondria from other tuberous tissue such as sweet potatoes, cauliflower buds, jerusalem artichokes, beetroots, and turnips.
[c] A Waring blender can be used in place of a juice extractor or hand-held homogenizer, but disruption must be done at low speed for a maximum of 2–5 sec. The blades should be routinely sharpened after several homogenizations to ensure that cutting rather than tearing of the tissue occurs.

Protocol 2. Isolation of mitochondria from green leaves

Equipment and reagents

- Plants: peas—use Feltham First, Little Marvel, or Progress varieties (spinach can be substituted for peas)
- Isolation medium (0.6–1 litre): 0.3 M sucrose, 3 mM 2-mercaptoethanol, 2 mM EDTA, 0.4% (w/v) defatted bovine serum albumin (BSA), 1% (w/v) polyvinylpyrrolidone (PVP-40, M_r 40 000), 1 mM $MgCl_2$, 40 mM MOPS–KOH pH 7.6
- Wash medium (50 ml): 0.3 M sucrose, 2 mM EDTA, 0.1% (w/v) BSA, 40 mM MOPS–KOH pH 7.6[a]
- Cheesecloth (or muslin)
- Purification solution A (50 ml): 0.3 M sucrose, 0.1% (w/v) BSA, 10 mM phosphate buffer pH 7.4, 10% (w/v) PVP-40, 28% (v/v) Percoll
- Purification solution B (50 ml): 0.3 M sucrose, 0.1% BSA, 10 mM phosphate buffer pH 7.4, 28% (v/v) Percoll
- A paint brush for resuspending the mitochondrial pellets
- A hand-held Moulinex or Braun mixer, a Polytron, Ultra-Turrax, or Waring blender
- A gradient maker

Protocol 2. *Continued*

Method

1. Grow pea plants from seed in vermiculite or Palmers pot bedding compost for 10–14 days at 20–25 °C under a 9–12 h photoperiod of warm white light from fluorescent tubes (light intensity greater than 50 Wm^{-2}). Harvest approx. 100 g of fully expanded leaves per 600 ml of isolation medium at 4 °C.[b]
2. Homogenize the leaves using a hand-held homogenizer for approx. 30–45 sec taking care to minimize frothing since this results in protein denaturation. This procedure results in only partial maceration of the tissue.
3. Complete the disruption procedure by transferring approx. 150 ml portions of the homogenate to a vessel suitable for use with a Polytron or Ultra-Turrax homogenizer and homogenize for approx. 5 sec at full speed.
4. Strain the homogenate through four layers of muslin (or cheesecloth) pre-wetted with isolation medium.
5. Transfer the filtrate into 500 ml bottles and centrifuge at 2960 *g* for 5 min.
6. Discard the pellet and centrifuge the supernatant at 17 700 *g* for 15 min.
7. Resuspend each of the pellets gently in approx. 3 ml of wash medium using a paint brush, and dilute them each to 30 ml with wash medium.
8. Transfer the resuspended pellets into 50 ml tubes and centrifuge at 1940 *g* for 10 min.
9. Centrifuge the supernatant for 12 100 *g* for 10 min to pellet the washed mitochondria. The pellets should be resuspended in a total volume of 2–3 ml of wash medium.
10. The procedure for the isolation of chlorophyll-free mitochondria from pea leaves employs a self-generating gradient of Percoll in combination with a linear gradient of PVP-40 (13). Place 17 ml of solution A to the chamber of the gradient mixer proximal to the exit tube and 17 ml of solution B to the distal chamber. Agitate solution A with a stirrer and immediately open the passageway between the two sides of the gradient maker and use a peristaltic pump to achieve a flow rate of approx. 5 ml/min. Layer the incoming solution into a 50 ml centrifuge tube. Repeat the operation to obtain a second Percoll/PVP-40 gradient.
11. Carefully layer 2–3 ml of the resuspended washed mitochondrial pellet on to each of the gradients and centrifuge at 39 200 *g* for 40 min.

Mitochondria are found as a tight light brown band near the bottom of the tube while thylakoids remain at the top of the tube.

12. Carefully remove the mitochondrial fraction, after aspiration of the upper part of the gradient, using a long-tipped Pasteur pipette, and dilute the fraction at least tenfold with wash medium.
13. Harvest purified mitochondria by centrifugation at 12 100 *g* for 10 min. Aspirate the supernatant taking care not to disturb the soft mitochondrial pellet. Add approx. 30 ml of wash medium, and centrifuge the resuspended pellet at 12 100 *g* for a further 10 min.
14. Remove the supernatant by aspiration and resuspend the purified mitochondrial pellet in minimum volume (0.5 ml) of its own juice or added wash medium using a paint brush to give a protein concentration of approx. 20–30 mg/ml based on a Lowry *et al.* (12) protein assay using BSA as a standard.[c]

[a] Glycine (2–5 mM) is often included in the wash medium for green leaf tissue to prevent inactivation of glycine decarboxylase.
[b] If spinach is substituted for pea leaves, the leaves should be deribbed prior to use, rinsed in ice-cold water, and excess water removed.
[c] Note that mitochondrial protein should be corrected for by the contribution by broken thylakoids by measuring the chlorophyll content (see *Protocol 7*) and assuming a thylakoid protein/chlorophyll ratio of 6.9 : 1 (14).

3. Chloroplasts

Photosynthesis may be simply defined as an energy transduction process whereby light energy is converted to chemical energy by plants (and certain bacteria), the whole process may be conveniently divided into two aspects. First there are those reactions which involve the harvesting of light energy and its subsequent transduction to chemical energy in the form of ATP and NADPH (both referred to as reducing power). This is achieved by a series of oxidation–reduction reactions. The second aspect of the whole process involves the reduction of CO_2 to the level of sugars using the reactions of the carbon reduction cycle (C3 cycle) which are driven by the light-generated reducing power.

Photosynthesis may be studied *in vitro* using isolated (and purified) chloroplasts from a variety of plant systems such as protoplasts (plant cells freed of their cellulose walls by enzymatic digestion) and leaf tissues. The chloroplast is bounded by a double membrane—the envelope—the outer component being permeable to substances up to an exclusion limit of 10 kDa, whilst the inner component contains a number of translocators and is the major permeability barrier to the entry and exit of metabolites from the interior of the organelle. The envelope contains some carotenoid pigment and is freely permeable to water and gases. Internal to the envelope is a gel-type matrix—the

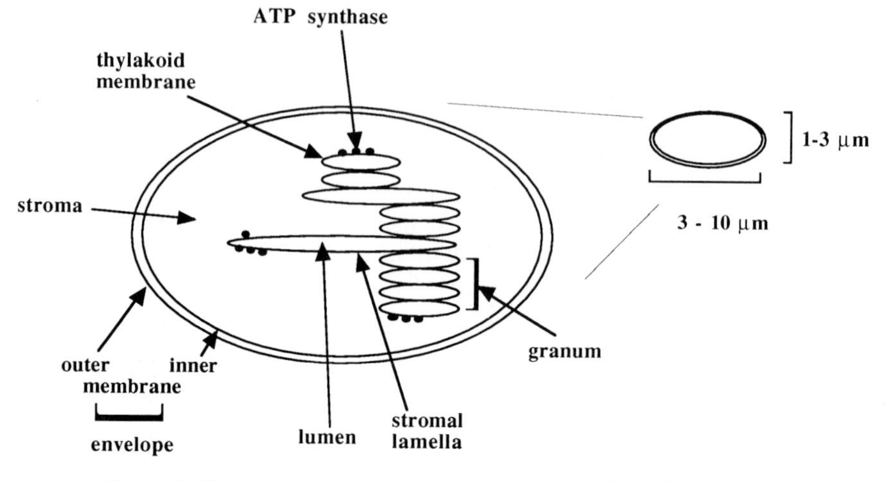

Figure 2. The major structural features of a higher plant chloroplast.

stroma—which contains the enzymes of the C3 cycle together with starch grains, lipid droplets (plastoglobuli), nucleic acids (the plastid genome is relatively small (120–270 kb) and apart from the housekeeping genes only encodes about 20 genes), ions, ribosomes, and cofactors such as ATP and NADP(H). Embedded in the stroma is an internal membrane system—the thylakoids—which are single membranes enclosing a lumen. The thylakoids may be stacked on top of each other to form grana or may run unstacked through the stroma (stromal lamellae). The thylakoid lumen is considered to be continuous throughout the entire thylakoid system. In/on these thylakoids are to be found the photosynthetic pigments, which exist as protein–pigment complexes (with probably no free pigment present), together with the electron carriers and the ATP synthase components (CF_0/CF_1) associated with the light harvesting and energy transduction processes. A diagram of the major features of the chloroplast structure is given in *Figure 2*.

The Z-scheme or photosynthetic electron flow chain (see *Figure 3*) describes the process of light-driven transfer of electrons from water to the terminal electron acceptor NADP. Quanta are absorbed by an assemblage of chlorophylls (both *a* and *b*) and carotenoid pigments arranged as light-harvesting or antennae systems (photosystems I and II). The photosystems transfer the absorbed energy to reaction centres or traps containing specialized chlorophyll *a* molecules, P680 or P700, both present in low concentrations (less than 1% of the total chlorophyll content) probably as dimers. It is from these traps that the electron flow reactions *per se* are initiated. The scheme is usually plotted against a redox potential scale and energy is provided by quanta to lift electrons from the traps against the thermodynamic gradient ('uphill') to acceptors which thus become reducing agents (phaeophytin and A_0 in the figure). As the electrons pass from these reducing agents

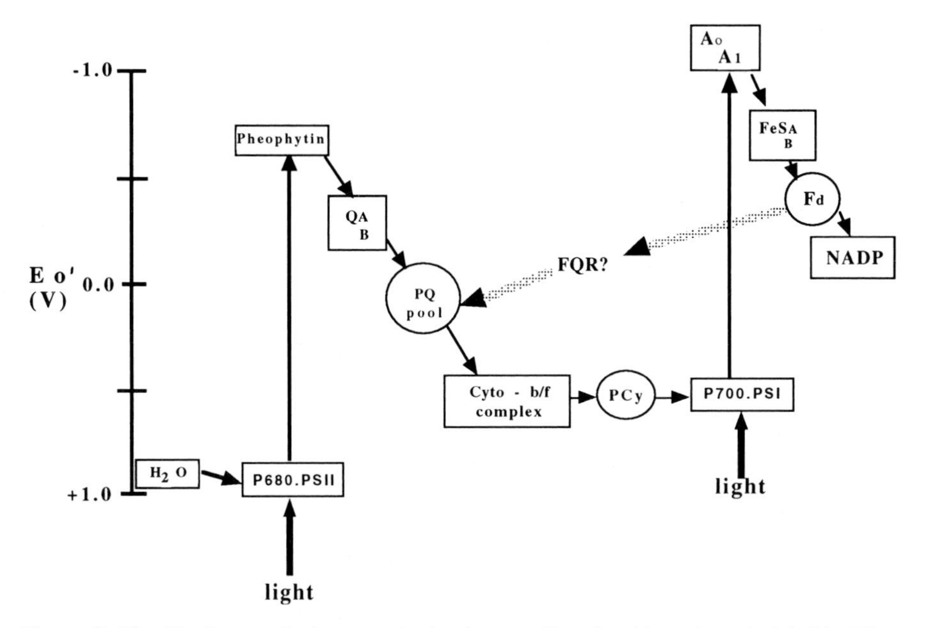

Figure 3. The Z-scheme of photosynthetic electron flow in chloroplast thylakoids. The encircled components are mobile carriers, the boxed components are membrane bound. The dotted line represents cyclic electron flow around photosystem I via a putative oxidoreductase. Abbreviations used: PSII, photosystem II and its associated trap P680; Q_A and Q_B are bound quinone molecules; PQ pool, plastoquinone pool; FQR, putative oxidoreductase; cyto-b/f complex, cytochrome *b/f* complex; PCy, plastocyanin; PSI, photosystem I and its associated trap P700; A_0 and A_1, iron–sulfur centres; Fd, ferredoxin.

('downhill') to electron carriers so energy is made available to drive the phosphorylation reaction. The electron flow depicted in *Figure 3* is a linear or non-cyclic flow wherein two photosystems participate, there is net oxidation–reduction (water is oxidized and NADP is reduced), and photophosphorylation occurs. Another mode of electron flow is recognized which involves just photosystem I and here there is a cyclic flow from P700 to ferredoxin thence to the cytochrome *b/f* complex back to photosystem I. This process (shown by the dotted line in *Figure 4*) does not involve any net oxidation–reduction reaction but there is an associated phosphorylation. The process of electron flow may be experimentally dissected by the use of a variety of electron donors, acceptors, and inhibitors and *Figure 4* illustrates the interaction of some of these agents with the electron transport chain. A full description and details of these dissection experiments can be found in ref. 15.

3.1 General isolation principles

Isolated chloroplasts may be 'intact' or 'whole' in which case they retain the envelope and their contents. Alternatively, isolation may produce 'broken'

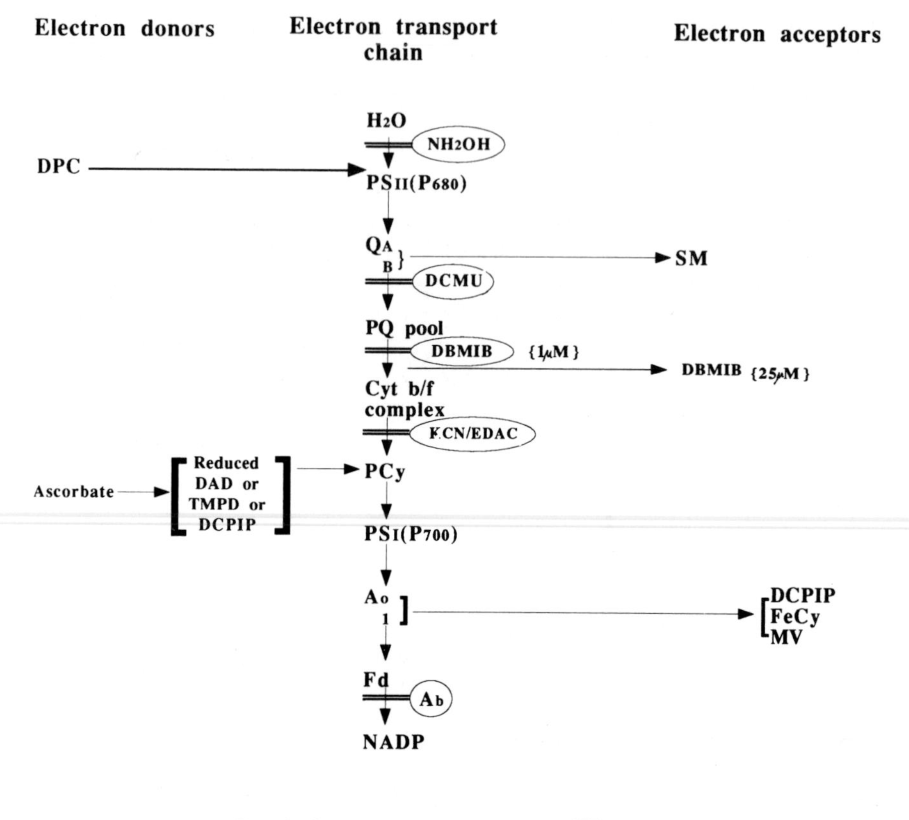

Figure 4. Photosynthetic electron transport interactions—electron acceptors and donors to the chain together with inhibitors of electron flow. Abbreviations used: Ab, antibody to ferredoxin; DAD, diaminodurene; DBMIB, 2,5-3-methyl-6-isopropyl-*p*-benzoquinone; DCMU, 3-(3,4-dichlorophenyl)-1,1-dimethylurea; DCPIP, dichlorophenyl indophenol; DPC, diphenylcarbazide; EDAC, carbodiimide; FeCy, ferricyanide (potassium salt); KCN, potassium cyanide; MV, methyl viologen; NH_2OH, hydroxylamine; SM, silicomolybdate; TMPD, tetramethyl-*p*-phenylene diamine; other abbreviations as in *Figure 3*.

chloroplasts where the envelope and stromal contents are lost (a more complete classification of chloroplast types may be found in the paper by Hall) (16). Intact chloroplasts can be competent in fixing CO_2 to differing degrees but intactness does not necessarily equate with maximal CO_2 fixation ability as a 'resealing' of ruptured envelope membranes can occur but the loss of stromal contents nevertheless curtails fixation. Broken chloroplasts, under suitable experimental conditions, are capable of catalysing a light-induced electron flow to added electron acceptors (a Hill reaction, see also *Protocol 8*) and the associated coupled reactions of phosphorylation. But breakage results in the loss of cofactors, e.g. ferredoxin, and these cofactors will need to be restored to the broken chloroplasts in order to achieve a physiological electron flow. The degree of chloroplast intactness can be assessed experimentally (see *Protocol 8*).

The investigation of such chloroplastic processes as carbon assimilation, electron flow and phosphorylation, metabolite transport, or protein targeting is best undertaken using intact chloroplasts as the starting material. The isolation of intact organelles means that the envelope can protect the stromal contents (including the thylakoids) from the destructive effects of degradative enzymes and phenolics released from the leaf tissue during the preparative procedure(s). This strategy also aids the retention of those proteins that are loosely bound to the thylakoid membranes, e.g. ferredoxin and which might be otherwise washed away during isolation. Furthermore intact organelles have better storage properties and retain their activity for longer periods than broken chloroplasts.

The principle of mechanical isolation involves:

- briefly homogenizing the leaf tissue (with minimal frothing) to produce a macerate
- quickly removing the cell debris, nuclei, and unbroken leaf tissue by filtering twice through muslin (thus obviating a centrifugation step)
- briefly centrifuging the filtrate
- finally gently resuspending the pelleted chloroplasts

All operations being conducted at 4 °C and as speedily as possible with the major objective being the preparation of good quality isolated organelles as opposed to obtaining the maximum quantity of chloroplasts from the tissue.

Isolation of intact chloroplasts can be achieved by using mechanical methods or may be derived from protoplasts. The choice of method is dependent, to a large extent, on the plant material available and whether the chloroplasts are for the study of photosynthesis itself or for the study of the photosynthetic properties of a particular plant. The enormous amount of literature on chloroplast photosynthesis has been derived mainly from studies using spinach chloroplasts (and to a lesser extent lettuce and pea chloroplasts) isolated by mechanical methods together with a smaller number of studies with these and other plants using chloroplasts derived from protoplast preparations of leaf material. It must be stressed from the outset that whatever method is adopted the use of poor or indifferent quality plant material will not yield intact chloroplasts with good photosynthetic activities. Further discussion and detailed accounts of the methods to be described can be found in refs 17–19.

The method for isolating chloroplasts from protoplasts (*Protocols 4* and *5*) is both expensive and time-consuming giving comparatively low yields of intact chloroplasts but it has the major advantage of being applicable to the leaves from a much wider range of plant species as compared to the mechanical method. The principle of preparation involves:

- digesting the chosen leaf material with a mixture of cell wall degrading enzymes (cellulase and pectinase)
- removing undigested leaf material by filtration
- harvesting the protoplasts by collecting after flotation in a gradient
- obtaining chloroplasts from the prepared protoplasts by disrupting the cell membrane to release the cellular contents and organelles (including the chloroplasts), which are then isolated by centrifugation

Protocol 3. Mechanically isolated chloroplasts

This protocol is generally restricted to species with soft leaves having a low content of oxalate, phenolics, and starch such as pea (*Pisum sativum*) or spinach (*Spinacea oleracea*) leaves and sometimes lettuce (*Lactuca sativum*) leaves. It is a fast, inexpensive method and gives a good yield of chloroplasts.

Equipment and reagents

- Plants: peas—use Feltham First, Little Marvel, or Progress varieties. For spinach use the varieties US hybrid 74 (also known as Yates hybrid 102) or Virtuosa. Do not use 'New Zealand' or 'Perpetual' spinach as these are beet cultivars and contain high levels of oxalate and phenolic compounds and do not give good chloroplasts.
- Sow the peas in vermiculite or potting compost to produce seedlings between 5–8 cm tall. Harvest the leaves after 10–14 days growth at 20–25 °C with 9–12 h light of intensity greater than 50 Wm^{-2}. Harvest the spinach leaves from actively growing, non-flowering plants kept in short days (8–8.5 h) under high light intensities.
- Isolation media. An effective medium (20) for spinach chloroplasts is: 330 mM sorbitol, 10 mM Na-pyrophosphate, 5 mM $MgCl_2$, 2 mM Na-isoascorbate adjusted to pH 6.5 with HCl. Media containing pyrophosphate are unsuitable for use with peas (in the absence of added adenylates) and may be replaced with media containing phosphate or zwitterionic buffers such as Hepes or Mes (50 mM). Tris buffer at higher pH values may also be used. Media low in cations may result in improved chloroplast intactness (possibly at the expense of storage stability). These 'low cation' media (21) enhance the separation of intact from broken chloroplasts: 330 mM sorbitol, 20 mM Mes, 0.2 mM $MgCl_2$ adjusted to pH 6.5 with Tris. A variation of the low cation medium without Tris comprises: 340 mM sorbitol. 2 mM Hepes, 0.4 mM KCl, 0.04 mM EDTA adjusted to pH 7.8 with KOH (22). Both of these low cation media are suitable for use with peas. Other media may be prepared by substituting KCl (200 mM) for the sugar/sugar alcohol osmoticum (23). All isolation media should be used as a semi-frozen slush at 4 °C.
- Resuspension media. A general resuspension medium frequently used is that described in (17): 330 mM sorbitol, 50 mM Hepes–KOH pH 7.6, 2 mM EDTA, 1 mM $MgCl_2$, 1 mM $MnCl_2$. If low cation isolation media have been used, use them also for resuspension media. Use these media at 4 °C and avoid the use of phosphate-containing media if experiments relating to the phosphorylation process are intended.
- Double strength resuspension media: media lacking osmoticum are required for the preparation of thylakoid membranes
- Double strength osmoticum: 660 mM sorbitol
- Wash media: use either the isolation or resuspension media as chilled solutions
- Homogenizer: use motor-driven homogenizers such as a kitchen blender, Polytron, Ultra-Turrax, or Waring blender (all are preferable to a mortar and pestle)—prechill prior to use
- Muslin and cotton wool to be used as filters

- A refrigerated low speed bench centrifuge with a swinging-bucket rotor is recommended—a non-refrigerated bench centrifuge can be used if the tubes and their holders are pre-chilled before immediate use
- Centrifuge tubes (50 ml): use new and/or unscratched centrifuge tubes as their walls are smooth and will not rupture the envelope during centrifugation
- Percoll™ (Pharmacia)
- Glass beakers (250 ml capacity) to hold the muslin filters
- Ice bowl/tray with ice to pre-chill all glassware and tubes prior to isolation
- Small soft paint brush for the resuspension of chloroplast pellets

Keep these tubes and glassware exclusively for chloroplast preparative work and clean/rinse them only with distilled water. Do not expose to detergent.

Method

1. Harvest leaves (approx. 50 g) at the start of the light period to avoid high levels of starch as these grains can rupture the chloroplast envelope during centrifugation. Derib the spinach leaves or use the whole aerial shoot of the peas (rather than picking individual leaves). Rinse the leaves in ice-cold water and remove excess water.
2. Chop the leaf material finely, place in a chilled blender with 150 ml of the chosen isolation medium, and briefly (within 5–10 sec) homogenize to a brei or coarse macerate with minimal frothing. Underhomogenize the tissue for best results.
3. Squeeze the ice-cold macerate through two layers of muslin into a 250 ml glass beaker kept on ice. This filtration removes coarse debris and unbroken tissue.
4. Pour, but do not squeeze, the filtrate through a pre-wetted (use ice-cold isolation medium) muslin cotton wool sandwich (four layers of muslin on either side of the cotton wool) into another 250 ml glass beaker on ice. This procedure helps to retain nuclei and broken chloroplasts and obviates the need for a short centrifugation step.
5. Divide the filtrate between the centrifuge tubes and centrifuge for 30–60 sec in the range 1500–6000 *g*. Exact times and values need to be determined for optimal results as these parameters can vary considerably for different plants.
6. Decant or aspirate away the supernatants, drain the tubes, and wipe any froth adhering to the sides with paper tissues.
7. Gently wash away the surface portion of the pellets (mostly broken thylakoids) with a small volume of the appropriate ice-cold medium and discard. Store the remaining compact pellets on ice to prevent leaching of ions until ready for resuspension.
8. Resuspend each pellet with 0.5–1.0 ml of resuspension medium by gently shaking each tube or by using the small paint brush.
9. Store the chloroplast suspension in the dark on ice.
10. To remove contaminants in the chloroplast suspension, e.g. mito-

Protocol 3. *Continued*

chondria or enzymes (from ruptured organelles) adsorbed to the chloroplasts, wash the pellets (or resuspend them) with 150 ml of ice-cold washing medium (use either isolation or resuspension media), and repeat the previous centrifugation and resuspension steps.

11. To increase the percentage of intact chloroplasts (at the expense of yield), layer 4–6 ml of chloroplast suspension on to a 6 ml cushion of buffered osmoticum containing 40% (v/v) Percoll. Centrifuge at 2000–3000 *g* for 1 min, decant the supernatant, and resuspend the pellet as before.
12. To prepare thylakoid membranes, resuspend the pelleted intact chloroplasts with ice-cold double strength media lacking osmoticum for 1 min, and then add an equal volume of double strength osmoticum. Optionally this stage may be followed by further repelleting and resuspension in full strength media. Alternatively lysis of the intact chloroplasts may be achieved in the experimental reaction vessel itself (see *Protocol 8*) by treating an aliquot of chloroplasts with double strength media lacking osmoticum for 1 min followed by the addition of an equal volume of double strength osmoticum.

Protocol 4. Preparation of protoplasts

Equipment and reagents

- Plants: leaves from actively growing and healthy plants (preferably less than two weeks old)
- Carborundum powder, single edged razor blades to abrade and cut the leaf surface
- Glass beaker (approx. 600 ml capacity) or shallow dish (20 cm diameter) to contain the digestion mixture
- Water-bath set to 25–30 °C for the incubation of tissue with digestion mixture.
- Light source: 60/100 W bulb and glass trough with water (to act as a heat filter)
- Nylon tea-strainer (500 μm mesh) and nylon meshes of 20, 80, and 200 μm mesh and 1 mm mesh
- Bench centrifuge preferably with a swinging-bucket rotor
- Glass centrifuge tubes (10–20 ml)—plastic tubes are not suitable since it is necessary to view the tube contents when making gradients
- Digestive enzymes: cellulase and pectinase
- 500 mM sorbitol
- Digestive mixture: no single digestion mixture is suitable for all tissues and the optimal mixture for a given tissue needs to be determined by trial and error. A guideline is to use cellulase in the range 1–3% (w/v), and pectinase in the range 0.1–5% (w/v). A good 'trial' digestion mixture to use is: 2% (w/v) cellulase, 0.3% (w/v) pectinase, 500 mM sorbitol, 5 mM Mes, 1 mM $CaCl_2$ adjusted to pH 6.0 with HCl. This has been used successfully with leaves from barley, peas, and wheat
- Medium I: 500 mM sorbitol, 5 mM Mes, 1 mM $CaCl_2$
- Medium II: 500 mM sucrose, 5 mM Mes, 1 mM $CaCl_2$
- Medium III: 400 mM sucrose, 100 mM sorbitol, 5 mM Mes, 1 mM $CaCl_2$ (media I, II, and III are best used chilled to 4 °C and all adjusted to pH 6.0 with HCl)
- Medium IV: 500 mM sorbitol, 1 mM $CaCl_2$, 50 mM Tricine–KOH chilled to 4 °C and adjusted to pH 7.8–8.0

- Medium V: 500 mM sucrose, 1 mM $CaCl_2$, 50 mM Tricine–KOH chilled to 4 °C and adjusted to pH 7.8–8.0
- Percoll™ (Pharmacia)

Keep all apparatus, glassware, and tubes used exclusively for protoplast preparative work, clean and rinse with distilled water only. Do not expose to detergent.

Method

1. Expose the leaf tissue cells to the digestion mixture as follows: abrade the leaf surface with the carborundum, or strip off the epidermis, or slice the leaf into transverse sections between 0.5–1 mm thick with a single edged razor blade. Change the blade frequently to ensure that cutting not tearing of the tissue occurs. If air enters the tissue during sectioning it may be necessary to cut the sections under 500 mM sorbitol and drain before the next stage or use vacuum infiltration to aid exposure of the cell walls to the digestive enzyme mixture.
2. Put 30 ml of the digestion mixture into a 600 ml capacity glass beaker or shallow dish.
3. Place the prepared leaf tissue (approx. 3–5 g) with the exposed leaf surfaces in contact with the digestion mixture.
4. Leave to digest at 25 °C for 2–3 h with illumination provided by the bulb approx. 20–25 cm above the digest with the heat filter between the digest and lamp.
5. Decant the digestion mixture and if it contains an appreciable number of protoplasts recover them by centrifugation at 100–150 *g* for 3–5 min.
6. Store the used digestion mixture in a freezer for future reclamation of the costly enzymes by ammonium sulfate precipitation and freeze-dry the resultant precipitate for storage.
7. Wash the digested tissue twice with approx. 20 ml of medium I to liberate the protoplasts and pour the washings first through a 500 μm mesh (a nylon tea-strainer may be used), then through a 200 μm mesh. This procedure helps to retain the large vascular strands.
8. Combine the washings and centrifuge them at 100–150 *g* for 3–5 min to yield a firm but not tightly packed pellet (capable of being gently resuspended). Discard the supernatant.
9. Gently resuspend each pellet in its own glass tube with approx. 1 ml of medium II and layer on to each suspension 2 ml of medium III to act as a wash layer.
10. Overlay these two layers within each tube with 1 ml of medium I to complete the gradient. Ensure that the three gradient layers do not mix.
11. Centrifuge the tubes at 100–150 *g* for 3–5 min: the intact protoplasts

Protocol 4. *Continued*

will float. Collect them at the interface between the two upper solutions using a Pasteur pipette.

12. Place the recovered protoplasts in medium I at a final total volume of approx. 5–10 ml.
13. A simpler gradient may be tried at step 9 by adding 10% (v/v) Percoll to the 1 ml of medium II containing the protoplasts. Overlay this with 2 ml of medium I and collect the protoplasts (after centrifugation) at the interface as before. Dilute these five- to tenfold with medium I for storage for up to a day in a refrigerator. (For prolonged storage replace the sorbitol with sucrose.)

Carry out all protoplast pipetting and handling techniques as gently as possible.

These procedures are suitable for most C3 plants but if C4 plants are used filter the initial crude digestion extract first through a 500 μm mesh or tea-strainer and then through an 80 μm mesh. This procedure will retain the bundle sheath protoplasts on the 80 μm mesh and these may be recovered and resuspended in medium I. Harvest the C4 mesophyll protoplasts that filter through the 80 μm mesh preferably using the Percoll gradient since these protoplasts are denser than the C3 protoplasts. If it is suspected that protoplasts are being pelleted in the gradient mixture they may be recovered by making the gradient denser by altering the tonicity up to 0.6 M (with respect to sucrose and sorbitol) or use Percoll as described for the gradient mixture.

Protocol 5. Preparation of protoplasts of CAM plants

Equipment and reagents

- See *Protocol 4*

Method

1. Follow *Protocol 4*, steps 1–6.
2. Wash the digested CAM tissue twice with medium IV to liberate the protoplasts and filter the washings through 1 mm and then 500 μm mesh. Combine these washings.
3. Centrifuge the combined washings at 100–150 *g* for 2–3 min and discard the supernatant.
4. Gently resuspend each pellet with 1–2 ml of medium IV. Because of their greater density CAM protoplasts may well settle out of the resuspension medium at this stage (5–10 min) and may be collected

without the need for a centrifugation step. Alternatively the CAM protoplasts may be harvested by layering the cells on to a cushion of medium V with Percoll added if necessary to increase the density.

5. Centrifuge at 100–150 *g* for 3–5 min and collect the protoplasts at the interface. If any protoplasts are observed in a pellet increase the density of the cushion and repeat the centrifugation step.

Protocol 6. Preparation of chloroplasts from the isolated protoplasts

Equipment and reagents

- Attach a 20 μm mesh (use a 30–50 μm mesh for CAM protoplasts) to the end of a disposable syringe with a plastic ring cut from a disposable pipette tip
- Rinse new syringes with alcohol to rid lubricant which is inhibitory to photosynthesis
- Cut short any pipette tips to be used with protoplasts to give a wider tip to accommodate the diameter of the protoplasts (between 30–40 μm for C3 and C4 protoplasts and up to 100 μm for CAM plant protoplasts)
- Resuspension medium 1: 330 mM sorbitol, 50 mM Hepes–KOH pH 7.6, 2 mM EDTA, 1 mM $MgCl_2$, 1 mM $MnCl_2$
- Resuspension medium 2: 330 mM sorbitol, 20 mM Hepes, 10 mM EDTA adjusted to pH 7.6
- Breaking medium: generally use the same medium as for the resuspension of chloroplasts

Often it will be necessary to increase the buffer strength (up to 250 mM) to combat acid release from the protoplast vacuoles and to add chelating agents (to 5 mM) with protective agents (soluble PVP to 1%, w/v) to the breaking media used when obtaining chloroplasts from CAM protoplasts. Trial and error will give the optimal composition of these media for individual CAM species.

Method

1. Pellet the protoplasts by centrifugation between 100–150 *g* for 2 min.
2. Discard the supernatant and resuspend the pellets in 0.5–1 ml of the required ice-cold breaking medium.
3. Draw up the protoplasts into the syringe and force them through the fine 20 μm mesh thus rupturing the cell membrane and releasing the internal contents.
4. Repeat this procedure once more and then centrifuge the resultant suspension at 200–300 *g* for approx. 1 min.
5. Discard the supernatant. The surface of the pellets at this point is often 'stringy' (especially at alkaline values), remove it with resuspension medium and discard.
6. Resuspend the remaining chloroplast pellets in the required ice-cold medium by gently shaking the tubes to give a highly intact chloroplast preparation (usually more than 95% intact).

4. Assays

4.1 Mitochondria

4.1.1 Respiratory activity measurements

One of the main functions of mitochondria is to convert the free energy of substrate oxidation into ATP (1,2,6,7). Therefore, a criterion for the isolation of functional mitochondria is the efficiency of this conversion. This can normally be estimated either from the ADP/O ratio or the respiratory control ratio. The ADP/O ratio is defined as the number of nanomoles of ADP phosphorylated divided by the amount of oxygen, in nanogram atoms, consumed during this process (6,7,10). This ratio is characteristic of the substrate undergoing oxidation and can have a value of approximately 2.5 for NAD^+-linked dehydrogenases and 1.5 for FAD-linked dehydrogenases (24). It is assumed that during the stimulated phase of oxygen consumption all oxygen uptake is coupled to ATP synthesis and that, upon return to the basal rate, all added ADP has been phosphorylated. The respiratory control ratio is the ratio of the state three respiratory rate (respiration in the presence of ADP) divided by the state four rate (the respiratory rate following exhaustion of added ADP) and its value gives a good indication of the quality of the mitochondrial preparation. A ratio of one is observed with uncoupled mitochondria, while ratios of four to six are usual with malate as substrate (6,7,10). No respiratory control is usually observed with mitochondria isolated from mature spadices of *Arum maculatum* or *Sauromatum guttatum* since any stimulation upon addition of ADP is masked by the very rapid alternative oxidase activity observed in these tissues (6,7). Isolated plant mitochondria are often characterized by displaying an increased state three and decreased state four rates following an initial state three to four transition, a phenomenon termed conditioning. It is therefore important in a determination of respiratory control or ADP/O ratios that values are taken after an initial state three to four cycle (see ref. 7).

Plant mitochondria readily oxidize externally added NADH (except some varieties of red beetroot) (25), and most citric acid cycle intermediates, with good respiratory control and ADP/O ratios close to theoretical. In general a counterion, particularly phosphate, is required for substrate uptake. Succinate oxidation normally requires the prior addition of ATP to activate succinate dehydrogenase and malate may require the addition of an oxaloacetate removing agent, such as glutamate (via glutamate–oxaloacetate transaminase) particularly under conditions when the rate of malate oxidation becomes progressively slower. Other citric acid cycle intermediates such as pyruvate, 2-oxoglutarate and glutamate are also oxidized by plant mitochondria but rates of oxidation and respiratory control ratios often tend to be lower than those observed when malate is the substrate. The oxidation of pyruvate is generally slow and normally requires both thiamine pyrophosphate (0.3 mM) and 'sparker' concentrations of malate (0.6 mM) and in some circumstances CoA (50 μM) (26). However, mitochondria isolated from the spadices of *Arum maculatum* and *Sauromatum guttatum* rapidly oxidize pyruvate without the need of added cofactors (27). Similarly 2-oxoglutarate (10 mM) normally requires thiamine pyrophosphate (0.3 mM) and malonate (3 mM), the latter being required to inhibit succinate oxidation which otherwise would result in lower measured ADP/O ratios. Mitochondria isolated from green leaf tissue of C3 plants also rapidly oxidize glycine (10 mM) with respiratory control and ADP/O ratios comparable to that observed with malate as substrate (see ref. 4).

Oxygen consumption is normally monitored in an oxygen electrode system (see ref. 28 for details). The system has two electrodes: one platinum and one silver, mounted in the base of a cell surrounded by a water jacket. Oxygen electrodes of this type can be obtained from Rank Bros., Bottisham, Cambridge, UK, or Hansatech Ltd, King's Lynn, UK. Air-saturated medium is placed in the cell and is separated from the electrodes by a thin Teflon membrane which is permeable to oxygen. Beneath this is a piece of tissue paper, to act as a spacer, soaked in saturated KCl which acts as an electrolyte. When a voltage is applied across the two electrodes, the platinum electrode polarized negative with respect to the Ag/AgCl electrode, oxygen undergoes an electrolytic reduction. With a polarization of −0.6 V, current flows from the Ag/ AgCl electrode to the platinum electrode. The current generated, at this voltage, is directly proportional to the concentration of the oxygen in solution. The amount of oxygen dissolved in solution can be calculated from its solubility coefficient, which depends on the temperature and electrolyte composition of the medium. The oxygen electrode should be thermostatted (22–25 °C) to circumvent any problems associated with lipid phase transitions in the membrane structure which may effect respiratory control and ADP/O values.

Protocol 7. Measurement of functional integrity of isolated mitochondria

Equipment and reagents

- Reaction medium (50 ml): 0.3 M mannitol, 5 mM $MgCl_2$ 10 mM KCl, 5 mM potassium phosphate, 10 mM Hepes–KOH pH 7.4
- Substrates (1 ml of each and kept frozen until required): 1 M sodium succinate, 1 M sodium malate, 0.1 M β-NADH (in alkaline solution)—substrates should be dissolved in a stock solution of 0.3 M mannitol and 10 mM Hepes–KOH pH 7.4
- Adenylates (0.5 ml of each kept frozen until required): 0.1 M ADP, 0.1 M ATP dissolved in 0.3 M mannitol and 10 mM Hepes–KOH pH 7.4
- A single pen recorder
- Sodium dithionite
- Inhibitors (0.2 ml of each, kept in freezer): 1 mM antimycin A or 1 mM myxothiazol, 0.1 M salicylhydroxamic acid (SHAM) dissolved in absolute ethanol
- An oxygen electrode, magnetic stirrer, and polarizing box (from Rank Bros., Bottisham, Cambridge, UK or Hansatech Ltd., King's Lynn, UK) or equivalent apparatus

Protocol 7. *Continued*

A. *Calibration of the oxygen electrode*

1. Pipette 2 ml of distilled water into the oxygen electrode chamber, add a few crystals of sodium dithionite. The oxygen concentration will rapidly decrease to zero, as indicated by the output of the electrode to the chart recorder which should be set to zero.
2. Wash out the electrode chamber several times with distilled water to ensure that no traces of dithionite remain. Pipette 2 ml of distilled water, equilibrated with air at 22–25 °C into the electrode chamber which itself is thermostatted at the same temperature.
3. Adjust the pen on the chart recorder to approx. 90% of its deflection using the oxygen electrode polarizing box. This level now corresponds to 254 μM dissolved oxygen in air-saturated distilled water (at 25 °C) or 237 μM in buffered media. As both the zero and maximum oxygen concentrations are known the electrode traces can be calibrated.
4. Since an accurate knowledge of the oxygen concentration is required for ADP/O estimates a convenient and rapid means of determining the oxygen content of the reaction medium is by using a limiting amount of NADH. The high affinity of the externally located NADH dehydrogenase for this substrate permits a stoichiometric titration of oxygen content. Pipette 2 ml of reaction medium into the chamber and add sufficient mitochondria to give a final concentration of 0.2–0.5 mg protein/ml. Add limiting final concentrations (100–200 μM) of spectrophotometrically standardized NADH to the chamber through the stopper and monitor the change in current on complete oxidation of the NADH.

B. *Measurement of respiratory activity*

1. Add 2 ml of reaction medium and sufficient mitochondria to give a final concentration of 0.2–0.5 mg protein/ml and following addition of stopper measure background rate for 1–2 min.
2. Add substrate. For malate, add 20 μl of the stock solution and for succinate, 10 μl in the presence of 250 μM ATP. For NADH or glycine, 20 μl of the stock solutions should be added. Measure the rate of oxygen consumption for 1–2 min.
3. Add 2.5 μl of the stock ADP solution and measure the oxygen uptake rate for approx. 5 min or until the rate decreases.
4. Repeat step 3.
5. The respiratory control ratio can be calculated by dividing the rate of oxygen consumption in state three by the rate in state four (i.e. the

rate of oxygen consumption following ADP exhaustion). The ADP/O ratio is calculated from the amount of ADP (in nanomoles) added divided by the amount of oxygen (in nanogram atoms) consumed during state three.[a]

6. The degree of alternative oxidase activity is estimated by the adding 2 μl of antimycin A or myxothiazol to the oxygen electrode chamber whilst mitochondria are respiring on any of the substrates in the presence of excess ADP (i.e. 1 mM ADP). Total respiratory inhibition should occur upon subsequent addition of 0.5–1 mM SHAM.

[a] Calculation of the ADP/O ratio assumes that all of the ADP is converted to ATP and state four respiratory activity does not continue during state three.

4.1.2 Membrane intactness

The ability of isolated mitochondria to transport and oxidize substrates, phosphorylate ADP, and maintain a membrane potential is obviously a function of the intactness of the membrane system which encloses the matrix. This latter compartment, containing the TCA cycle enzymes, is bounded by two membranes, an inner (IM) and an outer (OM), separated by an intermembrane space (IMS). Of the two membranes the OM is the more permeable, allowing free passage to molecules under M_r 12 000; the IM on the other hand represents the main barrier between matrix and cytoplasm. These permeability properties may therefore be used to assay the intactness of each membrane.

Substrate-dependent reduction of exogenous cytochrome *c* cannot occur in mitochondria which retain their OM. This is because, *in vivo*, cytochrome *c* (M_r = 12 400) is loosely associated with the outer face of the IM but is unable to cross the OM (27). Thus the intactness of mitochondria can be assessed from the degree of reduction of cytochrome *c*, since in fully intact mitochondria no reduction of cytochrome *c* occurs (6,10). Using succinate as reductant, it is possible to calculate OM intactness directly, by comparing rates of cytochrome *c* reduction in ruptured and intact mitochondria. If NADH is used as substrate, however, care must be taken in interpreting results. The presence of a flavoprotein and cytochrome b_{555} in the OM (29), or, in unpurified preparations, the cytochrome *b* found in microbodies, can also mediate reduction of cytochrome *c* by NADH. Fortunately, both such reductions are insensitive to antimycin A and inhibition of IM electron transport by this compound may be used to assay intactness.

As for any organelle, the latency of enzyme activity can be used to measure membrane intactness. In intact mitochondria there should be no reduction of exogenous NADH by oxaloacetate because malate dehydrogenase (MDH) is located in the matrix and NADH is unable to permeate the IM. A comparison, therefore, between intact and ruptured organelles will give a measure of intactness. It should be noted, however, that MDH is a ubiquitous cell protein and, unless highly purified, mitochondria may have some contamination by this enzyme. For this reason a further assay is often used relying on the impermeability of the IM to ferricyanide (29).

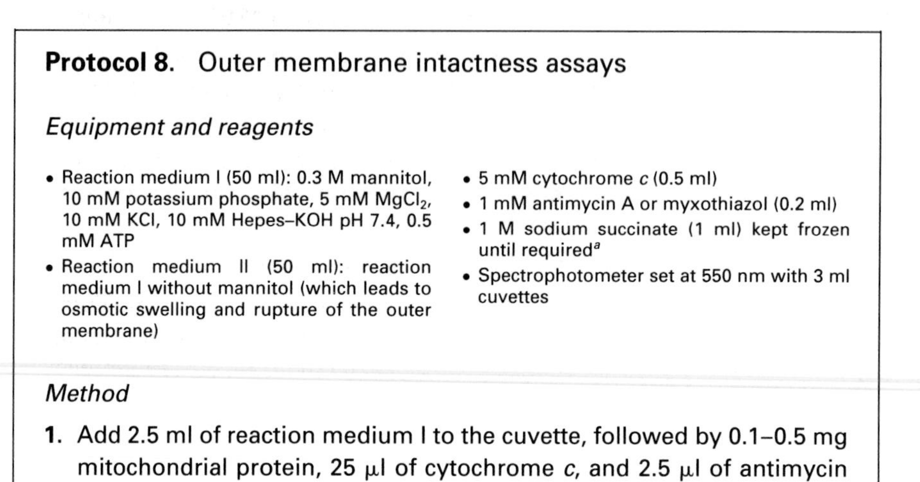

Protocol 8. Outer membrane intactness assays

Equipment and reagents

- Reaction medium I (50 ml): 0.3 M mannitol, 10 mM potassium phosphate, 5 mM $MgCl_2$, 10 mM KCl, 10 mM Hepes–KOH pH 7.4, 0.5 mM ATP
- Reaction medium II (50 ml): reaction medium I without mannitol (which leads to osmotic swelling and rupture of the outer membrane)
- 5 mM cytochrome *c* (0.5 ml)
- 1 mM antimycin A or myxothiazol (0.2 ml)
- 1 M sodium succinate (1 ml) kept frozen until required[a]
- Spectrophotometer set at 550 nm with 3 ml cuvettes

Method

1. Add 2.5 ml of reaction medium I to the cuvette, followed by 0.1–0.5 mg mitochondrial protein, 25 μl of cytochrome *c*, and 2.5 μl of antimycin A or myxothiazol.
2. Stir the contents of the cuvette and place in the spectrophotometer. Measure endogenous rate of cytochrome *c* reduction for 1–2 min.
3. Add 25 μl of succinate and measure rate for 2–5 min.
4. Repeat steps 1–4 using reaction medium II.
5. Comparison between the rate of cytochrome *c* reduction in reaction medium I and the rate observed after bursting the mitochondria in reaction medium II gives an estimate of the percentage of mitochondria with intact outer membranes.

[a] 100 mM NADH can be used as a substrate instead of succinate, but care must be taken in interpreting the results as outlined above.

Protocol 9. Inner membrane intactness assay

Equipment and reagents

- Reaction medium I (50 ml): 0.3 M mannitol, 10 mM potassium phosphate, 5 mM $MgCl_2$, 10 mM KCl, 10 mM Hepes–KOH pH 7.4, 0.5 mM ATP
- Reaction medium II (50 ml): reaction medium I without mannitol (which leads to osmotic swelling and rupture of the outer membrane)
- 100 mM KCN (0.5 ml)
- 100 mM NADH (0.5 ml)
- 100 mM oxaloacetate (0.2 ml) freshly prepared.
- Spectrophotometer set at 340 nm, with 3 ml cuvettes
- Microcentrifuge

Method

1. Dilute approx. 1 mg of mitochondrial protein with 1 ml of reaction medium I in a 1.5 ml Eppendorf tube and an identical aliquot with 1 ml of reaction medium II.
2. Centrifuge both samples in the microcentrifuge at 12 000 *g* for 2 min.
3. Aliquot 0.2–0.5 ml of the supernatants into the cuvettes and add the appropriate volume of reaction medium I to give a final volume of 2.5 ml.
4. Add 25 μl of KCN and 25 μl of NADH to each of the cuvettes. Stir the contents and place in the spectrophotometer. Measure the decrease in absorbance for 1–2 min.
5. Add 25 μl of oxaloacetate and measure rate for 2–5 min.
6. Comparison between the rate of NADH oxidation in reaction medium I and the rate observed after bursting the inner membrane in reaction medium II gives an estimate of the percentage of mitochondria with intact inner membranes.

4.2 Chloroplasts

In order to quantify most measurements of chloroplast photosynthesis it is usual to express results on a unit chlorophyll basis [(mg chlorophyll)$^{-1}$] and this entails the extraction of the chlorophyll from the chloroplast suspension with an organic solvent, usually 80% (v/v) acetone (since chlorophyll is not soluble in water). This procedure also results in the extraction of other photosynthetic pigments (the carotenoids) which absorb light in the 400–500 nm (blue) range of the spectrum and so the absorption of the extracted chlorophyll is determined in the red region (600–700 nm) of the spectrum. The following method (*Protocol 10*) is adapted from Bruinsma (30) and utilizes a single absorption measurement at 652 nm at which wavelength the extinction coefficients of chlorophylls *a* and *b* are the same and the total chlorophyll concentration is given by the expression: ($A_{652} \times 27.8$). This value gives the chlorophyll concentration in the acetone solution (mg/litre) and needs correction for the dilution of the original chloroplast suspension into the acetone solvent.

The degree of intactness of the chloroplast suspension may be assessed by comparing the rates of ferricyanide reduction (or its associated oxygen evolution) upon illumination both before and after osmotically shocking the chloroplast suspension (31). The principle of the assay is based upon the inability of the ferricyanide to cross the chloroplast envelope which therefore cannot react with the electron transport system in the thylakoid membranes present in intact chloroplast whereas with broken thylakoid systems the ferricyanide has access to the electron transport system. This estimation (*Protocol 11*) is carried out in an oxygen electrode and the equation below shows the stoichiometry of the process. The assay can also demonstrate the degree to which the chloroplasts are coupled.

$$4Fe(CN)_6^{3-} + 2H_2O \rightarrow 4Fe(CN)_6^{4-} + 4H^+ + O_2$$

Carefully prepared thylakoid systems are capable of catalysing coupled ATP synthesis. The synthesis (or hydrolysis) of ATP by illuminated chloroplasts is accompanied by the scalar consumption (or production) of protons. At pH 8.2 with excess Mg^{2+} present approx. one H^+ per ATP is involved (see ref. 32) according to the equation:

$$ADP^{3-} + PO_4^{2-} + H^+ \rightarrow ATP^{4-} + H_2O$$

Thus this reaction (*Protocol 12*) may be monitored by detecting the changes to pH of a weakly buffered suspension of chloroplasts illuminated in the presence of ADP and inorganic phosphate with excess Mg^{2+} present. The reaction may be conveniently carried out in an oxygen electrode using either ferricyanide, methyl viologen, or phenazine methosulfate (PMS) as the cofactor for electron transport.

Protocol 10. Estimation of chlorophyll

Equipment and reagents

- Acetone
- Spectrophotometer and cuvettes
- Bench centrifuge and plastic tubes (solvent resistant)

Method

1. Add sufficient water to 10–100 μl of the chloroplast suspension to give a final volume of 1 ml followed by 4 ml of acetone. Mix thoroughly.[a]
2. Centrifuge the acetone extract at approx. 3000 *g* for 2–3 min.
3. Retain the supernatant and read its absorbance at 652 nm (1 cm path length). Multiply this value by 27.8 to give micrograms of chlorophyll per millilitre of the acetone extract. Hence calculate the chlorophyll content of the chloroplast suspension.

[a] Avoid evaporation or exposure to bright light of the acetone extract as errors will be introduced by concentration or photo-oxidative effects.

Protoplast intactness may be assayed using a similar method. Assay glycollate oxidase (which is released from damaged protoplasts) before (A) and after disruption (B) by detergent (33). Use the oxygen electrode and the method described in ref. 31 to measure the rates (A) and (B) and use the intactness equation (below) to calculate the degree of intactness. Alternatively assess the intactness of protoplast preparations by their ability to exclude Evans blue (34) and/or accumulate fluorescein (35) and count the protoplasts in a haemocytometer.

Protocol 11. Estimation of the degree of chloroplast intactness (using a Hill reaction assay)

Equipment and reagents

- Double strength resuspending medium (see *Protocol 3*)
- Potassium ferricyanide (500 mM), make up fresh daily
- D L-glyceraldehyde (2 M) to inhibit CO_2 fixation
- NH_4Cl (500 mM) to act as uncoupler
- Slide projector or 150 W bulb and a glass water bottle to act as a heat filter
- Oxygen electrode and stirrer (see *Protocol 7* for details, operation, and calibration)

Method

1. Place the following mixture in an oxygen electrode. Add the reagents in the order given:
   i. 1 ml double strength resuspending medium
   ii. 1.0 ml water
   iii. 0.1 ml chloroplast suspension (50–100 g chlorophyll)
   iv. 10 μl potassium ferricyanide (500 mM)
   v. 10 μl D L-glyceraldehyde (2 M) (this reagent inhibits CO_2 fixation and prevents possible interference due to the evolution of oxygen associated with the C3 cycle)
2. Illuminate this mixture in the electrode (use a slide projector or 150 W bulb, with the heat filter). After approx. 1 min of illumination add 10 μl NH_4Cl (500 mM) to uncouple electron flow and give rates of oxygen evolution that are not restrained by the possible build-up of a proton gradient. The rate of oxygen evolution is used to calculate the rate of ferricyanide reduction (A) by the 'intact' chloroplast suspension.
3. Repeat the assay changing the order of addition of reagents to the electrode as follows: reagent (*ii*) followed by (*iii*), leave 1 min, then (*i*), (*iv*), and (*v*).
4. This osmotic shock results in the complete rupture of the chloroplast envelope thus allowing the ferricyanide to react at the now exposed thylakoid membranes.
5. Illuminate the broken chloroplast preparation and add uncoupler (10 μl NH_4Cl (500 mM) as above. This rate of oxygen evolution will allow the calculation of the rate of ferricyanide reduction (B) by the fully broken chloroplasts.
6. Calculate the percentage intactness of the original suspension as follows:

$$\text{Intactness \%} = [(B - A)/B] = 100.$$

Protocol 12. Estimation of photophosphorylation

Equipment and reagents

- Lightly buffered assay medium: 2 mM Tricine containing 10 mM $MgCl_2$, 4 mM sodium phosphate, 50 mM KCl, 1 mM ADP, 0.06 mM PMS, adjust the pH to 8.2[a]
- 660 mM sorbitol
- Standard HCl (1 and 10 mM) for calibration purposes
- Standard KOH (1 and 10 mM) for calibration purposes
- Microsyringes (10 and 25 μl volume)
- Oxygen electrode with stirrer
- Micro- or semi-micro combination electrode with pH meter (expanded scale) with 'buffer control' (or a similar device) to back off the recorder output (see also ref. 36)
- Chart recorder (full scale deflection 1–10 mV)
- Slide projector or 150 W bulb and a glass water bottle to act as a heat filter
- Red Perspex sheet to act as a light filter

Method

1. Place 1.5 ml of the assay buffer medium in the well of the oxygen electrode and add chloroplast suspension equivalent to 50–100 μg chlorophyll. Leave for 1 min to shock the chloroplasts osmotically.
2. Add the sorbitol solution to a final volume of 3.0 ml and insert the pH electrode. Check and adjust the pH of the mixture to 8.2 with HCl or KOH. Ensure the mixture is well stirred.
3. Adjust the pH meter and chart recorder so that pH changes of 0.05 units or less can be monitored. Record the pH drift.
4. Illuminate and monitor the uptake of protons. The initial pH change is due to proton uptake into the thylakoid lumen and is followed by a slower but more sustained uptake representing the phosphorylative process. (If the pH electrode is sensitive to white light then a red filter should be interposed between it and the light source.)
5. Turn off the light and once the pH drift is constant, calibrate the proton uptake by adding 5–10 μl amounts of the standard HCl. Calculate the amount of ATP formed after making corrections for the drift of the pH electrode.

[a] Adenylate kinase activity can be inhibited by the inclusion of 2 μm Ap_5A (P^1, P^5-di(adenosine-5') pentaphosphate) in the assay medium.

Acknowledgements

Work in the authors' laboratory is supported by the BBSRC, British Council, and The Royal Society.

References

1. Moore, A. L., Siedow, J. N., Fricaud, A-C., Vojnikov, V., Walters, A. J., and Whitehouse, D. G. (1992). In *Plant organelles, Society for Experimental Biology*

Seminar Series (ed. A. K. Tobin), Vol. 50, p. 188. Cambridge University Press, Cambridge.

2. Whitehouse, D. G. and Moore, A. L. (1995). In *Advances in cellular and molecular biology of plants: molecular biology of the mitochondria* (ed. C. S. Levings and I. K. Vasil). Vol. 2, p. 313. Kluwer Academic Publishers.
3. Moore, A. L., Wood, C. K., and Watts, F. Z. (1994). *Annu. Rev. Plant Physiol. Plant Mol. Biol.*, **45,** 545.
4. Douce, R. and Neuburger, M. (1989). *Annu. Rev. Plant Physiol.*, **40,** 371.
5. Moore, A. L., Dry, I. B., and Wiskich, J. T. (1991). *Plant Physiol.*, **95,** 34.
6. Douce, R., Bourguignon, J., Brouquisse, R., and Neuburger, M. (1987). In *Methods in enzymology* (ed. L. Packer and R. Douce), Vol. 18, p. 403. Academic Press, London.
7. Moore, A. L. and Proudlove, M. O. (1983). In *Isolation of membranes and organelles from plant cells* (ed. J. L. Hall and A. L. Moore), p. 153. Academic Press, New York.
8. Moore, A. L. and Proudlove, M. O. (1987). In *Methods in enzymology* (ed. L. Packer and R. Douce), Vol. 18, p. 415. Academic Press, London.
9. Edwards, G. E. and Gardestrom, P. (1987). In *Methods in enzymology* (ed. L. Packer and R. Douce) Vol. 18, p. 421. Academic Press, London.
10. Neuburger, M. (1985). In *Encyclopaedia of plant physiology: higher plant cell respiration* (ed. R. Douce and D. A. Day), Vol. 18, p. 7. Springer-Verlag, Berlin.
11. Moore, A. L., Fricaud, A-C., Walters, A. J., and Whitehouse, D. G. (1993). In *Methods in molecular biology* (ed. J. M. Graham and J. A. Higgins), Vol. 19, p. 133. Humana Press Inc., Totowa.
12. Lowry, O. H., Rosebrough, N. J., Farr, A. L., and Randall, R. J. (1951). *J. Biol. Chem.,* **193,** 265.
13. Day, D. A., Neuburger, M., and Douce, R. (1976). *Aust. J. Plant Physiol.*, **12,** 119.
14. Nash, D. and Wiskich, J. T. (1982). *Aust. J. Plant Physiol.*, **9,** 715.
15. Allen, J. F. and Holmes, N. G. (1986). In *Photosynthesis energy transduction: a practical approach* (ed. M. F. Hipkins and N. R. Baker), p. 103. IRL Press, Oxford.
16. Hall, D. O. (1972). *Nature New Biol.*, **235,** 125.
17. Walker, D. A. (1980). In *Methods in enzymology* (ed. A. San Pietro) Vol. 69, p. 94. Academic Press, London.
18. Leegood, R. C. and Walker, D. A. (1983). In *Isolation of membranes and organelles from plant cells* (ed. J. L. Hall and A. L. Moore), p. 185. Academic Press, New York.
19. Whitehouse, D. G. and Moore, A. L. (1993). In *Methods in molecular biology* (ed. J. M. Graham and J. A. Higgins), Vol. 19, p. 123. Humana Press Inc., Totowa.
20. Cockburn, W., Walker, D. A., and Baldry, C. W. (1968). *Biochem. J.*, **107,** 89.
21. Nakatani, H. Y. and Barber, J. (1977). *Biochim. Biophys. Acta*, **461,** 510.
22. Cerovic, Z. G. and Plesnicar, M. (1984). *Biochem. J.*, **223,** 543.
23. Robinson, S. P. (1986). *Photosynthesis Res.*, **10,** 93.
24. Hinkle, P. C., Kumar, M. A., Resetar, A., and Harris, D. L. (1991). *Biochemistry*, **30,** 3576.
25. Day, D. A., Rayner, J. R., and Wiskich, J. T. (1976). *Plant Physiol.*, **58,** 38.
26. Day, D. A. and Hanson, J. B. (1977). *Plant Sci. Lett.,* **11,** 99.
27. Proudlove, M. O., Beechey, R. B., and Moore, A. L. (1987). *Biochem. J.*, **247,** 441.

28. Rickwood, D., Wilson, M. T., and Darley-Usmar, V. M. (1987). In *Mitochondria: a practical approach* (ed. V. M. Darley-Usmar, D. Rickwood, and M. T. Wilson), p. 1. IRL Press, Oxford.
29. Douce, R., Christensen, E. L., and Bonner, W. D. (1972). *Biochim. Biophys. Acta*, **275,** 148.
30. Bruinsma, J. (1961). *Biochim. Biophys. Acta*, **52,** 576.
31. Lilley, R. McC., Fitzgerald, M. P., Reinits, K. G., and Walker, D. A. (1975). *New Phytol.*, **75,** 1.
32. Chance, B. and Nishimura, M. (1967). In *Methods in enzymology* (ed. R. W. Estabrook and M. E. Pullman), Vol. 10, p. 641. Academic Press, New York and London.
33. Nishimura, M., Hara-Nishimura, I., and Robinson, S. P. (1984). *Plant Sci. Lett.*, **37,** 171.
34. Nagata, H. and Takebe, I. (1970). *Protoplasma*, **64,** 460.
35. Power, J. B. and Chapman, J. V. (1985). In *Plant cell culture: a practical approach* (ed. R. A. Dixon), p. 37. IRL Press, Oxford.
36. Hind, G. (1993). In *Photosynthesis and production in a changing environment: a field and laboratory manual* (ed. D. O. Hall, J. M. O. Scurlock, H. R. Bolhar-Nordenkampf, R. C. Leegood, and S. P. Long), p. 283. Chapman and Hall, London.

9

Ribosomes and polysomes

ULRICH A. BOMMER, NILS BURKHARDT, RALF JÜNEMANN, CHRISTIAN M. T. SPAHN, FRANCISCO J. TRIANA-ALONSO, and KNUD H. NIERHAUS

1. Introduction

Isolation of ribosomes and assay systems for testing the translational apparatus have been part of the experimental routine of many laboratories for more than two decades, and excellent collections of the methods have been published previously, including two books in this series. However, a number of methods have been gradually and continuously improved and optimized. Major developments have included the improvement of ribosome isolation procedures, the establishment of highly efficient assay systems, and the application of heteropolymeric mRNA. The protocols in this chapter are concerned with the isolation of polysomes, ribosomes, and ribosomal subunits from prokaryotic and eukaryotic sources, as well as test systems for both total protein synthesis and single ribosomal functions, taking into account the recent developments.

2. Isolation of ribosomes and polysomes

2.1 Prokaryotic ribosomes

The methods described in this section incorporate recent optimizations (1) for the isolation of polysomes, highly active tightly-coupled ribosomes, and ribosomal subunits. The tightly-coupled ribosomes represent run-off ribosomes. Ribosomes that have been dissociated and reassociated *in vitro*, differ in their features for tRNA binding (2) and are not considered here. These ribosomes are sometimes incorrectly termed 'tightly-coupled' ribosomes.

Special precautions must be taken for the isolation of ribosomes and their subunits. They are aimed at preventing nucleolytic and proteolytic degradation of ribosomal components:

- all plasticware and glassware as well as water and most buffers should be sterilized

- chemicals of the highest quality (RNase-free, if possible) should be used throughout
- all steps should be performed at 4 °C and as quickly as possible
- in some cases, the application of ribonuclease inhibitors and protease inhibitors (e.g. phenylmethylsulfonyl fluoride, PMSF) is recommended

Earlier protocols often recommend diethylpyrocarbonate (DEPC) treatment of water and buffers. In our experience, the use of sterilized high quality water (e.g. from a Milli-Q apparatus) is sufficient.

2.1.1 Isolation and characterization of prokaryotic polysomes

Since most ribosomes are engaged in protein synthesis in exponentially growing cells, such cells are a rich source of polysomes. To isolate polysomes the cells are harvested and gently lysed. The polysomes are then separated from run-off ribosomes and subunits using sucrose gradient centrifugation.

The protocols described here are modified from refs 3 and 4. *Protocol 1* describes the basic procedure to isolate polysomes. In *Protocol 2* a more extended method is described, where the nascent peptide chains are labelled and polysomes are assayed by reaction with puromycin.

Protocol 1. Isolation of polysomes from *E. coli*

For the efficient preparation of polysomes it is very important to work always at low temperature (≤ 4 °C) and as fast as possible.

Reagents

- LB medium (Luria–Bertani medium): 10 g/litre bactotryptone (Difco, USA), 5 g/litre yeast extract (Difco, USA), 10 g/litre NaCl—after sterilization, add 1/100 vol. of sterilized 20% (w/v) glucose
- Buffer 1: 20 mM Hepes–KOH (pH 7.8 at 0 °C), 6 mM $MgCl_2$, 100 mM NaCl, 16% (w/v) sucrose
- Buffer 2: 20 mM Hepes–KOH (pH 7.8 at 0 °C), 6 mM $MgCl_2$, 100 mM NaCl
- Buffer 3: 20 mM Hepes–KOH (pH 7.5 at 0 °C), 6 mM $MgCl_2$, 150 mM NH_4Cl, 2 mM spermidine, 0.05 mM spermine, 4 mM 2-mercaptoethanol
- Lysozyme

Method

1. Grow *E. coli* cells in 100 ml of rich medium (e.g. LB medium) to an absorbance of 0.5–0.6 A_{600} units/ml.[a] If no functional activity of the polysomes is needed, addition of chloramphenicol to a final concentration of 1 mM just before harvesting will increase the yield.
2. For a fast cooling of the cell suspension put 100 g crushed ice in 250 ml centrifuge tubes (e.g. GSA rotor, Sorvall); cool to −20 °C and crush ice again with a spatula. Pour the cells over this crushed ice.
3. Pellet cells at 2500 *g* for 5 min, resuspend them in 1 ml buffer 1, and transfer them to a 1.5 ml Eppendorf tube. Take care that the pH is not below 7.5 (0 °C), otherwise lysozyme will not work properly.

4. Add 15 μl lysozyme (50 mg/ml in water) and incubate on ice for 2 min.
5. Store cells at −80 °C for at least 1 h.
6. After thawing the cells in an ice–water bath centrifuge at 32 000 *g* for 30 min at 2 °C.
7. Dilute cleared lysate threefold with buffer 2 and layer 5–10 A_{260} units on to a 10–40% (w/v) sucrose gradient in buffer 3. The total yield of cleared lysate should be around 100 A_{260} units. Centrifuge in a swinging-bucket rotor, e.g. Beckman SW40, at 80 000 *g* for 7 h at 2 °C and collect in 0.8 ml fractions. A typical absorbance profile is shown in *Figure 1*.

[a]Throughout this chapter A_x means absorbance at x nm (e.g. A_{260}, absorbance at 260 nm); A_x unit means the amount of material, which gives an absorbance of 1.0 at x nm when dissolved in 1 ml solvent and when the light path is 1 cm.

Protocol 2. Isolation and functional analysis of polysomes

Reagents

- LB medium (see *Protocol 1*)
- Buffer 1: 20 mM Hepes–KOH (pH 7.5 at 0 °C), 6 mM $MgCl_2$, 150 mM NH_4Cl, 2 mM spermidine, 0.05 mM spermine, 4 mM 2-mercaptoethanol
- Buffer 2: 20 mM Hepes–KOH (pH 7.8 at 0 °C), 6 mM $MgCl_2$, 100 mM NaCl, 16% (w/v) sucrose
- Lysozyme and puromycin

Method

1. Grow cells in 30 ml of rich medium (e.g. LB medium) to an absorbance of about 0.5 A_{600} units/ml, pour them over crushed ice. Pellet and resuspend them in 1 ml of buffer 1 containing 0.4 mCi ^{3}H-labelled amino acid mixture (Amersham, TRK.550).
2. Incubate the cells for 3 min at 37 °C and cool them down quickly to 0 °C. During this incubation the nascent polypeptide chains will be preferentially labelled.
3. Pellet the cells and resuspend them in 1 ml of buffer 2. Add 15 μl lysozyme (50 mg/ml) and incubate on ice for 2 min.
4. Pellet cells for 2 min in a microcentrifuge and resuspend them in 400 μl of buffer 1. The cells should not be pelleted too long in order to resuspend them easily.
5. Divide the cell suspension into two 200 μl aliquots and store them at −80 °C.
6. After thawing in an ice–water bath centrifuge at 32 000 *g* for 10 min.
7. Add puromycin made up in buffer 1 (adjust pH to about 7.5) to one

Protocol 2. *Continued*

aliquot (final concentration 1 mM) and the same volume of buffer 1 to the other aliquot; leave on ice for 10 min.

8. Put the aliquots on to a sucrose gradient (10–40% (w/v) in buffer 1) and centrifuge in a swinging-bucket rotor, e.g. Beckman SW40, at 80 000 *g* for 7 h at 2 °C and collect fractions of 0.8 ml.
9. Precipitate 100 μl of the fractions with 2 ml of 10% (w/v) hot trichloroacetic acid (TCA) as described in *Protocol 11*. The difference in counts between the two preparations divided by the counts of the sample without puromycin, will give the fraction of polysomes in the *post*-translocational state (see *Figure 1*).

2.1.2 Isolation of tightly-coupled ribosomes and ribosomal subunits from prokaryotes

Protocols 3–5 describe reliable and well tried methods for the isolation of tightly-coupled ribosomes and ribosomal subunits from *E. coli*. For successful preparation of highly active ribosomal particles it is important to select an *E. coli* strain that is deficient in RNases, e.g. *E. coli* K12, strain A19 or D10, CAN20-19E, or MRE600, to minimize rRNA degradation (see also ref. 5). After fermentation in rich medium it is optimal to harvest the cells at an absorbance of about 0.5 A_{650} units/ml, i.e. in the early mid-log phase, since it has been observed that at later growth stages an RNase activity tends to stick to the ribosomes (6). In order to obtain a crude ribosomal fraction the cells are disrupted and the resulting homogenate is fractionated in a sequence of centrifugation steps. Depending on whether tightly-coupled ribosomes or ribosomal subunits are required, the final purification via sucrose gradient centrifugation (rate-zonal centrifugation) is performed under association conditions, i.e. 6 mM Mg^{2+}, 30 mM NH_4^+, or dissociation conditions, i.e. 1 mM Mg^{2+}, 200 mM NH_4^+, respectively. In contrast to other protocols we do not recommend washing the ribosomes with 0.5 M (or even higher concentrations of) NH_4Cl, because a partial loss of ribosomal proteins (S1, S5, S6, S8, S16, L1, L3, L6, L10, L11, L7/12, L24, L29, L30, L32/33; U. Fehner and K. H. Nierhaus, unpublished data) has been observed under these ionic conditions. In the sucrose gradient 30S and 50S subunits as well as 70S ribosomes migrate in distinct peaks well separated from slower sedimenting material, e.g. tRNAs, synthetases, and factors (*Figure 2*). If fractions are carefully pooled as indicated in *Figure 2* one rate-zonal centrifugation is enough to yield homogeneous 30S and 50S or 70S particles, respectively, with minimal cross-contamination.

From the same bacterial cells (*Protocol 3*), the post-ribosomal supernatant (S-100) can be prepared (*Protocols 4* and *6*). The S-100 is required for functional assays (Sections 3.1.1 and 3.2.2).

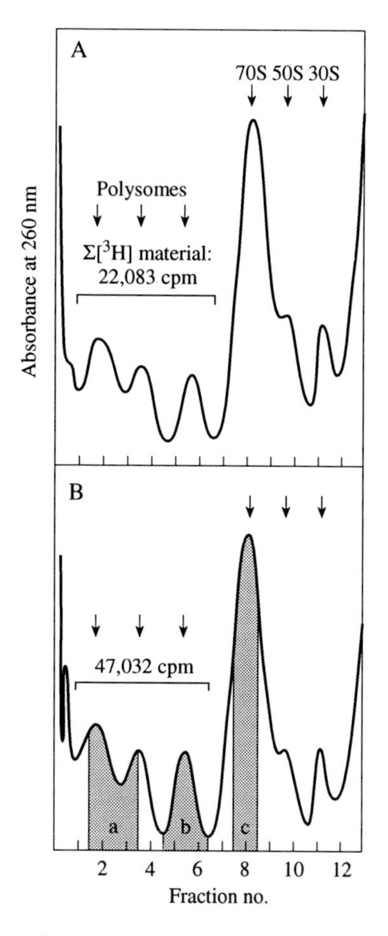

Figure 1. Size distribution of ribosomal particles in the cytosol from *E. coli* cells (A) with and (B) without puromycin treatment according to *Protocol 2* (taken from ref. 4). A_{260} absorption profiles of 10–40% sucrose gradients are shown. (B) A total of 47 032 c.p.m. [^{3}H] is incorporated in the nascent polypeptide chains. (A) 22 083 c.p.m. of this is remaining after puromycin treatment. This means that 47% of polypeptide chains are not reactive with puromycin and are therefore in the pre-translocational state, whereas 53% of polysomes are in the post-translocational state. The shaded areas indicate (a) tri- and tetrasomes, (b) disomes, and (c) run-off 70S ribosomes. Rate of sedimentation increases right to left.

Protocol 3. Large scale fermentation (100 litres) of *E. coli* cells

Reagent

- Cultivation medium: 1 kg bactotryptone (Difco, USA), 0.5 kg yeast extract (Difco, USA), 0.5 kg NaCl, and 1 litre of a 20% (w/v) glucose solution per 100 litre medium

Protocol 3. *Continued*

Method

1. Prepare 100 litres of cultivation medium in, for example, a Bioengineering-B98 fermentor and sterilize *in situ*.
2. Inoculate with a 2.5 litres overnight culture of *E. coli*, e.g. K12-D10.
3. Ferment under continuous aeration at 37 °C.
4. Harvest the cells at an absorbance of about 0.5 A_{650} units/ml by continuous flow centrifugation (e.g. using a Padberg centrifuge No. 41). 100–150 g wet cells are usually obtained from a 100 litre culture.
5. Shock-freeze the cells and store in portions of 100 g at −80 °C.

Protocol 4. Isolation of tightly-coupled ribosomes

Reagents

- Buffer 1 (association conditions): 20 mM Hepes–KOH (pH 7.6 at 0 °C), 6 mM $MgCl_2$, 30 mM NH_4Cl, 4 mM 2-mercaptoethanol
- Alumina (Alcoa A305, Serva, Heidelberg)
- RNase-free sucrose ultrapure (BRL, UK)

All steps should be performed with sterilized equipment on ice.

Method

1. Take, for example, 300 g of frozen cells and suspend them in 600 ml of buffer 1.
2. Centrifuge at 10 000 *g* for 15 min; discard the supernatant and mix the pelleted cells with 600 g of alumina.
3. Grind the cells for 30–45 min in a Retsch mill (for small amounts of cells, i.e. below 150 g, grind for only 25–30 min); a viscous cell paste is generated. Add 450 ml of buffer 1 and continue to homogenize for 10 min.
4. Centrifuge the suspension at 16 000 *g* for 10 min to remove the alumina.
5. Discard the pellet and centrifuge the supernatant at 32 000 *g* for 60 min in order to remove the cell debris, which appears as a loose, brownish pellet. Decant the supernatant carefully avoiding any contamination with the debris.
6. Centrifuge at 70 000 *g* for 17 h. Decant the supernatant which can be used as a source of enzymes and factors (see *Protocol 6* and Section 3.1.1). Remove the viscous, yellowish material which overlays the brownish ribosome pellet by rinsing with buffer 1.
7. Add 30–40 ml buffer 1, break the pellet to pieces with a spatula, and resuspend using a magnetic stirrer within 1–2 h.
8. Clarify the crude ribosome fraction by low speed centrifugation, determine the absorbance at 260 nm, shock freeze in portions of 6000–8000 A_{260} units, and store at −80 °C.

9. Subject portions of 6000–8000 A_{260} units to rate-zonal centrifugation at 21 000 r.p.m. (Beckman Ti15 rotor) for 16 h on a 6–40% (w/v) sucrose gradient made up in buffer 1.
10. Fractionate the gradient, and in order to minimize the cross-contamination with 50S subunits and loose coupled ribosomes, pool just two-thirds of the 70S peak as indicated in *Figure 2A*.
11. Pellet the tightly-coupled ribosomes by ultracentrifugation at 70 000 *g* for 24 h. Do not apply centrifugation fields higher than 100 000 *g* in order to avoid pressure induced dissociation (7). The 70S ribosomes appear as a clear, colourless pellet. Resuspend in buffer 1 and adjust the concentration to 300–700 A_{260} units/ml.
12. Clarify the suspension by low speed centrifugation, shock-freeze, and store in small portions at −80 °C.
13. 70S tightly-coupled ribosomes are routinely checked for homogeneity by analytical sucrose gradient centrifugation, for their tRNA binding capacities (Section 3.2.2), and for their activity in poly(U)-dependent poly(Phe) synthesis (Section 3.1.1). For a number of experiments it is useful to confirm the intactness of the rRNA (*Figure 3A*) and the completeness of the protein moiety (*Figure 4*).

Protocol 5. Isolation of ribosomal subunits

Reagents

- Buffer 1 (association conditions) (see *Protocol 4*)
- Buffer 2 (dissociation conditions): 20 mM Hepes–KOH (pH 7.6 at 0 °C), 1 mM $MgCl_2$, 200 mM NH_4Cl, 4 mM 2-mercaptoethanol

Method

1. Follow *Protocol 4*, steps 1–6. For isolating ribosomal subunits resuspend the pellet in buffer 2 (dissociation condition) in the same manner as described above.
2. Subject portions of 6000–8000 A_{260} units to zonal centrifugation at 24 000 r.p.m. (Beckman Ti15 rotor) for 18 h using a 6–40% (w/v) sucrose gradient in buffer 2.
3. Pool carefully, i.e. avoiding cross-contamination, the fractions containing 30S and 50S subunits, respectively (*Figure 2B*). Collect the subunits by ultracentrifugation at 140 000 *g* for 20 h. 50S subunits yield a clear, colourless pellet, whereas the 30S pellet is generally light yellowish.
4. Resuspend in buffer 1 and adjust the concentration to 300–500 A_{260} units/ml.
5. Clarify by low speed centrifugation, shock-freeze, and store at −80 °C.

Protocol 5. *Continued*

6. Ribosomal subunits should be routinely checked for their homogeneity, for their poly(U)-dependent poly(Phe) synthesis activity (Section 3.1.1), and for the intactness of their rRNA (*Figure 3B* and *C*). Sometimes it might be necessary to check the protein moiety for completeness in addition.

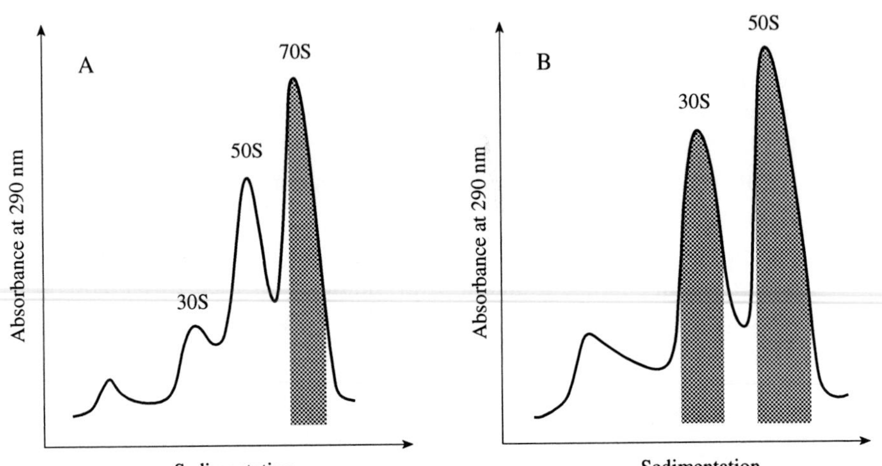

Figure 2. Preparation of (A) tightly-coupled 70S ribosomes (association ionic conditions) and (B) ribosomal subunits (dissociation ionic conditions) by rate-zonal centrifugation (*Protocol 4*). The A_{290} absorption profiles of 6000 A_{260} units of crude 70S ribosomes applied to 6–40% sucrose gradients are shown. Shaded areas of the gradients were pooled in order to obtain 70S ribosomes or 30S and 50S subunits, respectively.

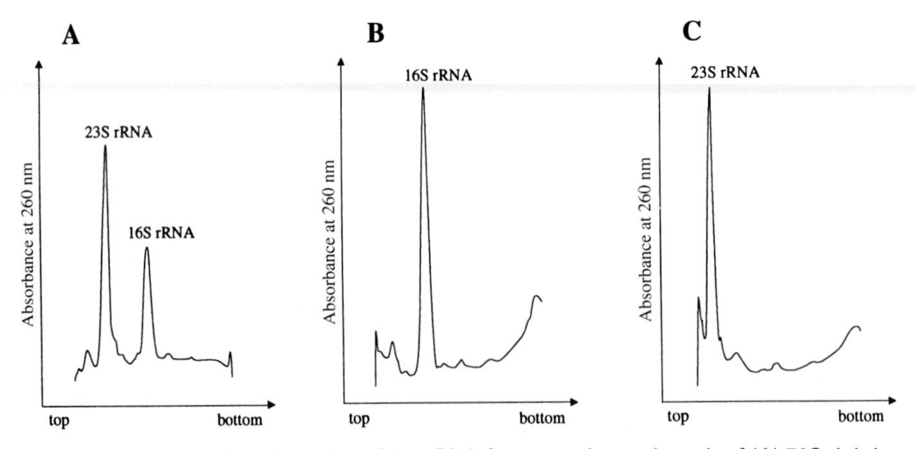

Figure 3. Analysis of the integrity of the rRNA from one A_{260} unit each of (A) 70S tightly-coupled ribosomes, (B) 30S subunits, and (C) 50S subunits. Extraction of the rRNA from the ribosomal particles and analysis on SDS–RNA gels was performed as described in ref. 5. The A_{260} absorption profiles of the gels are shown.

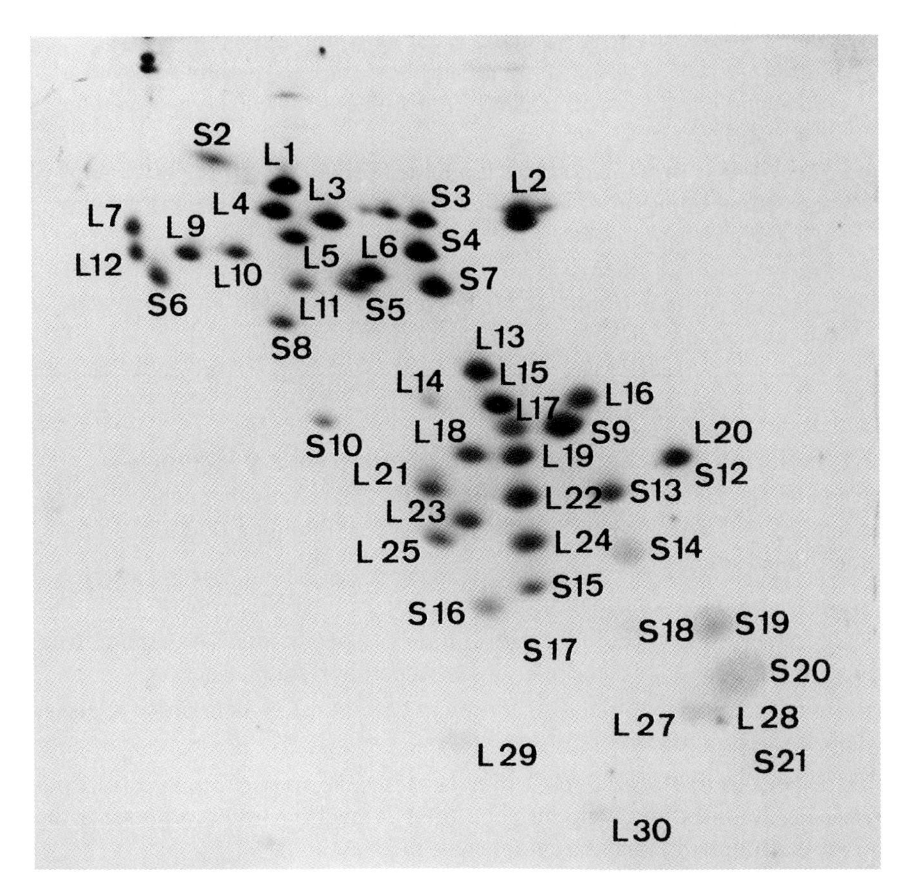

Figure 4. Two-dimensional gel electrophoresis of ribosomal proteins. Protein extraction and gel electrophoresis were performed as described in ref. 19. The protein from 3 A_{260} units of 70S tightly-coupled ribosomes was applied to the gel. L1, L2 . . . mark the proteins of the large ribosomal subunit, S1, S2 . . ., the proteins of the small ribosomal sub-

Protocol 6. Preparation of post-ribosomal supernatant (S-100)

Reagent

- Buffer 1 (association conditions) (see *Protocol 4*)

Method

1. After pelleting the ribosomes by ultracentrifugation (*Protocol 4*, step 6) decant the supernatant into sterile tubes.
2. Centrifuge again at 100 000 *g* for 4 h at 4 °C to remove remaining ribosomal particles completely.
3. Dialyse the upper 3/4 of the supernatant against a 100-fold volume of

Protocol 6. *Continued*

buffer 1 three times for 30 min using Spectrapore membranes with a molecular weight cut-off of 6000. Free amino acids will be removed by this dialysis.

4. Shock-freeze in small portions (approx. 1 ml) and store at −80 °C. Use each aliquot only once; avoid thawing and refreezing since the activity will decrease significantly.
5. Determine the optimal amount of S-100 to be used in the poly(Phe) synthesis system (*Protocol 11*) by varying the S-100 input between 5 μl and 30 μl.

2.2 Eukaryotic ribosomes

2.2.1 Isolation and characterization of eukaryotic polysomes

Polysomes are the subcellular fraction of ribosomes actively engaged in protein synthesis. There may be various reasons for obtaining polysomal preparations, for example:

- as source of active ribosomes for fractionated translation systems
- to investigate the subcellular distribution of specific mRNAs among free, membrane-bound, or cytoskeleton-associated polysomal fractions
- to study the subpolysomal distribution of certain mRNAs in order to assess their translational efficiency

With respect to the latter goal, it may be desirable to perform size fractionation of polysomes involving also the mRNP fraction which represents the pool of (temporarily) inactive cytoplasmic mRNAs.

Special care has to be taken in order to prevent RNA degradation (see Section 2.1). When preparing ribosomes or polysomes from eukaryotic tissues, it is particularly important to perform the homogenization procedure as gently as possible (see *Protocol 9*) in order to avoid disruption of lysosomes, and the application of RNase inhibitors is highly recommended.

Here we provide protocols for the preparation and size fractionation of total polysomes and for the separation of the different polysomal compartments. *Protocol 7* is adapted from ref. 8.

Protocol 7. Preparation and size fractionation of total polysomes from eukaryotic cells[a]

Reagents

- Buffer 1: 20 mM Tris–HCl pH 7.5, 5 mM $MgCl_2$, 100 mM KCl, 5 mM 2-mercaptoethanol, 250 mM sucrose
- Buffer 2: 50 mM Tris–HCl pH 7.5, 10 mM $MgCl_2$, 500 mM KCl, 5 mM 2-mercaptoethanol, 250 mM sucrose, 25% glycerol
- Buffer 3: 20 mM Tris–HCl pH 7.5, 3 mM $MgCl_2$, 100 mM KCl
- Triton X100

Method

1. Suspend tissue or cells in 3 vol. of ice-cold buffer 1 (homogenization buffer), homogenize in a chilled Teflon–glass homogenizer (five to six strokes).
2. Remove debris, nuclei, and mitochondria by centrifugation at 12 000 *g* for 2 × 10 min.
3. Add Triton X-100 (2% (w/w) final concentration[b]) to the post-mitochondrial supernatant and mix gently. Layer the mixture over an equal volume of 1.0 M sucrose made up in buffer 1, and centrifuge at 260 000 *g* for 2 h at 4 °C.
4. Discard the supernatant, gently rinse the centrifuge tube with buffer 1, and drain the polysome pellet for a few minutes over tissue paper.
5. Factor-free polysomes for translation assays. Resuspend the polysome pellet in buffer 2 by gentle homogenization. Centrifuge the polysomes through a cushion of 1 M sucrose in buffer 2, as in step 3. The polysomes can be stored as pellets at −70 °C.
6. Size fractionation of polysomes. Suspend the crude polysomes (step 4) in cold buffer 1 (0.5 ml/g of tissue) by gentle homogenization. Alternatively, use directly the detergent-treated post-mitochondrial supernatant.[c]
7. Layer the suspension on to sucrose gradients made up in buffer 3. Sucrose gradients may be purpose designed. Crude post-mitochondrial supernatants should be analysed on 20–50% gradients. Otherwise, 10–40% or 15–45% gradients may also be used. Make the volume of the sample as small as possible. A typical load is 3–10 A_{260} units of polysomes per 10 ml of gradient. Centrifuge at about 150 000 *g* for 1 h at 4 °C.
8. Fractionate the gradients while monitoring the A_{260} in a recording spectrophotometer at a suitable flow rate. The actual flow rate will depend on rotor size, type of flow cell, etc. Collect 20–30 fractions per gradient for subsequent analysis.

[a] The method described is suitable for tissue such as liver. For application to other tissues/cells, the conditions for homogenization including the buffer may be varied (see also *Protocol 9*). For many types of cultured mammalian cells, an alternative procedure to step 1 may be used, e.g. hypotonic or detergent lysis.

[b] This dissolves the ER membranes, thereby releasing the polysomes. The Triton X-100 concentration should be chosen depending on the tissue. In the case of liver, some authors use 1% sodium deoxycholate in addition to 2% Triton. On the other hand, in the case of cultured cells, which contain considerably less ER membranes, 1% Triton will be sufficient. No detergent has to be added at this step if the cell lysis was performed in the presence of detergent.

[c] If the mRNP fraction is required for the analysis, apply the post-mitochondrial supernatant directly. However, for better purity of the polysomes, especially from tissues like liver, steps 3 and 4 should be included.

Up to the early 1980s, polysomes were thought to occur mainly as two distinct populations, i.e. in a free and membrane-bound form, depending on the destination of the protein being synthesized (reviewed in ref. 9). Accordingly, protocols have been described to separate these two polysomal subpopulations (e.g. Table 9 in ref. 8). Meanwhile, evidence has accumulated that polysomes can actually be found in three subcellular compartments: free, membrane-bound, and associated with the cytoskeleton, and methods have been developed to obtain polysomes from all these fractions (compiled in ref. 10). The following protocol, kindly provided by Dr J. E. Hesketh (Rowett Research Institute, Bucksburn, Aberdeen AB2 9SB, UK), describes the isolation of these three polysomal fractions including mRNPs from 3T3 fibroblasts.

Protocol 8. Preparation of free, cytoskeleton-bound, and membrane-bound polysomes from 3T3 fibroblast cells

Reagents

- Emetine–phosphate-buffered saline (PBS): emetine inhibits translational elongation, thus preventing run-off of ribosomes. Add 100 ng emetine/ml of PBS. Dissolve emetine in ethanol at 5 mg/ml, dilute 1:50 with PBS (phosphate-buffered saline: 0.24 g/litre KH_2PO_4, 1.44 g/litre Na_2HPO_4, 0.2 g/litre KCl, 8 g/litre NaCl pH 7.4) to give 0.1 mg/ml then add 1 μl/ml medium.
- Buffer 1: 10 mM Tris–HCl pH 7.6, 5 mM $MgCl_2$, 0.5 mM $CaCl_2$, 25 mM KCl, 250 mM sucrose
- Buffer 2: 10 mM Tris–HCl pH 7.6, 5 mM $MgCl_2$, 0.5 mM $CaCl_2$, 130 mM KCl, 250 mM sucrose
- Stock solutions for addition to buffers RNase inhibitor (Promega): 50 U/μl (add 4 μl/1.5 ml); 10 mg/ml heparin (add 10 μl/ml); 10% (w/v) Nonidet P-40; 10% (w/v) Na deoxycholate

Method

1. If compatible with the experiment, pre-treat the cells with emetine 30 min before harvesting.
2. Wash cells three times with emetine–PBS, scrape into ice-cold buffer 1 and centrifuge at 200 *g* for 5 min at 4 °C.
3. Resuspend pellet in 1.5 ml buffer 1 containing 4 μl RNase inhibitor, 0.1 mg/ml heparin, 0.05% Nonidet P-40. Leave on ice for 10 min. Centrifuge as in step 2. The supernatant is fraction A (free polysomes).
4. Resuspend the pellet in 5 ml buffer 1 containing 0.1 mg/ml heparin, centrifuge as above and discard the supernatant.
5. Resuspend the pellet in 1.5 ml buffer 2 containing additions as in step 3. Leave on ice for 10 min. Centrifuge at 800 *g* for 10 min at 4 °C. The supernatant is fraction B (cytoskeleton-bound polysomes).
6. Resuspend the pellet in 1.5 ml buffer 2 containing 4 μl RNase inhibitor, 0.1 mg/ml heparin, 0.5% (w/v) Nonidet P-40, 0.5% (w/v) sodium deoxy-

cholate. Leave on ice for 10 min. Centrifuge at 1800 *g* for 10 min at 4 °C. The supernatant is fraction C (membrane-bound polysomes).

7. To collect the polysomes from fractions A, B, and C, layer the respective fraction (1.5 ml) on top of 15 ml 40% (w/v) sucrose made up in buffer 2. Centrifuge at 25 000 *g* for 17 h at 4 °C. The polysomes are in the pellet and the mRNPs are at the top of the cushion.

Recently, even more gentle methods involving *in situ* extraction have been developed for the isolation of cytoskeleton-associated polysomes (see e.g. ref. 11).

Polysomes may be analysed in suitable cell-free translation systems (see Section 3.1.3) or subjected to mRNA distribution analysis after size fractionation (*Protocol 7*). For extraction of mRNA from polysome fractions, extreme care should be taken to prevent RNA degradation. Usually, proteinase K plus SDS treatment is applied prior to phenol/chloroform extraction. (A detailed protocol is given in Table 11 of ref. 8.) For quantification of mRNA species in the polysome fractions, essentially three different methods have been applied: Northern blot analysis (12) nuclease S1 protection assay (13), and reverse transcription in combination with PCR methods (14). There is no space to provide protocols for these techniques here. However, examples of their application can be found in the cited references. The choice of the procedure is very much dependent on the abundance of the mRNA species in question.

2.2.2 Isolation of ribosomes and ribosomal subunits from eukaryotic cells

In this section, a standard protocol will be described for preparation of total ribosomes and ribosomal subunits from mammalian liver, which is still the most common source for animal 80S ribosomes and their subunits. For further details and preparation of mitochondrial, chloroplast, and plant cytoplasmic ribosomes, the reader is referred to Chapter 1 of ref. 15, as well as to earlier reviews and method collections (9, 16). The preparation of yeast ribosomes will be described at the end of this section (*Protocol 10*), as this is rarely covered in such reviews, and as yeast protein synthesis is attracting increasing attention.

The following procedure has been successfully applied in our previous work. Although such protocols differ slightly in some details (15, 16), they all comprise the following main steps:

- preparation of the post-mitochondrial supernatant
- release of the ribosomes from the ER membranes and collection of the crude ribosomes by centrifugation
- ribosome dissociation by puromycin treatment in the presence of high salt
- separation of the ribosomal subunits by sucrose gradient centrifugation

Protocol 9. Preparation of rat liver ribosomal subunits

Reagents

- Buffer 1 (homogenization buffer): 50 mM Tris–HCl pH 7.5, 5 mM $MgCl_2$, 100 mM KCl, 0.2 mM EDTA, 10 mM 2-mercaptoethanol, 250 mM sucrose
- Buffer 2: 50 mM Tris–HCl pH 7.5, 5 mM $MgCl_2$, 100 mM KCl, 10 mM 2-mercaptoethanol
- Buffer 3: 5 mM Tris–HCl pH 7.5, 1.5 mM $MgCl_2$, 50 mM KCl, 10 mM 2-mercaptoethanol
- Buffer 4: 20 mM Tris–HCl pH 7.5, 2 mM $MgCl_2$, 100 mM KCl, 10 mM 2-mercaptoethanol
- Puromycin stock: dissolve puromycin at about 2.5 mg/ml in water, add buffer 3 containing 0.5 M KCl to yield a concentration of 1 mg/ml, adjust pH to 7.5 (20 °C) using 0.5 M KOH
- Buffer 5: 20 mM Tris–HCl pH 7.5, 2 mM $MgCl_2$, 50 mM KCl, 10 mM 2-mercaptoethanol, 0.2 M sucrose
- Buffer 6: 5 mM Tris–HCl pH 7.6 at 20 °C, 5 mM $MgCl_2$, 0.5 M KCl, 10 mM 2-mercaptoethanol
- Buffer 7: 50 mM Tris–HCl pH 7.5, 30 mM $MgCl_2$, 10 mM 2-mercaptoethanol

Method

1. Sacrifice overnight starved rats (200–250 g body weight) by decapitation, remove the livers, and place them in an excess of ice-cold buffer 1 (homogenization buffer).
2. Weigh the livers and homogenize (2 ml homogenization buffer 1 per gram of tissue) using a loosely-fitting Teflon–glass homogenizer. Do not exceed five or six strokes using a motor-driven Teflon pestle.[a]
3. Remove nuclei and mitochondria by centrifugation at 12 000 *g* for 20 min. Carefully collect the upper 2/3 of the supernatant and filter through four layers of cheesecloth.
4. For release of the ribosomes from the microsomal membranes, add 1/10 vol. of freshly prepared 10% sodium deoxycholate and mix gently.
5. Layer the post-mitochondrial supernatant over a sucrose pad (about 0.6 vol.) consisting of 1 M sucrose in buffer 2 and centrifuge overnight at 130 000 *g* (e.g. in a Beckman Ti60 rotor).
6. Carefully remove the supernatant, rinse the clear ribosomal pellet with buffer 3, and resuspend the pellet gently in the same buffer using a glass rod followed by a small Teflon–glass homogenizer rotating at low speed. The ribosome concentration should be about 20 mg/ml as judged by the A_{260} reading.[b]
7. Adjust the KCl concentration to 0.5 M. Add a freshly prepared puromycin solution to give a ratio of 1 mg puromycin per 100 mg of ribosomes. Adjust to a final concentration of 10 mg ribosomes/ml maintaining 0.5 M KCl. Incubate 30 min on ice and 15 min at 37 °C. Clarify the solution by centrifugation for 5 min at 10 000 *g*.
8. For collection of 80S ribosomes. Centrifuge ribosomes at 100 000 *g*

for 4 h through a 0.5 M sucrose cushion in buffer 4. Resuspend pellet in buffer 5 to about 15 mg/ml. Store in liquid nitrogen.

9. For separation of ribosomal subunits. Layer dissociated ribosomes (from step 7) on 15–30% linear sucrose gradients in buffer 6. Centrifuge in a swinging-bucket rotor (e.g. Beckman SW27 rotor) at 30 000 *g* for 14 h at 20 °C. Fractionate the gradients while monitoring the absorbance at 260 nm. Pool the main fractions of the 40S and 60S peaks. Avoid the overlapping area.
10. Dilute the solutions by adding about 0.6 vol. of buffer 7. This will lower the sucrose density and increase the Mg^{2+} concentration and thereby improve the recovery especially of the 40S ribosomal subunits. Centrifuge at 140 000 *g* for 18 h (e.g. in a Beckman Ti60 rotor). Resuspend the subunits in buffer 5 and store them in aliquots in liquid nitrogen[c].
11. To remove contamination of the 60S ribosomal subunits by dimerized 40S subunits,[d] dialyse the pooled 60S subunits against buffer 5 without sucrose, then layer them on 15–30% (w/v) sucrose gradients made up in the same buffer. Centrifuge at 240 000 *g* for 4.5 h at 20 °C. Fractionate gradients and collect 60S subunits as in steps 9 and 10.

[a] For even more gentle homogenization, the livers may be pressed through a pre-chilled tissue press in order to remove the connective tissue. The liver tissue can be homogenized at a lower speed of the Teflon pestle.
[b] 13 A_{260} units/ml corresponds to a ribosome concentration of 1 mg/ml.
[c] Concentration of 40S subunits by ultrafiltration in Centrikon-100 tubes (Amicon) is more gentle and yields better subunits (T. Pestova and C. Hellen personal comm.).
[d] The contaminating 40S subunits will associate with 60S subunits which allows separation of excess 60S subunits from the 80S ribosomes which are formed.

The procedure described above can also be used for ribosome preparation from cultured cells. Typically, cells are lysed in lysis buffer consisting of 20 mM Tris–HCl pH 7.5, 5 mM $MgCl_2$, 100 mM KCl, 0.5% Nonidet-P40, 10 mM 2-mercaptoethanol, and protease inhibitors. A compilation of dissociation conditions for ribosomes from various eukaryotic sources is given in ref. 9. As the ribosome content of cultured cells is considerably lower compared to the liver, it is often difficult to obtain ribosomes from such sources. A procedure for a high recovery of ribosomes from 3T3 cells has been described more recently (17). However, this procedure is mainly designed for analysis of the phosphorylation state of ribosomal protein S6.

For investigation of initiation reactions, it may be desirable to prepare native 40S ribosomal subunits ($40S_N$) which represent intermediate initiation complexes carrying certain factors and Met-$tRNA_f$. A procedure for the isolation of $40S_N$ from rabbit reticulocytes and rat liver can be found in ref. 18. Yeast ribosomes prepared by the following protocol have been successfully used for the investigation of some peculiarities of yeast translation.

Protocol 10. Preparation of yeast ribosomes

Reagents

- YPD medium: 20 g/litre glucose, 20 g/litre peptone, 10 g/litre yeast extract
- Buffer 1: 60 mM Tris–acetate (pH 7.0 at 4 °C), 5 mM Mg(acetate)$_2$, 50 mM NH_4Cl, 2 mM spermidine, 0.1 mM EDTA, 10 mM 2-mercaptoethanol, 10% (v/v) glycerol, 0.1 mM PMSF
- Buffer 2: 50 mM Tris–acetate (pH 7.0 at 4 °C), 5 mM Mg (acetate)$_2$, 500 mM NH_4Cl, 0.1 mM EDTA, 10 mM 2-mercaptoethanol, 10% (v/v) glycerol, 0.1 mM PMSF
- Buffer 3: 50 mM Tris–acetate (pH 7.0 at 4 °C), 5 mM Mg(acetate)$_2$, 50 mM NH_4Cl, 2 mM spermidine, 5 mM 2-mercaptoethanol, 0.1 mM DTE, 25% (v/v) glycerol, 0.1 mM PMSF
- Puromycin
- GTP

Method

1. Grow wild-type *Saccharomyces cerevisiae* (A_1 protease$^-$ strain EJ926) in a fermentor (30–100 litres) at 30 °C in YPD medium. Harvest the culture at an absorbance of 0.4–0.5 A_{650} units/ml, i.e. in the early log phase. Collect the cells by continuous flow centrifugation, wash them once with 0.9% (w/v) KCl and store at −80 °C.
2. For preparing cell-free extracts resuspend the cells in 1.5 vol. of buffer 1 and grind them with 5 vol. of acid-washed glass beads in a Waring blender (10 × 30 sec homogenization with 30 sec pause under continuous cooling).
3. Centrifuge the suspension at 20 000 *g* for 30 min in order to remove the glass beads and the cell debris. The supernatant (S-20) is used for the preparation of 80S ribosomes and the S-100 fraction.
4. In order to obtain crude 80S ribosomes adjust the S-20 to 0.4 M KCl and centrifuge at 65 000 *g* for 5 h.
5. Resuspend the ribosome pellet in buffer 2 and store in small portions at −80 °C. If the ribosomes have to be stripped from endogenous peptidyl tRNA, mRNA, and soluble factors, which may interfere with the functional analyses, include the puromycin treatment (15) described in the next steps.
6. Preserving the composition of buffer 2, add to the crude 80S fraction obtained in step 5 puromycin and GTP to a final concentration of 1 mM each. The final concentration of ribosomes should be 100–500 A_{260} units/ml.
7. Incubate the mix at 30 °C for 30 min.
8. Pellet the ribosomes through a 25% (v/v) glycerol cushion (about 1/10 of the tube volume) in buffer 2 at 65 000 *g* for 12 h using a swinging-bucket rotor.

9. Resuspend the ribosomes in buffer 2 and pellet again as in step 8.
10. Resuspend finally in buffer 3, adjust the ribosome concentration to 200–500 A_{260} units/ml, and store in small portions at −80 °C.

3. Assay systems

In the first part of this section, two protein synthesis systems for testing the overall activities of ribosomes will be described. The poly(U)-dependent poly(Phe) synthesis assay, as here described for *Escherichia coli* ribosomes (*Protocol 11*), is the standard system to test the activity of both prokaryotic and eukaryotic ribosomes. However, as the highly artificial poly(U) template is used as mRNA, this system is restricted to assaying only the elongation activity of ribosomes. If testing of both initiation and elongation capacity is required, other protein synthesis systems utilizing natural mRNA have to be chosen. Phage MS2 RNA is often used as natural template for bacterial ribosomes (19). As an example of an eukaryotic protein synthesis system capable of assaying ribosomal activities with natural mRNAs, a recently developed highly active fractionated translation system based on the reticulocyte lysate will be described. The second part contains assay systems for single reactions of the whole translation process. Protocols for the test of three initiation reactions with eukaryotic ribosomes will be provided, as well as an assay which allows the step by step analysis of the elongation cycle. The latter one is described for the *Escherichia coli* system (*Protocol 16*).

3.1 Protein synthesis systems

3.1.1 Poly(U)-dependent poly(Phe) synthesis

The activity in the poly(U)-dependent poly(Phe) synthesis system is an important criterion for estimating the quality of either a 70S or a 50S/30S bacterial ribosome preparation. *Protocol 11* describes an assay system of simple composition which is easy to handle and can be performed with large numbers of samples in parallel. Since the ionic conditions (6 mM Mg^{2+}, 150 mM NH_4^+, 2 mM spermidine, 0.05 mM spermine) are the same as in the tRNA binding system (Section 3.2.2), both systems are compatible and may be combined easily, e.g. to perform *N*-acetyl-Phe-$tRNA^{Phe}$ primed poly(Phe) synthesis.

In the system described here, the post-ribosomal supernatant (S-100) is used as the source of enzymes, i.e. elongation factors EF-Tu, EF-G, and EF-Ts, phenylalanyl-tRNA-synthetase, and nucleoside diphosphate kinase. The S-100 is prepared according to *Protocols 4* and *6*.

Protocol 11. Poly(U)-dependent poly(Phe) synthesis system

Reagents

- Reaction mixture (final concentrations are given): 20 mM Hepes–KOH (pH 7.6 at 0 °C), 6 mM Mg(acetate)$_2$, 150 mM NH_4Cl, 4 mM 2-mercaptoethanol, 1 mg/ml poly(U), 0.1 mM [^{14}C]Phe (specific activity 10 d.p.m./pmol), 1.5 mM ATP, 50 μM GTP, 5 mM phosphoenolpyruvate, 30 μg/ml pyruvate kinase, 400 μg/ml tRNA^{bulk} (*E. coli* MRE600), 50 μM spermine, 2 mM spermidine
- [^{14}C]Phe poly(U)mix. Add to 1.2 ml of a 5 mM phenylalanine solution 0.6 ml [^{14}C]Phe (50 μCi/ml, Amersham) and 0.6 ml of a poly(U) solution (100 mg/ml, Boehringer). Add 0.6 ml distilled water and store at −20 °C in 20 × 150 μl portions.
- Buffer 1 (association condition): 20 mM Hepes–KOH (pH 7.6 at 0 °C), 6 mM $MgCl_2$, 30 mM NH_4Cl, 4 mM 2-mercaptoethanol
- Energy mix. Take 0.8 ml of 1 M Hepes–KOH (pH 7.6 at 0 °C), 140 μl of 1 M Mg(acetate)$_2$, 3.3 ml of 4 M NH_4Cl, and 11.2 μl of 2-mercaptoethanol (14.2 M), 50 μl of 100 mM spermine (Fluka), and 400 μl of 500 mM spermidine (Fluka). Add 1.5 ml of 100 mM ATP-Na_2 (Boehringer, 605.2 mg in 10 ml H_2O), 833 μl of 6 mM GTP-Li_2 (Boehringer, 32.1 mg in 10 ml H_2O), and 1.25 ml of 400 mM phospho*enol*pyruvate (Boehringer, 828.4 mg in 10 ml H_2O). Mix well and adjust pH to 7.6 (0 °C) with 1 M KOH (about 1 ml will be needed). Add distilled water to 10 ml and store at −20 °C in 33 × 300 μl portions.
- 1% (w/v) BSA
- 5% (w/v) TCA
- Diethyl ether/ethanol (1:1)

Method

1. Prepare an assay pre-mix for a suitable number of samples. The values for 30 single reactions are given (the corresponding values for one reaction are presented in parentheses). Take 600 μl (20 μl) distilled water[a] and mix with 300 μl (10 μl) of the energy mix, 90 μl (3 μl) pyruvate kinase (1 mg/ml in water), and 60 μl (2 μl) tRNA^{bulk} from *E. coli* MRE600 (Boehringer, 20 mg/ml). Add the optimized amount of S-100 and buffer 1 to give a total volume of 1950 μl (65 μl); S-100 and buffer 1 together should amount to 900 μl (30 μl). Finally, 150 μl (5 μl) of the [^{14}C]Phe/poly(U) mix is added to the mixture.
2. Distribute 70 μl of the assay pre-mix to glass vials (approx. 5 ml volume) for each assay.
3. Add the ribosomal subunit to be tested in limiting amounts. Therefore, establish the following stoichiometric ratios per assay:[b]
   - 50S to be tested: 1 A_{260} unit of 50S and 1 A_{260} unit of 30S
   - 30S to be tested. 2 A_{260} units of 50S and 0.5 A_{260} units of 30S

   Alternatively, 1.5 A_{260} units of 70S ribosomes may be added. Include a minus ribosomes control as well as a control with either 30S or 50S alone to estimate the cross-contamination. Complete with buffer 1 to 100 μl per assay (ribosomes and buffer 1 together should amount to 30 μl).
4. Mix and incubate at 30 °C for 60 min.
5. Add one drop of a 1% (w/v) bovine serum albumin solution and 2 ml of 5% (w/v) trichloroacetic acid (TCA). Mix the sample.

6. Incubate for 15 min at 90 °C and cool down on ice afterwards[c].
7. Pass sample through a glass fibre filter (Schleicher and Schüll, No. 6, 2.3 cm diameter).
8. Wash the filter three times with 2 ml of 5% (w/v) TCA and once with 2 ml diethylether/ethanol (1:1)[d].
9. Add an appropriate scintillation cocktail and determine the radioactivity retained on the filter. Calculate the phenylalanine incorporation per ribosome[b] (10 d.p.m. [^{14}C]Phe correspond to 1 pmol).

[a] Other components dissolved in water, e.g. antibiotics, may be added at this stage.
[b] 1 A_{260} unit 30S subunits corresponds to 72 pmol, 1 A_{260} unit 50S subunits to 36 pmol, and 1 A_{260} unit 70S ribosomes to 24 pmol.
[c] This step deacylates the [^{14}C]Phe-tRNA which coprecipitates with the protein.
[d] This removes the strongly quenching TCA.

3.1.2 Eukaryotic cell-free translation systems

Several cell-free protein synthesis systems have been established mainly aimed at assaying the efficiency of natural as well as artificial mRNA. By far the most popular systems are the wheat germ extract (20) and the rabbit reticulocyte lysate (21). The main advantage of these cellular extracts is that they maintain high translational efficiency which closely resembles the *in vivo* situation. However, they are not suitable for studying certain aspects of translational regulation, e.g. during virus infection of animal cells. For such purposes, a series of translation systems based on 10 000–30 000 *g* supernatants of extracts from cultured cells, such as mouse ascites tumour cells, HeLa cells, Chinese hamster ovary cells, as well as from yeast have been developed. An excellent and comprehensive description of all these systems including general guidelines for cell-free translation work is given in a previous book of this series (22). In addition, refs 21, 23, and 24 provide detailed protocols which we commend. However, as these systems are mainly designed to assess translational efficiency in general, rather than to test the ribosomal activities *per se*, we will focus on the assays described below.

3.1.3 Fractionated cell-free translation systems

Highly fractionated protein synthesis systems mainly based on the rabbit reticulocyte lysate have been developed in the late 1970s (24,25). They are composed of ribosomal subunits, mRNA(s), suitable sources of aminoacyl-tRNA synthetases and tRNAs, and purified initiation and elongation factors. These assay systems were primarily designed in order to explore the activity of single eukaryotic initiation factors. They are also suitable for assaying other components of the translational machinery. However, as the preparation of the individual initiation factors is very laborious, it is evident that such translation assays can only be handled in specialized laboratories. Another drawback of highly fractionated systems is that they retain considerably less efficiency compared to the parental lysate. Thus, when employing a fractionated protein synthesis system, it is always advisable to stay as close as possible to the 'native' conditions and to avoid any unnecessary fractionation step.

For these reasons, partially fractionated protein synthesis systems have been developed which usually consist of the following components (26):

(a) Salt-washed ribosomes: depending on the requirements of the researcher, the total ribosome fraction, polysomes, or purified ribosomal subunits plus exogenous mRNA will be employed. Stripping the attached translation factors off the ribosome fraction is usually achieved by washing the ribosomes with a buffer containing 0.5 M KCl.

(b) The resulting ribosomal salt wash fraction can be either further purified or directly utilized as a source of initiation factors.

(c) The post-ribosomal supernatant (S-100), or rat liver 'pH 5 enzymes' are used to supplement the system with elongation factors, aminoacyl-tRNA synthetases, and tRNA.

(d) All necessary low molecular weight components including buffer, ATP, GTP, radioactive and non-radioactive amino acids, and an energy-regenerating system are usually added as a mix.

The polysomal system has been successfully used in order to measure differences in the total translation factor activity in the cytosolic fraction of different tissue samples, e.g. during liver regeneration (27). This system has often been criticized because it measures mainly elongation by pre-existing polysomes. But the fact that it responds to the addition of initiation factors (27) indicates that it is capable of reinitiation. However, if it is preferable to use an initiation-dependent translation assay, a more fractionated system utilizing ribosomal subunits plus exogenous mRNA (26) is the right choice. More recently, this system has been considerably improved by employing a rapid centrifugation method using the Beckman table-top ultracentrifuge TL100 (28). This allows a drastic reduction of the centrifugation time needed to collect the ribosomes, which is the most critical step during fractionation of cell lysates. Thereby, it is possible to retain as much as 70% of the efficiency of the original rabbit reticulocyte lysate. The following protocol is adapted from ref. 28.

Protocol 12. A highly efficient fractionated translation system from reticulocyte lysate

Reagents

- Buffer 1: 20 mM Hepes–KOH pH 7.2, 6.1 mM $Mg(acetate)_2$, 500 mM KCl, 10 mM NaCl, 0.1 mM EDTA, 0.1 mM DTT
- Buffer 2: 20 mM Hepes–KOH pH 7.2, 1.1 mM $Mg(acetate)_2$, 25 mM KCl, 10 mM NaCl, 0.1 mM EDTA, 0.5 mM DTT, 10% glycerol

A. *Preparation of the subcellular fractions*

1. Crude ribosomes and post-ribosomal supernatant. Supplement the reticulocyte lysate[a] with 20 μM haemin and 2 mM GTP/$MgCl_2$[b] and centrifuge in a Beckman TL100.2 rotor at 250 000 *g* for 20 min at 4 °C. Remove the upper two-thirds of the supernatant. This is the S-100 fraction. Shock-freeze in aliquots and store at −70 °C. Discard the lower third of the supernatant. Rinse the crude ribosome pellet and proceed as in step 2.
2. Ribosomal salt wash and washed ribosomes. Resuspend the ribosome pellet in buffer 1 (one-tenth the volume of the original lysate) and layer the suspension on to a 100 μl cushion of buffer 1 containing 10% (v/v) glycerol. After centrifugation as in step 1, remove the upper two-thirds of the supernatant, dialyse for 4 h against buffer 2, and freeze in aliquots at −70 °C. This is the ribosomal salt wash (RSW). Discard the lower third of the supernatant inclusive of the glycerol cushion. Resuspend the ribosome pellet (washed ribosomes) in buffer 2 (one-tenth the volume of the original lysate) and freeze at −70 °C.
3. Fractionation of the ribosomal salt wash (optional[c]). Add to the RSW solid ammonium sulfate up to 45% saturation. Stir for 30 min at 4 °C. Collect the precipitate by centrifugation (20 min, 10 000 *g*). This is fraction A. To the supernatant add further ammonium sulfate up to 70% saturation. Stir and collect the precipitate as above (fraction B). Dissolve both precipitates in buffer 2 (one-tenth the volume of the original lysate), dialyse for 4 h, and store in aliquots at −70 °C.

B. *Reaction mixture*

1. Master mix. Make up a master mix (2.5-fold concentrated) to give a final concentration of the following components in the assay: 120 mM KCl, 2 mM $Mg(acetate)_2$, 0.5 mM spermidine, 0.5 mM ATP, and 0.1 mM GTP supplementary to the endogenous pools, 10 mM creatine phosphate, 50 μg/ml creatine kinase, 0.2 mM glucose-6-phosphate, 0.5 mM DTT, amino acids (except methionine) 100 μM each, methionine to give a final concentration of 25 μM (inclusive of the radioactive methionine).
2. Assay mixture. Make up 20 μl reaction mixtures of the following typical composition: 8 μl S-100 fraction, 2 μl ribosomes, 2 μl RSW, 8 μl master mix containing [^{35}S]methionine to give a final radioactive concentration of 50–80 μCi/ml. For controls, replace the respective fraction by buffer 2.
3. Alternative variants:
   (a) For translation of exogenous mRNAs, use fractions obtained from nuclease-treated lysates.[a]

Protocol 12. *Continued*

(b) For testing different ribosome preparations (e.g. 40S subunits phosphorylated on protein S6), start from nuclease-treated lysate[a] and replace the endogenous ribosome fraction by the ribosomal subunits (see *Protocol 9*) or washed polysomes (*Protocol 7*) to be tested. In the case of ribosomal subunits, add either globin mRNA or any other mRNA.

(c) For testing initiation factor preparations, use fractions A or B of the RSW (part A, step 3) and supplement with suitable initiation factors. For details see ref. 28.

C. *Sample processing*

1. Incubate reaction mixtures for 30–60 min[d] at 30 °C. Remove 5 μl aliquots and determine [^{35}S]methionine incorporation by trichloroacetic acid precipitation and scintillation counting (for details see refs 21 and 22). Alternatively, analyse translation products of 5 μl samples by SDS gel electrophoresis followed by fluorography.

[a]For this translation system, either untreated or micrococcal nuclease-treated rabbit reticulocyte lysates (21) may be used. Here, the protocol is given for the untreated lysate. In this case, the endogenous globin mRNA is utilized. For translation of exogenous mRNA, the nuclease-treated lysate should be used instead.
[b]This is necessary in order to prevent pressure-activation of the haemin-controlled repressor (HCR) during centrifugation. Active HCR is a protein kinase specifically phosphorylating the α-subunit of eIF-2 thereby rendering the lysate inactive. For haemin treatment see ref. 21.
[c]This step is only necessary if you wish to test certain initiation factors. (See ref. 28 for details.) Some RSW preparations contain inhibiting activities. In this case it is advisable to ammonium sulfate precipitate the RSW (70% saturation) and to carry out a DEAE–cellulose chromatography step (26).
[d]Determine optimal incubation time experimentally. Usually, amino acid incorporation is linear up to 1 h.

3.2 Testing of partial reactions of the ribosome cycle

3.2.1 Initiation reactions of eukaryotic ribosomes

The process of polypeptide chain initiation comprises all steps leading to the assembly of the initiation complex consisting of initiator tRNA, mRNA, and both ribosomal subunits. In both eukaryotic and prokaryotic cells the complex is formed on the small ribosomal subunit and is finally joined by the large subunit. However, there are two main differences in the mechanisms of initiation between these cell types:

(a) There are many more initiation factors involved in eukaryotic polypeptide chain initiation.

(b) Whereas in prokaryotes mRNA and initiator tRNA are bound to the 30S ribosomal subunit in 'concerted action' involving all three initiation factors, in eukaryotes the whole process occurs in the following distinct sequential manner:

i. eIF-2 and GTP-dependent binding of Met-$tRNA_f$ to the 40S ribosomal subunit

ii. mRNA binding to the [40S·Met-$tRNA_f$·factor] complex

iii. GTP hydrolysis, factor release, and subunit joining (for details see refs. 29, 30).

A wide variety of assay systems have been developed in order to monitor the different initiation reactions. Protocols for most of these assays as well as for the preparation of individual initiation factors are given in previous methods collections (24, 25, 30). Here we have selected those assays specifically involving ribosomal activities. We have chosen to describe the eukaryotic system because the intermediate complexes are fairly stable and, therefore, the single reactions can be easily assayed. The assays described below are mainly based on the initiation factor-dependent binding of radiolabelled tRNA and/or mRNA to either the 40S ribosomal subunit or to the 80S ribosome. The complexes are usually analysed by sucrose gradient centrifugation.

Protocol 13 describes the initiation factor eIF-2 dependent initiator tRNA binding to the 40S subunit. This assay has been successfully used to test the activity of 40S ribosomal subunits modified by pre-incubation with antibodies (31).

Protocol 13. Met-$tRNA_f$ binding to 40S ribosomal subunits

1. Preparation of [^{35}S]Met-$tRNA_f$.[a] Incubate 2 mg calf liver tRNA in a 1 ml incubation mixture containing 50 mM Hepes–KOH pH 7.5, 10 mM ATP, 15 mM $MgCl_2$, 0.1 mM EDTA, 2 mM DTT, 1 mM CTP, 10 μM total methionine (100 μCi [^{35}S]methionine)[b] and appropriate amounts of crude *E. coli* synthetases[c] for 20 min at 37 °C. Extract RNA by phenol extraction in the presence of 50 mM sodium acetate buffer pH 5. Dialyse or gel filter the RNA in the same buffer. Ethanol precipitate and collect the RNA, dissolve in water, and count radioactivity. Typically, the extent of aminoacylation is 150 pmol methionine/mg of tRNA.
2. Formation of the ternary initiation complex [eIF-2·GMPPCP·Met-$tRNA_f$]. Incubate 10 μg of purified eIF-2[d] and 10 pmol [^{35}S]Met-$tRNA_f$ in a 50 μl incubation mixture containing 20 mM Tris–HCl pH 7.5, 100 mM KCl, 2 mM 2-mercaptoethanol, and 0.5 mM GMPPCP[e] for 10 min at 37 °C.
3. Formation and analysis of the quaternary initiation complex [eIF-2·GMPPCP·Met-$tRNA_f$·40S subunit]. Add 12.5 μg of 40S ribosomal subunit and $MgCl_2$ to a final concentration of 7 mM,[f] and incubate another 5 min at 37 °C. Analyse incubation mixtures by centrifugation

Protocol 13. *Continued*

in 10–30% (w/v) sucrose gradients made up in 20 mM Tris–HCl pH 7.5, 100 mM KCl, 5 mM $MgCl_2$, for 80 min at 400 000 *g* and 2 °C in a Beckman SW60 rotor.[g] Fractionate gradients while monitoring A_{260}. Count fractions (or aliquots thereof) by scintillation counting. Under these conditions, a maximal binding of 0.3 mol Met-$tRNA_f$/mol of 40S subunit was observed (31).

[a] Aminoacylation of eukaryotic tRNA using bacterial aminoacyl-tRNA synthetases and methionine results in the specific charging of the eukaryotic initiator tRNA (32).
[b] May be varied as desired.
[c] Prepare crude *E. coli* aminoacyl-tRNA synthetases from *E. coli* extracts in 20 mM Tris–HCl pH 7.5, 2 mM $MgCl_2$, 2 mM DTT, 5% (v/v) glycerol by gel filtration on Sephadex G25 and subsequent DEAE–cellulose chromatography (elution at 0.25 M KCl) (32). Concentrate by ultrafiltration, dialyse against the above buffer, and store in aliquots in liquid nitrogen. Determine optimal amount of synthetase for aminoacylation experimentally.
[d] Initiation factor eIF-2 may be purified from the ribosomal salt wash (*Protocol 12*) as described (33). The crude salt wash may work as well. However, partial purification is advisable, as the crude ribosomal wash may contain an inhibiting fraction (26).
[e] It is important to use a non-hydrolysable GTP analogue, as traces of eIF-5 present in eIF-2 preparations may cause GTP hydrolysis resulting in inhibition of eIF-2.
[f] Mg^{2+} is partially inhibitory to ternary complex formation, but is required for stabilizing the 40S initiation complex.
[g] For resolution of 80S, 60S, and 40S ribosomal particles, the SW40 rotor is preferable.

The 'shift assay' is used to measure the mRNA-dependent shift of radiolabelled Met-$tRNA_f$ from the 40S subunit to the 80S initiation complex. As the shift reaction comprises a whole series of events (mRNA binding to the Met-$tRNA_f$·40S complexes, release of initiation factors from the 40S complex, and joining of the 60S ribosomal subunit) which involve several initiation factors, it is most easily performed with the unfractionated rabbit reticulocyte lysate as described in *Protocol 14* (adapted from ref. 34). The assay has also been used in a highly fractionated system in order to explore the function of individual initiation factors (24). Thus it could be adapted to a partially fractionated system prepared as in *Protocol 12*.

Protocol 14. Met-$tRNA_f$ shift assay

1. A 1 ml reaction mixture contains 0.8 ml crude reticulocyte lysate (21) and the following additions: 100 mM KCl, 1 mM $MgCl_2$, 1 mM ATP, 0.2 mM GTP, 5.5 mM creatine phosphate, 100 μg creatine kinase, 0.1 mM each of 20 amino acids, 0.5 mM DTT, 30 μM haemin.
2. Incubate 60 μl aliquots of the above reaction mixture in the presence of 0.1 mM sparsomycin[a] for 5 min at 30 °C. Add about 200 000 c.p.m. of [^{35}S]Met-$tRNA_f$ (*Protocol 13*) and the mRNA to be tested (e.g. 2 μg globin mRNA) to each aliquot. Leave one aliquot without mRNA as a control. Incubate for another 2 min, dilute sixfold with ice-cold gradient buffer (step 3), and layer on to sucrose gradients.

3. Analyse reaction mixtures on 15–30% (w/v) sucrose gradients made up in the following buffer: 20 mM Tris–HCl pH 7.5, 25 mM KCl, 10 mM NaCl, 1 mM $MgCl_2$, 0.25 mM DTT. Centrifuge for 3 h at 200 000 *g* in a Beckman SW40 rotor. Fractionate the gradients while monitoring the A_{260}, and count the radioactivity of the fractions by liquid scintillation counting as described in *Protocol 13*.

[a] Sparsomycin inhibits polypeptide chain elongation, thereby preventing a further shift of the [^{35}S]methionine into polysomes.

In the presence of mRNA, the Met-tRNA migrates with the 80S ribosomes, whereas in the control assay without mRNA, it remains with the 40S subunits (*Figure 5*).

An assay which measures the functional activity of 80S initiation complexes by methionyl-puromycin formation, is described in *Protocol 15*. Puromycin binds to the ribosomal A site and is able to accept the aminoacyl moiety of the P site bound tRNA by means of the peptidyltransferase reaction. The proper binding of [^{35}S]Met-tRNA$_f$ into the ribosomal P site during initiation can be assessed by the puromycin reaction. The formed [^{35}S]methionyl-puromycin is extracted from the incubation mixture by ethyl acetate. The protocol is adapted from ref. 35.

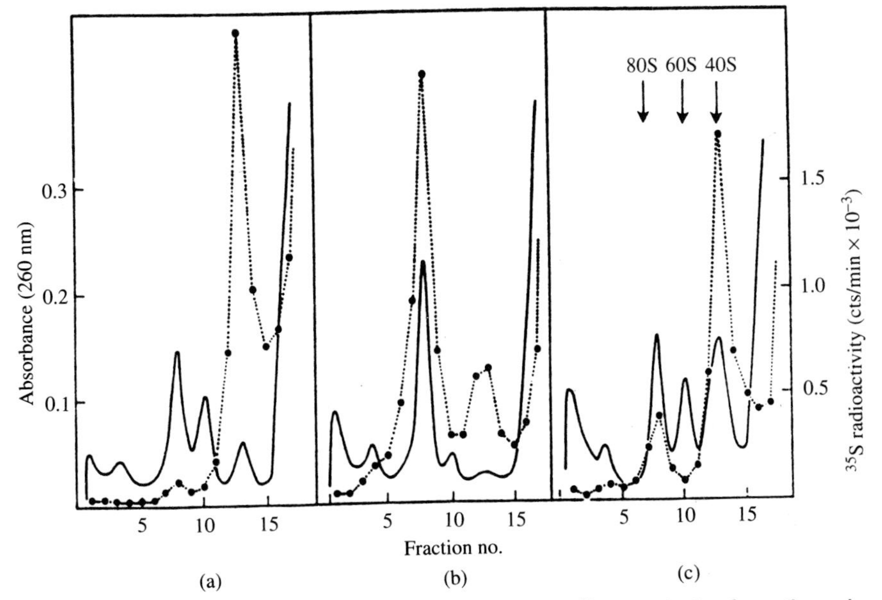

Figure 5. Met-tRNA$_f$ shift assay (*Protocol 14*). Sucrose gradient analysis of reaction mixtures containing (a) no mRNA, (b) 2 μg globin mRNA, and (c) 5 μg bacteriophage f2 RNA (—— absorbance, ●------● radioactivity). Rate of sedimentation increases right to left. Reprinted from ref. 34 with permission.

Protocol 15. Methionyl-puromycin formation

1. The 50 μl reaction mixtures each contain: 20 mM Tris–HCl pH 7.5, 100 mM KCl, 3 mM $MgCl_2$, 1 mM DTT, 0.5 mM GTP, 0.5 mM puromycin,[a] 0.3 A_{260} units AUG trinucleotide, 1 A_{260} unit 80S ribosomes (or the equivalent amounts of 40S and 60S subunits), 15 pmol [^{35}S]Met-$tRNA_f$, and crude initiation factors (e.g. ribosomal salt wash, *Protocol 12*) or purified factors eIF-2, -4C, -4D, and -5.
2. Incubate for 20 min at 30 °C.
3. Terminate the reaction by addition of 0.9 ml of 100 mM protamin phosphate pH 8.0. Extract with 1 ml ethyl acetate (for details see *Protocol 17*). Count 700 μl of each extract.

[a] For handling of the puromycin stock see *Protocol 17*.

3.2.2 Single reactions of the elongation cycle with heteropolymeric mRNA

The partial reactions of the ribosomal elongation cycle as well as the functional competence of ribosomes and translation factors can be studied using an experimental approach based on the procedure described by Watanabe in 1972 (36). This method allows a controlled stepwise execution of a complete elongation cycle. In the first step a 70S·mRNA·tRNA complex (or 80S·mRNA·tRNA in the case of the eukaryotic system) is formed, in which the tRNA is located in the ribosomal P site. In the second step the A or the E site can be filled with the corresponding cognate tRNA, and in the third step the complexes containing tRNAs in P and A sites are translocated to the E and P sites, respectively, upon addition of elongation factor EF-G (EF-2) and GTP. The efficiency of the translocation reaction and the binding state of the tRNAs can be determined in a fourth step by means of the puromycin reaction: the P site bound acyl-tRNA will give a positive reaction, while that in the A site will not react (37, 38). In summary, the puromycin reactivity of a P site bound aminoacyl- or peptidyl-tRNA should not be affected by the translocation factor (EF-G or EF-2), while an A site bound species should show a factor dependent puromycin reaction. The reproducibility and physiological meaning of the experimental data obtained by this method depends greatly on the following experimental prerequisites:

(a) Near physiological conditions with respect to the concentrations of the major ions. The concentrations of Mg^{2+} and other bioactive components like polyamines have a strong influence on the tRNA binding features in all ribosomal systems.

(b) Homologous system with respect to the source of ribosomes and tRNA, in order to avoid ambiguities concerning tRNA affinity for ribosomal binding sites.

(c) Use of heteropolymeric mRNAs containing unique codons in order to ensure unequivocal assignments of the bound tRNA to the various ribosomal sites, as already shown with *E. coli* ribosomes (39) and yeast ribosomes (40). The first steps in the characterization of the system studied can be conveniently performed with homopolymeric mRNAs like poly(U) or poly(A).

Violation of these criteria enlarges the probability of artefacts. Indeed, most of the conflicting findings concerning ribosome function can be traced back to a violation of at least one of these rules The ionic conditions can be optimized using the information available in the literature concerning the physiological concentrations of Mg^{2+} and polyamines. The tRNAs should be purified from the same source as the ribosomes. Finally an appropriate heteropolymeric model mRNA should be chosen. The following criteria should be used for the design of a model mRNA (see, for example, ref. 39):

(a) The sequence should contain unique codons which are frequently used in the organism studied.
(b) Rare combinations of adjacent codons should be excluded.
(c) The cognate tRNAs should be available in purified form. Otherwise the species most easy to isolate with high specific activity should be selected.
(d) A minimum of predictable secondary structure is recommended. The computer programs FOLD (41) or MFOLD (42) offer a convenient way to examine this aspect.

Once a sequence is designed the synthesis of the model mRNA can be performed using phosphoramidite chemistry (43) when the number of nucleotides is below 60 and the number of experiments is well defined. For large quantities or when the model mRNA is longer than 60 nucleotides, the best approach for synthesis is the run-off transcription *in vitro* using bacteriophage RNA polymerases (44–46).

Protocol 16. Synthesis of model mRNAs via *in vitro* transcription with T7 RNA polymerase[a]

1. The standard transcription mix contains 40 mM Tris–HCl (pH 8.0 at 37 °C), 22 mM $MgCl_2$, 1 mM spermidine, 5 mM dithioerythritol, 100 μg/ml of BSA (RNase and DNase-free), 1 U/μl ribonuclease inhibitor (RNasin), 5 U/ml of inorganic pyrophosphatase, 20 nM of linearized plasmid DNA template (or 40 nM for single-stranded DNA templates), 3.75 mM (or the indicated values) each of ATP, GTP, and CTP, and 0.8 mM UTP (see ref. 46), and 40 μg/ml of purified T7 RNA polymerase.
2. Incubate 3–4 h at 37 °C.

Protocol 16. *Continued*

3. Stop the reaction by adding 0.5 M EDTA pH 8.0 to a final concentration of 25 mM.
4. Extract the transcription mix with 1 vol. of phenol/chloroform/isoamyl alcohol (25:24:1) and precipitate the nucleic acids with ethanol.
5. Recover the precipitated RNA by centrifugation (20 min at 10 000 *g*), dissolve in RNA denaturing sample buffer (10 mM Tris–HCl pH 7.5, 1 mM EDTA, 8 M urea, xylene cyanol 0.05%, bromophenol blue 0.05%; 10 μl of sample buffer/pmol of DNA template used in the transcription assay is normally enough), and purify the transcript via gel electrophoresis under denaturing conditions.

[a] A typical preparative incubation is done in 1–5 ml and yields 20–100 nmol of purified mRNA after preparative electrophoresis purification under denaturing conditions.

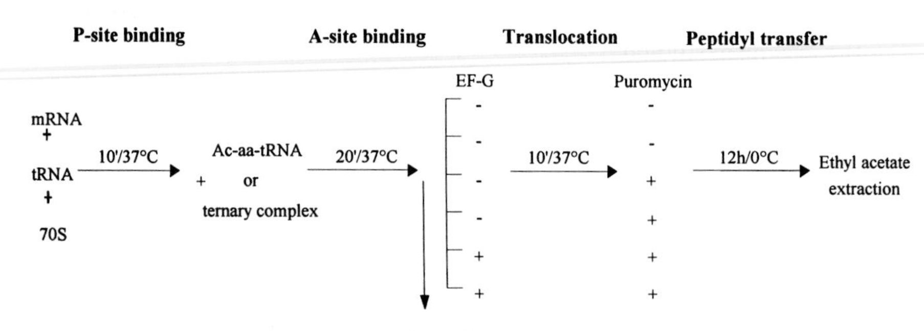

Figure 6. Flow scheme of the stepwise execution of the single reactions of the ribosomal elongation cycle (see *Protocol 17*).

Protocol 17. Single reactions of the elongation cycle (see *Figure 6*)

Reagent

- Buffer 1: 20 mM Hepes–KOH (pH 7.5 at 0 °C), 6 mM $MgCl_2$, 150 mM NH_4Cl, 2 mM spermidine, 0.05 mM spermine, 4 mM 2-mercaptoethanol
- Buffer 2: 20 mM Hepes–KOH (pH 7.5 at 0°C, 6 mM $MgCl_2$, 150 mM KCl, 1 mM DTE, 0.01 mM GDP, 10% (v/v) glycerol
- For the *E. coli* system all the steps are performed in buffer 1

Method

1. P site binding. Incubate 90 pmol of 70S ribosomes (10 pmol per single determination) in 225 μl of binding mix containing 900 μg of poly(U) or a five- to ten-fold molar excess of heteropolymeric mRNA over ribosomes,[a] with a cognate deacylated tRNA in an 1.2–1.5-fold molar ratio to ribosomes for 10 min at 37 °C.[b]

2. A site binding. Increase the volume of mix to 450 μl while adding labelled *N*-acetylaminoacyl-tRNA or ternary complex[c] in a 0.8 to threefold molar ratio to ribosomes. Incubate the mix for 20 min at 37 °C. Take two samples of 50 μl of binding mix to measure the binding by nitrocellulose filtration.[d]
3. Translocation. Transfer six samples of 50 μl of binding mix to 1.5 ml Eppendorf tubes. To two of them add EF-G (0.3–0.6 pmol/pmol 70S) and GTP (final concentration 0.5 mM) in 10 μl of buffer 2. Add 10 μl of buffer 2 and GTP to the four remaining samples. Incubate for 10 min at 37 °C.
4. Puromycin reaction (peptidyl-transferase activity). Add 5 μl of puromycin stock solution (10 mM in buffer 1, final concentration 0.7 mM)[e] to the samples containing EF-G and to two of the samples without factor. Add 5 μl of buffer 1 to the remaining samples. Incubate at 0 °C for 4–18 h and stop the reaction by adding 65 μl of 0.3 M sodium acetate pH 5.5, saturated with $MgSO_4$. Extract with 1 ml of ethyl acetate (1 min mixing), incubate 10 min at 0 °C, centrifuge briefly in a microcentrifuge, and measure the radioactivity contained in 700 μl of the ethyl acetate phase. The radioactivity extracted in the controls (minus puromycin) is subtracted from the plus puromycin determination in order to calculate the amount of acyl-puromycin formed.[f]

[a]The optimal amount of heteropolymeric mRNA is determined according to its ability to stimulate specific tRNA binding.
[b]In experiments where the tRNA or the mRNA are labelled, the binding can be determined at this stage by performing a nitrocellulose filtration assay (47).
[c]The ternary complex (EF-Tu·GTP·aminoacyl-tRNA) is formed immediately before its addition to the assay. One aliquot containing 10–20 pmol of aminoacyl-tRNA (1–2 pmol per pmol of 70S ribosomes after addition to the assay), 0.5 mM GTP and 12–24 pmol of EF-Tu (1.2 pmol/pmol of aminoacyl-tRNA) in a volume of 5–10 μl in the same ionic conditions as the binding assays is pre-incubated for 2 min at 37 °C and used immediately. The stability of the ternary complex can be checked by cold TCA precipitation. Samples of the preparation are precipitated at different times (at the beginning and at the end of the binding assay at least) and compared to samples of the aminoacyl-tRNA in the same conditions (without factor). More than 90% of the initial precipitable radioactivity is preserved after a 2 h assay when EF-Tu is present while a significant reduction (20–40%) is detected when the factor is not added.
[d]Include binding incubations without ribosomes as standard controls in order to determine the background of radioactivity adsorbed to the filters. This background is normally low (below 10% of the binding signal) and directly proportional to the concentration of the radioactive component in the assay. Controls without mRNA can also be included (e.g. to test a new preparation of heteropolymeric mRNA).
[e]A successful puromycin reaction depends critically on the way in which the puromycin stock solution is prepared and handled. Two basic rules for the preparation of a puromycin stock solution with maximal activity must be observed:

(a) The pH of the solution must be neutral. Since the puromycin is obtained commercially as a hydrochloride, the pH of the solution has to be neutralized by adding 1 M KOH (approx. 1/100 of the volume).

(b) The puromycin stock solution must be maintained at room temperature (otherwise it precipitates, lowering the effective concentration). Under these conditions the stock solution retains its maximum activity for about 1 h.

[f]This protocol has also been applied successfully with yeast 80S ribosomes using the following ionic conditions in all steps: 30 mM Hepes–KOH (pH 7.5 at 0 °C), 5 mM Mg(acetate)$_2$, 50 mM NH_4Cl, 2 mM spermidine, 5 mM 2-mercaptoethanol, 9% (v/v) glycerol (40).

Acknowledgements

The authors wish to thank Prof. M. J. Clemens for help and advice during preparation of the manuscript, Drs J. E. Hesketh and H. B. J. Jefferies for providing yet unpublished protocols, and G. Diedrich for help in preparing some of the figures.

References

1. Rühl, M. (1989). Diploma Thesis at the Technical University Berlin.
2. Triana-Alonso, F., Chakraburtty, K., and Nierhaus, K. H. (1995). *J. Biol. Chem.*, **270**, 20473.
3. Tai, P.-C. and Davis, B. D. (1979). In *Methods in enzymology* (ed. K. Moldave and L. Grossmann) Vol. 59, p. 362. Academic Press, London.
4. Remme, J., Margus, T., Villems, R., and Nierhaus, K. H. (1989). *Eur. J. Biochem.*, **183,** 281.
5. Nierhaus, K. H. (1990). In *Ribosomes and protein synthesis: a practical approach* (ed. G. Spedding), p. 161. IRL Press, Oxford.
6. Dohme, F. and Nierhaus, K. H. (1976). *J. Mol. Biol.*, **107,** 585.
7. Infante, A. A. and Baierlein, R. (1971). *Proc. Natl. Acad. Sci. USA*, **68,** 1780.
8. Clemens, M. J. (1984). In *Transcription and translation: a practical approach* (ed. B. D. Hames and S. J. Higgins), p. 211. IRL Press, Oxford.
9. Bielka, H. (ed.) (1982). *The eukaryotic ribosome.* Springer-Verlag, Berlin, Heidelberg, New York.
10. Hesketh, J. E. and Pryme, I. F. (1991). *Biochem J.*, **277,** 1.
11. Biegel, D. and Pachter, J. S. (1992). *J. Cell. Biochem.*, **48,** 98.
12. Kaspar, R. L., Kakegawa, T., Cranston, H., Morris, D. R., and White, M. W. (1992). *J. Biol. Chem.*, **267,** 508.
13. Jefferies, H. B. J., Thomas, G., and Thomas, G. (1994). *J. Biol. Chem.*, **269,** 4367.
14. Fagan, R. J., Lazaris-Karatzas. A., Sonenberg, N., and Rozen, R. (1991) *J. Biol. Chem.*, **266,** 16518.
15. Spedding, G. (ed.) (1990). *Ribosomes and protein synthesis: a practical approach.* IRL Press, Oxford.
16. Moldave, K. and Grossmann, L. (ed.) (1979). *Methods in enzymology*, Vol. 59. Academic Press, London.
17. Krieg, J., Olivier, A. R., and Thomas, G. (1988). In *Methods in enzymology* (ed. H. F. Noller and K. Moldave), Vol. 164, p. 575. Academic Press, London.
18. Lutsch, G., Benndorf, R., Westermann, P., Bommer, U. A., and Bielka, H. (1986). *Eur. J. Cell. Biol.*, **40,** 257.
19. Funatsu, G., Nierhaus, K. H., and Wittmann-Liebold, B. (1972). *J. Mol. Biol.*, **64,** 201.
20. Erickson, A. H. and Blobel, G. (1983). In *Methods in enzymology* (ed. S. Fleischer and B. Fleischer), Vol. 96, p. 38. Academic Press, London.
21. Jackson, R. J. and Hunt, T. (1983). In *Methods in enzymology* (ed. S. Fleischer and B. Fleischer), Vol. 96, p. 50. Academic Press, London.

22. Clemens, M. J. (1984). In *Transcription and translation: a practical approach* (ed. B. D. Hames and S. J. Higgins), p. 231. IRL Press, Oxford.
23. Wu, R., Grossman, L., and Moldave, K. (ed.) (1983). *Methods in enzymology*, Vol. 101. Academic Press, London.
24. Trachsel, H., Erni, B., Schreier, M. H., and Staehelin, T. (1977). *J. Mol. Biol.*, **116,** 755.
25. Safer, B., Jagus, R., and Kemper, W. M. (1979). In *Methods in enzymology* (ed. K. Moldave and L. Grossmann), Vol. 60, p. 61. Academic Press, London.
26. Schreier, M. H. and Staehelin, T. (1973). *J. Mol. Biol.*, **73,** 329.
27. Bommer, U. A., Junghahn, I., and Bielka, H. (1987). *Biol. Chem. Hoppe Seyler*, **368,** 445.
28. Morley, S. J. and Hershey, J. W. B. (1990). *Biochimie*, **72,** 259.
29. Merrick, W. C. (1992). *Microbiol. Rev.*, **56,** 291.
30. Wahba, A. J., Woodley, C. L., and Dholakia, J. N. (1990). In *Ribosomes and protein synthesis: a practical approach* (ed. G. Spedding), p. 31. IRL Press, Oxford.
31. Bommer, U. A., Stahl, J., Henske, A., Lutsch, G., and Bielka, H. (1988). *FEBS Lett.*, **233,** 114.
32. Stanley, W. M. Jr. (1974). In *Methods in enzymology* (ed. L. Grossmann and K. Moldave), Vol. 29, p. 530. Academic Press, London.
33. Oldfield, S. and Proud, C. G. (1992). *Eur. J. Biochem.*, **208,** 73.
34. Darnbrough, C., Legon, S., Hunt, T., and Jackson, R. (1973). *J. Mol. Biol.*, **76,** 379.
35. Merrick, W. C. (1979). In *Methods in enzymology* (ed. K. Moldave and L. Grossmann), Vol. 60, p. 108. Academic Press, London.
36. Watanabe, S. (1972). *J. Mol. Biol.*, **67,** 443.
37. Allen, D. W. and Zamecnik, P. C. (1962). *Biochim. Biophys. Acta*, **55,** 865.
38. Traut, R. R. and Monro, R. E. (1964). *J. Mol. Biol.*, **10,** 63.
39. Gnirke, A., Geigenmüller, U., Rheinberger, H.-J., and Nierhaus, K. H. (1989). *J. Biol. Chem.*, **264,** 7291.
40. Triana, F., Nierhaus, K. H., Ziehler, J., and Chakraburtty, K. (1993). In *The translational apparatus* (ed. K. H. Nierhaus, F. Franceschi, A. Subramanian, V. Erdmann, and B. Wittmann-Liebold), p. 327. Plenum Publishing Corp., New York.
41. Zuker, M. and Stiegler, P. (1981). *Nucleic Acids Res.*, **9,** 133.
42. Zuker, M. (1989). *Science*, **244,** 48.
43. Scaringe, S. A., Franklyn, C., and Usman, N. (1990). *Nucleic Acids Res.*, **18,** 5433.
44. Milligan, F. and Uhlenbeck, O. C. (1989). In *Methods in enzymology* (ed. J. E. Dahlberg and J. N. Abelson), Vol. 180, p. 51. Academic Press, London.
45. Weitzmann, C. J., Cunningham, P. R., and Ofengand, J. (1990). *Nucleic Acids Res.*, **18,** 3515.
46. Triana, F., Dabrowski, M., Wadzack, J., and Nierhaus, K. H. (1995). *J. Biol. Chem.*, **270,** 6298.
47. Nirenberg, M. and Leder, P. (1964). *Science,* **145,** 1399.

10

Electron microscopy of organelles

NIGEL JAMES

1. Introduction

The electron microscope has made major contributions to our understanding of tissues, cells, and organelles and how they function. Electron microscopes have become widespread in their use and are a reliable means of examining biological structures. The process of producing sections using a microtome (microtomy) is also reliable provided that the samples are adequately prepared and embedded. Electron microscopy enables these samples to be examined with a high degree of resolution of approx. 1 nm which is much greater than that of optical microscopy which is limited by the intrinsic properties of light. A further fundamental difference between optical and all forms of electron microscopy is that optical microscopical techniques for examining large quantities of potentially living specimens exist whilst the electron microscope is capable of examining only relatively small quantities of exclusively dead and artefactually transformed specimens.

Both electron and optical microscopy techniques are ultimately limited by the nature of the specimen examined and the techniques of preparation. The stages involved in examining specimens by electron microscopy involve fixation (preservation), embedding (in a specimen to provide support), microtomy (sectioning), staining (to visualize the important contents), and the electron microscope operations. Of all these stages, the fixation–embedding–staining triad is vital and of these, fixation and embedding are the most important since they cannot be repeated on an original specimen. If these two stages have been performed satisfactorily then the production of sections by subsequent microtomy is usually successful. In the present account, since local conditions are so variable, both microtomy and the electron microscope operation are best avoided in specific detail due to instrumental peculiarities and differences in operation.

There are fundamental differences between the preparation of specimens for electron microscopy from fractionated specimens compared to examination *in situ*. Intact tissue preparations have to be fixed either by immersion of small tissue fragments arbitrarily selected from a larger sampling mass, or in vertebrate studies, prepared by perfusion through the cardiovascular system.

Both techniques have their deficiencies. In the case of isolated fractionated specimens many of the technical problems of fixation are circumvented though all surrounding structural and spatial information in two and three dimensions is destroyed. No technique is really adequate for both good universal fixation with retention of structural and spatial identities of both organelle preparations and intact specimens.

Ultimately, a fundamental skill in electron microscopy is to obtain the final image which is a representative sample of the original specimen such that it is also 'acceptable' for analysis. However good the ideas of the investigator, the presence of poorly-fixed tissues or organelles often mitigates against prompt acceptance of papers by journal editors or by peer review. Some knowledge of the way in which the electron microscope yields its images is of paramount importance in achieving both meaningful and acceptable quality images. Knowledge is also important in understanding the increasing variety of electron microscope techniques.

Finally, no electron microscopy study can be regarded as adequate, as in other scientific endeavours, without some quantitative measurement. Quantitative measurement transforms the subjective impression incapable of statistical analysis to a rigorous objective measurement suitable for numerical analysis. Specific numerical techniques constitute a form of 'mathematical' fractionation.

Electron microscopical investigations have been carried out for 50 years and the volume of practical and theoretical knowledge generated is enormous (1–7). Some subjects within electron microscopy, for example, staining for enzyme activity even by the 1970s had generated multivolume texts. The present chapter cannot aspire to more than indicate in most general terms the potential range of activities with a few sample practical procedures. It is intended as an introduction to the appropriate journals and classical reference texts of the subject so that potential investigators may find all the more readily the methodology and other sources appropriate to their investigations. The thousands of electron microscope procedures cannot be encompassed in a short article. Of much more benefit is an understanding of the general principles which guide laboratory practice and the access to the vast databases of skill and knowledge.

2. Fixation

2.1 Importance of fixation

Good fixation is the key to successful electron microscopy. Poor fixation results in inadequate results which represent additional distortions beyond those induced by the preservation process of fixation. Poor fixation may preclude reproducibility which is essential both for objective and quantitative science.

Fixation of biological material is at a relatively primitive stage of development. Many advances will undoubtedly derive from the use of alternative fixatives. How much organic material is lost from biological specimens due to their lack of reactivity with a fixative presently in use is unknown. All material that is fixed is highly unlikely to have retained all the compounds normally present in cells. A sensible approach to the interpretation of the results obtained from experimental tissue subject to prior fixation is to regard the specimen as an artefact of high precision but relatively low accuracy compared with its *in vivo* state. The single most effective advice to be given by the present author is that specimens that are poorly fixed must be discarded in favour of better fixed material.

2.2 Chemical fixation

Fixation is used to preserve biological specimens for electron microscopy with the aim of causing as little alteration as possible in their structure. Ultimately, however, since some structures are visible only with the electron microscope, the unfixed living nature of specimens can only be guessed at and real changes are unknown quantities. The aim of fixation is to preserve the original sizes and, in the case of intact tissues, spatial relationships, of cells and organelles whilst allowing further investigative manipulations, such as dehydration and embedding procedures (8). They may also have to be sufficiently innocuous as to allow the retention of certain enzyme activities.

Unlike the fixation of structures of cellular dimensions or larger where diffusion processes markedly influence the time taken for fixative to diffuse into the cells, with organelle fractions their much smaller dimensions and their ability to undergo mixing with fixative solutions removes some of the problems of fixation artefacts.

A variety of fixatives are in common use, most frequently glutaraldehyde (glutaric acid dialdehyde) and osmium tetroxide are used, though many others, such as formaldehyde and acrolein are available for special purposes. The use of a chemical fixative will be determined by the specific aim of the study.

Fixation, for example with glutaraldehyde, is achieved by the combined processes of diffusion and chemical reaction. As the glutaraldehyde diffuses into the specimen it acts to cross-link proteins etc., which in turn reduces the diffusion rate of further glutaraldehyde into the specimen.

The most important determining factor in the diffusion process is the penetration rate (penetration depth per unit time) as indicated by the diffusion constant, D. A comparison of the diffusion constants of a small molecule formaldehyde (D = 2.0 mm/h) and a larger functionally similar molecule, glutaraldehyde (D = 0.34 mm/h) are instructive. They are reactive mono- and dialdehyde reagents and a considerable volume of investigation in to their chemical effects has been carried out. Much of their diffusion properties however remains unknown. In the first hour of diffusion into intact tissue formaldehyde, as indicated by the 'wave front' of the diffusant, potentially penetrates to a depth of 2.0 mm. Glutaraldehyde penetrates to a lesser depth of about 340 μm. Clearly, in cell and organelle preparations the dimensions are much smaller and penetration may be achieved in seconds or minutes depending on organelle size. Mixtures of formaldehyde and glutaraldehyde are utilized in tissue work so as to benefit from the more rapid formaldehyde diffusion compared with slower diffusion of glutaraldehyde.

The numerical aspects of the process of diffusion can be of practical value in designing an experiment. The relation between chemical diffusant concentration, and depth to which a diffusant penetrates a specimen and its diffusion coefficient is well known. Further, the effects of the rate of chemical reaction occurring simultaneously with diffusion are known.

$$C_{x,t} = C_0 \, erfc \, \frac{x}{2\sqrt{Dt}} \qquad [1]$$

Equation 1 for example, illustrates the relation between the initial concentration, C_0, the concentration at time t, $C_{x,t}$, x the depth of penetration, and D the diffusion coefficient for diffusion into a semi-infinite medium and it illustrates fundamental aspects of the diffusion process. The error function complement (*erfc*) term $x/2\sqrt{Dt}$ determines the nature of the penetration. Important consequences of Equation 1 are:

- the distance to which the wave front penetrates is proportional to $\sqrt{t}$
- the amount of diffusant varies as $\sqrt{t}$
- the time taken for any point to reach a given concentration is proportional to the square of the distance from the surface
- the speed at which a wave front of a diffusant penetrates a specimen is independent of the external concentration of diffusant

Increasing the concentration of a fixative does not alter the penetration velocity of the wave front but only increases the concentration at any point behind the wave front more rapidly.

A greater concentration may increase the reaction rate of fixative with the specimen. This may variously be an advantage or a disadvantage. Increased concentrations may generate increased hardness for microtomy or severely inhibit enzymic activities. The advantage of small organelle preparations is that the independence of wave front penetration speed and fixative concentration potentially permits lowered fixative concentration compared with intact tissues thus allowing, for example, more representative enzyme assays. The numerical relationships should give some insights into the minimal effective fixation times consistent with achieving the required degree of fixation. For spherical specimens, other relationships hold which can readily be determined from standard texts on diffusion processes. More complicated relations apply where specimens are heterogeneous by particulate inclusions or laminated in composition.

The relation between the sizes and diffusion (mass transfer) coefficients, *D*, of for example, different aldehydes is given in approximate and elementary form by the well known relation of Einstein (Equation 2):

$$D = \frac{kT}{6\pi r\eta} \qquad [2]$$

where *k* is the Boltzman constant, *T* the absolute temperature, (K^0), *r* the radius of an equivalent spherical particle, and the equivalent viscosity. This enables at least a rough estimate to be made as to possible changes of diffusion behaviour due to changing the size of fixation molecule. There is, however, no allowance for the rate of chemical reactions. Accurate estimates are impossible where the biological specimens have unknown properties and chemical reactivities.

Each aldehyde reacts differently with chemical groupings in the preparations and results in the retention of different materials within these preparations. It is highly likely that some structural elements are lost from cells since they do not react with the fixatives or in sufficient amounts to ensure their retention.

Both aldehydes inactivate different enzymes to different amounts so the choice is dependent on the aims of the investigation. Some benefit may occur by fixation in the presence of the substrate (either natural or artificial) which occupies an enzyme site and may therefore serve to protect it during fixation. It must however, be able to relinquish its site during the staining procedure.

Aldehydes cannot be compared with other fixatives as the reaction processes which occur simultaneously with diffusion can significantly alter their diffusion coefficients. Further, the cross-linking abilities of glutaraldehyde are such that the molecules behind the diffusion wave front reduce the rate of diffusion of the fixative further by restricting the passage of further molecules.

The retention of macromolecular materials is partially dependent on the nature of any adjacent structure. For example, lipids may readily be lost due to failure to bind with the structural elements of the cell. Artefactual alterations may also ensue with the failure to bind lipids adequately. A major factor in determining whether fixation is 'adequate' is its subjective appearance in a micrograph. The presence of empty artefactual 'holes' in mitochondria can be interpreted as 'poor' fixation. Fixation is the single topic within microscopy which retains an element of being almost an 'art form' in addition to being purely science.

2.3 Osmolarity of fixative solutions

Osmolarity is defined in terms of the solution molarity (mole per litre of solution). For non-electrolytes which do not ionize the molar and osmolar concentrations are more or less equivalent but dissociating electrolytes exert greater osmotic pressure than their molar concentrations would indicate. An isotonic solution is one which exerts an osmotic pressure identical to that normally experienced by the organelle such that there is neither net inflow nor outflow of water from the organelle such that it appears neither to have become swollen nor shrunk. Hypertonic fixative will result in shrinkage whilst hypotonic fixative results in swelling. One aim of the investigator is to decide what osmolarity of fixative is important and yet allows the specific aims of the experiment or analysis to be carried out.

Techniques of fixation and the osmolarity of fixative are not related to blood composition. In the mammalian body cells do not lie in contact with blood but (with the exception of its own cells) are isolated from it by the blood vessel endothelium. The cells lie in contact with interstitial tissue fluid (lymph) which has a unique composition in each organ. Some organs do not have tissue fluid and lymphatic systems (e.g. brain and placenta). Blood does not have a fixed composition (Starling's hypothesis) but is dynamically changing in capillaries and has a fixed composition only in the test-tube.

Each tissue and cell and each organelle within cells possess their own unique osmotic characteristics and consequently each has to be fixed according to the cellular or organelle species studied. That cells in tissues have markedly different overall osmotic pressures, for example, in the kidney, is well known. The organelles contained within them are also heterogeneous with respect to osmotic pressure. Mitochondria, nuclei, sarcoplasmic reticulum, transport vesicles, and myofibrils in skeletal muscle all possess different osmotic environments, each requiring its own optimal osmolarity of fixative. The osmolarity also alters with the stage of development, for example, embryonic tissue is different from mature adult tissue, this is particularly so with nervous tissue.

The osmolarity of a fixative solution is dependent upon both the amount of chemical fixative and the medium (including any buffer). Approximate osmolarities of both fixatives and common media are well established and a few examples are given in *Table 1*.

A simple 'rule of thumb' relating the effective osmotic pressure (EOP) to the osmotic pressure of glutaraldehyde (Mg) and medium (Mv) is given in Equation 3.

$$EOP = \frac{Mg}{2} + Mv \qquad [3]$$

It is useful to remember that for pure glutaraldehyde a 1% solution is equivalent to 100 mOsm/litre.

2.4 Temperature, pH, and other factors

Chemical reactions of fixation are markedly influenced by both pH and temperature as well as by time. The aim of the buffer containing the fixative

Table 1. Osmolarity and pH of glutaraldehyde solutions in phosphate buffer

Chemical content	Osmolarity (mOsm)	pH
0.1 M phosphate buffer	230	7.4
1.2% (w/v) glutaraldehyde	370	7.2–7.3
2.3% (w/v) glutaraldehyde	490	7.2–7.3
4.0% (w/v) glutaraldehyde	685	7.2–7.3

is to maintain optimal fixation conditions both for reactions and retention of biological macromolecules. The buffer also contributes markedly to the osmolarity of the fixative.

The speed of reaction, both of fixative and intrinsic autolytic and other processes alter with temperature. Diffusion speed may also be affected though this is not of much practical importance. The structure of some organelles may be themselves temperature dependent for example, microtubules (neurotubules) in nerve axons. Their morphology and size and consequently the properties of the resulting image may well be dependent on fixation temperature.

Buffers are important not only for provision of an appropriate osmolarity but also for ensuring the appropriate pH for preservation reactions with fixative and for the preservation of the integrity of the organelles. The integrity of neurotubules and possible interconversion with neurofilaments for example may be buffer as well as temperature dependent. Care should be taken over temperature of fixation to ensure it is actually carried out at the optimal temperature claimed.

Fixation of organelle fractions is relatively easier due to homogeneity of the specimen. With isolated organelles, an important difference from intact tissue is that the order of fixation and chemical manipulation may be altered more readily than with intact tissue due to the ease both of access of substrates and of fixatives to the specimens. There is no universal fixative or panacea for the lack of good fixation. Each investigation requires its own tailor-made fixation schedule which is designed to optimize the retention of morphology and/or function relevant to the particular study.

Some useful fixation mixtures are given below. Mixtures of glutaraldehyde and formaldehyde which result in better structural preservation include the following formulations:

(a) Fixative 1
- glutaraldehyde (25%, w/v) 5 ml
- paraformaldehyde (10%, w/v) 10 ml
- cacodylate buffer (0.2 mol/litre) 25 ml
- distilled water to a final volume 50 ml

- osmolarity adjusted with sucrose or NaCl
- pH may be adjusted as appropriate

An alternative but equally effective fixative is given by:

(b) Fixative 2

• glutaraldehyde (50%, w/v)	2 ml
• formaldehyde (40%, w/v)	10 ml
• $NaH_2PO_4.H_2O$	1.16 g
• NaOH	0.27 g
• distilled water	88 ml

Both mixtures are widely used general-type fixatives for both animal and plant specimens. As with all glutaraldehyde preparations the use of freshly prepared fixative solutions from purified glutaraldehyde is essential. A glutaraldehyde–hydrogen peroxide mixture has been suggested for better enzyme preservation and which may result from the supply of oxygen required for greater cross-linking with protein groups:

(c) Fixative 3

• cacodylate buffer (0.1 mol/litre)	25 ml
• glutaraldehyde (25%, w/v)	5 ml
• hydrogen peroxide (15%, w/v) dropwise addition and stirring	5–25 drops

- after several hours transfer to 3–5% glutaraldehyde at room temperature
- wash in buffer and post-fix with 2% (w/v) OsO_4 for 2 h at room temperature

Avoid attempting to prepare mixtures of hydrogen peroxide and formaldehyde as potentially explosive compounds may be synthesized. Note also that many of these reagents, cacodylate, glutaraldehyde, formaldehyde, OsO_4, are extremely toxic. If you are unfamiliar with their handling procedures please consult your safety officer.

2.5 Cryofixation

Cryofixation is a form of physical fixation which has the same general aims as that of chemical fixation. It is achieved by extremely rapid freezing of very small specimens to low temperatures of −190 °C using liquid nitrogen as the cooling agent. The advantage of cryofixation is that, correctly performed, it is extremely rapid and intracellular movements of macromolecules and other substances can be inhibited, or at least impeded, within milliseconds. Cryofixation however represents only a small part of the relatively extensive subject of cryotechnology (9,10).

Theoretically the aim is to generate vitrification of the specimen with the formation of amorphous ice. In intact tissues this is impossible due to the formation of ice crystals within the superficial layers more than approx. 500 μm in thickness. With isolated organelle preparations it may be possible either to utilize particulate specimens within the superficial regions or, alternatively use a spray-freeze method of material already stained or incubated for specific chemical reactions. Under the general heading of cryofixation are such directly related techniques as cryoultramicrotomy, high-pressure freeze methods, freeze-drying, freeze-substitution, and the more specialized techniques, particularly for membranes, of freeze-fracturing and freeze-etching. All the techniques have in common the need to keep specimens sufficiently and continuously cold to prevent potential thawing which would result in diffusion of materials within specimens.

The advantages of physical cryofixation is the avoidance of chemical influence on sensitive biological materials, for example, reactive sites of enzymes and the formation of chemically induced artefacts. Probably the majority of chemical events consequent upon fixation are completely unknown. Antigenicity is not so severely masked by cryofixation as by chemical fixatives and the retention of small molecules and other cell constituents is potentially greater. Cisternae of endoplasmic reticulum, Golgi cisternae, vesicles, and the matrix of mitochondria usually appear much denser with cryofixation than with chemical fixation due to the retention of greater amounts of biological materials, probably proteins.

Cryofixation, due to its rapidity, may allow the detection of processes such as the contact of vesicles with membranes where chemical fixation techniques may be too slow to capture such events. It allows *in situ* localization of diffusible cell components and may be combined with chemical fixation as a final stage rather than as a preliminary procedure. Cryoprotectants (such as the low molecular weight membrane penetrating glycerol or the higher molecular weight non-penetrating sucrose) may be used to reduce the damage caused by the formation of ice crystals.

Specimens to be frozen such as a centrifugation pellet can be treated as a small tissue specimen and frozen by direct immersion in a cooling fluid chilled by liquid nitrogen. Some support may be necessary to avoid smearing of the pellet contents during microtomy. The cooling fluid, such as isopentane, must remain liquid at relatively high temperatures to prevent the formation of an insulating gas bubble surrounding the specimen. For this reason liquid nitrogen cannot be used directly. The use of environmentally unfriendly liquids of the fluorocarbon Freon^{R} series are now discontinued.

2.6 Cryoultramicrotomy

The maintenance of very low temperatures is essential with cryoultramicrotomy in which sectioning is carried out in an environment of −120°C using tungsten coated glass knives. Debate still exists as to whether pressure induced thawing or simple mechanical fracture occurs at the knife edge though thawing seems unlikely at −180 °C. Interpretation of results must allow for the possibility of thawing. Sections of approx. 120 nm thickness are now achievable with cryoultramicrotomy. Sections can be freeze-dried and subjected to fixation if required for staining purposes.

2.7 Freeze-fracture preparations

In principle, there is little difference between the process of cryoultramicrotomy and freeze-fracture preparations. In freeze-fracture preparations, specimens are subjected to a fracturing process followed by the preparation of a metal–carbon replica of the fractured surfaces. Membranes are split along their central hydrophobic plane exposing intramembranous structures such as intramembranous particles (10). Subsequently, staining with colloidal gold or other electron dense-containing materials can be carried out.

The fractured surface is first replicated with a heavy metal and a covering supporting layer of carbon is deposited. Exact technical details are readily available in standard texts though will be subject to local operating conditions of the available hardware.

3. Embedding

Apart from fixation, the process of embedding is the next most important technical stage of preparing specimens for electron microscopy. At worst, unsatisfactory embedding ensures that the specimen cannot be sectioned for electron microscopy. However, with reasonably well fixed and embedded materials, errors and lack of initial skills can often be remedied on resectioning or restaining or re-examining tissue in the electron microscope. With either poor fixation or poor embedding discarding of the specimen is inevitably necessary.

Specimens for electron microscopy are always too small for easy manipulation and are necessarily embedded in a suitable medium. The embedding medium provides support and protection not only for the specimen but for the thin section subsequently prepared for further staining and microscopical manipulations. The embedding medium usually consists of a resin, most frequently water immiscible for structural analysis, though increasingly the medium is water miscible for histochemical and other functional analyses. Following fixation, excess fixative must be removed by washing followed by dehydration prior to infiltrating with resin. Dehydration is important since penetration of the embedding medium will be incomplete in the presence of water. The raw resin is liquid (though viscous) and contains additional compounds to harden (by polymerization) it so as to facilitate sectioning. Other compounds may also be included in the embedding medium, such as modifiers and catalysts which influence the rate of hardening and the final degree of hardness. A major aim of embedding is to ensure that the hardness of medium allows adequate sections to be cut and where the hardness of the specimen and surrounding embedding medium (matrix) are similar.

In general, few substances are suitable for use as embedding media for electron microscopy. Those substances (resins) that are available must be chosen with some care according to the aim of the investigation.

There are three types of water immiscible resins, epoxy, polyester, and methacrylate resins. There are several water miscible resins of importance (Durcupan^{R}) and various water miscible methacrylates such as glycol methacrylate and acrylate-methacrylate (Lowicryl^{R}) mixtures. Acrylic resins are particularly useful for enzyme histochemistry and immunocytochemistry.

In addition, some embedding media can be formulated which are mixtures of resins since the majority of methacrylates, polyesters, and epoxy resins are miscible with each other. The aim of using a mixed embedding medium is to derive benefit of the optimal properties of each of its components. For example, a mixture may produce a resin with the thermal stability of Araldite^{R} and with the picture contrast properties of Epon^{R}. A mixed embedding media should produce good tissue structure preservation, homogeneous penetration, and ease of sectioning. The selection of the embedding medium, as the fixative, depends on the nature of the particular specimens, the aims of the study, the staining techniques to be used, together with the maintenance of any antigenic or enzymic activity and the size of the specimens. In the case of organelle fractions specimen size is of least importance for embedding procedures.

The suitability of an embedding medium is determined by a number of deciding factors. For example, there must be good structural preservation with little or no extraction of or damage to cellular components. The viscosity of the monomer (unpolymerized) embedding medium, which is a function of mean molecular chain length, must not be so low that tissue deformation (shrinkage) occurs during polymerization. On the other hand the viscosity must not be too high as to prevent its uniform diffusion into the biological specimens. The medium has to polymerize evenly throughout the specimen so as not to induce differential properties in different domains of the specimen. There should be no significant swelling of the medium during polymerization sufficient to distort structural integrity or spatial relationships particularly if any subsequent form of quantitative assay is contemplated.

The embedding medium has to withstand the forces and stresses of sectioning without imparting distortion to the specimens. The forces at the extremely fine cutting edge of a knife generate enormous pressures and temperatures in the microenvironment of the section plane. The medium must have sufficient hardness and heat resistance yet have sufficient elasticity and plasticity to yield even thickness consistent with the required properties of sections without specimen damage.

At the staining stage the properties of the embedding media must allow the appropriate staining procedures to be carried out. These may range from stain penetration to achieve sufficient electron density for structural identity and contrasting or binding with specific staining reagents such as monoclonal antibodies. In the vacuum of the electron microscope, the medium must withstand electron bombardment and be stable to potential deformation induced by embedding medium sublimation.

The media must be readily available and consistent in reproducibility of results. As with fixatives, no medium is perfect and at best only a compromise of best subjective benefits is attainable. X-ray diffraction studies which provide information on the ordered structures of a molecule give an index of the size of these structures and have been used to indicate optimal embedding media. Araldite, with the smallest values of ordered structure is considered by many to be the most ideal medium possibly due to its minimal rigid structure. Ultimately, however, only the investigator can make a choice. N.B. An important safety note is that as with the handling of all chemicals in the laboratory care should be taken to avoid exposure either by cutaneous contact or inhalation. Contact dermatitis and allergic phenomena are becoming more common with many experimental procedures.

3.1 Hardeners, modifiers, and catalysts

The incorporation of hardeners in the embedding medium, such as dodecenyl succinic anhydride (DDSA) results in block hardening, though with DDSA the degree of hardening is minimal. A range of alternative hardeners is available for varying the hardness of blocks as required.

The hardness of the final block can be reduced by the use of a plasticizer or flexibilizer. Dibutyl phthallate (DP) in varying amounts can be used to control block hardness. Tertiary amine catalysts are used to accelerate polymerization of resins. A typical example is benzyldimethylamine (BDMA).

4. Microtomy

Microtomy is the process whereby sections of the appropriate thickness are obtained for microscopy. Semi-thin sections approx. 0.5 μm in thickness and stained with a suitable dye such as alkaline 1% toluidine blue (in borax) for 1 min at approx. 90°C may be needed for basic orientation or identification of the specimen. Most important is the ability to provide thin sections approx. 70 nm for electron microscopy. Thin sectioning has become standardized and is capable of yielding consistently reproducible results. Sections of between 30–90 nm in thickness are required for clarity of image formation with conventional 100 kV electron microscopes. Thicker sections, up to approx. 5 μm, may be visualized with the high voltage electron microscope though these dimensions are greater than those required for organelle preparations.

Clearly, the choice of appropriate fixative and embedding medium and adequacy of specimen block consistency is relevant to the process of microtomy. Sections are prepared using a microtome bearing, most commonly, a simple glass knife though occasionally a diamond knife may be used.

Glass is a convenient material for the manufacture of knives. It is cheap (particularly compared with a diamond knife) and readily available in batches of consistent mechanical properties. It is strong, up to two million psi stress factor, and does not undergo significant distortion or compression. A glass edge is capable of being smoother and sharper than any steel edge and can be readily replaced by rapid glass breaking methods—most often using a special glass knife making machine, for example, LKB Knifemaker[R]. Glass knives are capable of producing larger sections than those made of diamond.

Not all glass knives made with a knife making machine will be adequate and some will need to be discarded upon manufacture but the cost of discarding knives is negligible. The number of high quality sections available from a knife will depend on the specimen composition, hardness, size, and thickness of the sections required. The harder the specimen the shorter the life of the knife edge. Since glass is a supercooled liquid and undergoes some reaction with air, even unused knives deteriorate, though slowly, and are best prepared fresh. Coating with tungsten may improve cutting qualities.

Diamond knives have some advantages despite their cost. For example they can cut sections of less than 30 nm (down to 10 nm) and can cut harder tissue though this is not often encountered with organelle preparations. Diamond knives come ready prepared for immediate use and show more clear images of relatively smaller structures. They may need resharpening at some considerable cost. Only a perfect knife produces perfect sections. Optimally fixed and embedded specimens can be cut with an adequate knife but non-optimally embedded specimens require exceptionally good knives.

The higher the resolution required the thinner must be the section. However, as the thickness of the section is decreased so is the contrast of the image. The smallest detail resolved is usually approx. one-tenth of the section thickness. Hence staining is required for image contrast.

Some index of section thickness can be estimated by observing their interference colours. Grey sections are approx. 25–35 nm in thickness whilst silver are 50–65 nm and gold 90–120 nm. The thickness of a section prepared for electron microscopy is best estimated from its interference colour and is relevant to the object of study. Thickness of a section must be adequate for the identification of small structures. For example, some collagen fibrils can be visualized using gold sections but not easily using grey sections, whilst for other structures, such as most subcellular organelles, grey sections are mandatory. Section thickness is variable and is commonly $\geq 10\%$.

During microtomy sections are floated on to a liquid surface (often distilled water) partly to allow them to restore shape comparable with that of the parent block under the influence of surface tension and other mechanical forces. The reader will appreciate that deformation of the sections and hence of the structures contained within it, is one of the many important contributions to measurement errors in estimating specimen sizes accurately.

Details of the sectioning procedure will depend on the local operating conditions and type of microtome used. Sections are picked up on an electron microscope grid (often Formvar coated to provide additional support strength in the electron beam) bearing apertures which allow the passage of electrons. The sections for electron microscopy can subsequently be stained, often by the grid being placed section downwards on to the surface of a drop of appropriate staining solution. Grids of different slot sizes and made of different materials to avoid possible unwanted reactions with reagents in the staining solution, are readily available commercially.

Defects in the microtomy process are numerous. Some, due to knife defects which cause scratches, wrinkling, irregular folds, or holes can be remedied by using a new knife. Others may be more difficult to rectify. Crumpling of sections and knife chatter may be due to imperfect embedding procedures. Each fault needs to be diagnosed accurately prior to attempting a remedy.

5. Staining

5.1 The electron microscope family

All electron microscopes have in common three essential components. First, an electron gun to generate a stream of primary electrons. Second, a system of wire coils through which controlling currents are passed to manipulate the electron beam and which act, by analogy with the optical microscope, as lenses. Third, a signal detector for detecting the nature of the interactions between electron beam and specimen. With variations this basic concept can be modified to generate an entire family of electron microscopes, each member of which can perform unique tasks.

The most widely used forms of electron microscopy are the transmission electron microscope (TEM) and the scanning electron microscope (SEM) (11,12). The TEM utilizes sections of biological specimens whilst the scanning electron microscope examines prepared surfaces of specimens. The TEM needs objective and projector lenses to view specimens on a viewing screen whilst the SEM requires scanning coils and an electron detector. With the addition of an X-ray detector the SEM becomes a microprobe analyser (EPMA) or in a scanning transmission electron microscope (STEM) it is moved to lie below the specimen. The high voltage electron microscope (HVEM) may use voltages of approx. 1 mV to achieve greater penetration of much thicker specimens than conventional TEM.

5.2 Image contrast

In the transmission electron microscope image contrast is due to selective electron scattering by the atoms both of the biological specimen and the stain which are present in the section. Basically molecules consist of positively charged atoms lying within nuclei of varying size. The atomic number (denoted by Z) is an indication of their size. Surrounding the central nucleus is a shell of negatively charged electrons. An electron beam passing through a section is influenced by the nuclei of the specimen molecules. The electrons are deflected as they approach a positively charged atomic nucleus and follow an altered trajectory around that nucleus. The greater atomic number of the nucleus the larger the positive charge and the greater the electron deflection. Such deflections or scattering may either be elastic or inelastic.

Elastic scattering can be defined as a process in which the direction of the electron path is altered without detectable alteration in its energy (ΔE). ΔE would need to be significantly more than approx. 0.1 eV before it could be detected. Such (Rutherford) scattering is due to electrostatic interactions between the primary electron beam and the nuclei. Where the energy of a primary electron is E_0 then the probability $\Pr(\theta)$ of it being scattered through an angle θ is given by Equation 4:

$$\Pr(\theta) \propto \frac{1}{E_0^2 \sin^4\theta} \qquad [4]$$

The probability of an electron suffering n random events whilst travelling through a specimen distance x, is given by the Poisson distribution (Equation 5)

$$\Pr(n) = \left(\frac{1}{n!}\right)\left(\frac{x}{\lambda}\right)^n \exp\left(-\frac{x}{\lambda}\right) \qquad [5]$$

where λ is its mean free path. Inelastic scattering is a general purpose description involving any process causing primary electrons to lose energy (ΔE).

Low atomic weight molecules typical of biological organic molecules cause inelastic scattering of electrons and the resulting aberrations generate low resolution, blurred images with relatively low spatial contrast. High atomic weight molecules in the section cause the greatest and more elastic scattering. Scattering is approximately a linear function of atomic number.

Image contrast is therefore proportional to atomic number provided all other microscopical and operational variables are constant. The aim of a staining procedure in electron microscopy is, therefore, to induce electron scattering differentially in biological specimens by using a selective electron dense stain. The efficiency of an electron stain is therefore determined by the mass of the heavy element in a section stain. The most effective stains have large atomic numbers and are usually heavy metals.

Regional loss of electrons from the electron beam due to the increased local scattering induced by staining is detected by a lighter local image generated in the halide emulsion of the photographic negative film held in the microscope. In images, those regions of greatest scattering are termed 'electron dense' whilst those local regions of the image where there is minimal scattering are termed 'electron lucent'. In positive prints of the exposed negative the electron dense regions appear dark whilst electron lucent regions appear lighter and yield appropriate levels of image contrast.

Staining for electron microscopy is either positive in which the organelles of interest are stained (13) or negative in which the surrounding matrix is stained for contrast (14–16). Positive staining is carried out with a specific aim of increasing electron density either to demonstrate structure, or increasingly, to demonstrate a process or manipulation of the biological specimen. Specific biological structures may be stained or the stain may demonstrate a histochemical reaction to test for the presence of a specific chemical substance of the site of enzymic activity. More recently, one of the most exciting developments in electron microscopy has been the ability to produce pure monoclonal antibodies which can be combined (bound or conjugated) with electron dense stains to demonstrate the sites of specific antigens. An alternative approach is to use a negative staining procedure though presently it appears to have less potential and be less widely used than positive staining.

5.3 Structural staining

Heavy metals variously used in electron microscopy include iron (Z = 26), copper (Z = 29), silver (Z = 47), lanthanum (Z = 57), praseodymium (Z = 59), neodymium (Z = 60), gadolinium (Z = 64), tungsten (Z = 74), osmium (Z = 76), gold (Z = 79), lead (Z = 82), and uranium (Z = 92). The most commonly used stains are lead which is used as lead citrate and uranium as uranyl acetate. Osmium occupies a semi-anomalous position in the list since it may also be used in fixation procedures and its use may also be described routinely as a post-fixative. Some methods use iron and gold in colloidal form. Gold may be bound to specific proteins which, for example, may be monoclonal antibodies used to identify specific antigens. Copper is used in the form of a polyvalent basic dye, most commonly, alcian blue 8G (which has a molecular diameter of approx. 3 nm and a weight of 1300 daltons). Each metal preparation has specific properties which yield individual staining patterns and each also acts as a stain by different binding mechanisms. Clearly the mechanism of binding and staining may additionally be affected by the fixation and embedding procedures so careful experimental design in the choice of fixation and embedding is always required.

The most widely used stain in electron microscopy for morphological study is lead citrate. The staining solution is prepared by dissolving 1.33 g of lead nitrate ($Pb(NO_3)_2$) in 30 ml of CO_2-free distilled water containing 1.76 g of sodium citrate ($Na_3(C_6H_5)_7 2H_2O$). The citrate acts as a chelating agent and sufficient is present to prevent the formation of undesirable lead carbonate. Lead citrate may also be used as a double stain with uranyl acetate which is an excellent general stain and is the most widely used counterstain. Its binding properties depend markedly upon its concentration and its pH. Uranyl acetate penetrates the entire thickness of the section and components are consequently stained throughout the section. Uranyl acetate increases overall contrast and stabilizes membranous and nucleic acid-containing structures. Concentrations of solutions range from 1–4% (w/v) and may be contained either in distilled water or in media containing up to 95% ethanol. Staining times and temperatures are variable according to section and other local conditions. Optimal staining procedures may require trial and error modification by individual investigators and what is regarded as adequate staining is often either purely subjective or directly related to the investigation being pursued. For example, staining of intranuclear (rather than isolated) chromatin can be more densely stained by the incorporation of relatively high concentrations of methanol (25%) in staining solutions.

5.4 Antibody staining

The almost infinite variety of antigens all of which will ultimately require localization precludes anything other than the most general indication of principles. The staining of antigen sites is one of the major topics of biology and medicine (17,18). For example, the structure of membranes and the spatial mapping of membrane components using antibody–electron microscopical techniques is one of the most important of all topics ranking almost with that of DNA base sequencing procedures. Immunoantibody staining can utilize colloidal gold staining techniques or the use of antibodies conjugated with enzymes such as horse-radish peroxidase and which are then stained (*vide infra*). Consequently there is often appreciable overlap in the technologies of these procedures.

5.5 Colloidal gold

Gold in colloidal form is sufficiently dense to impart electron density to biological macromolecules such as primary antibody molecules. The binding of colloidal gold to biological molecules is not based on covalent bonding so biological activity of the macromolecules is usually retained. Colloidal gold (spherical) particles of different sizes are available so multiple staining procedures can be undertaken in which the sizes of the gold particles (2–150 nm) are used to identify specific staining sites. The affinity of colloidal gold to biological substances is high and withstands much electron microscope preparation without significant loss. It is readily available from commercial sources or relatively easy to prepare in a well equipped laboratory (19,20).

The sizes of gold particles can readily be measured by attaching colloidal gold to Formvar coated grids. Essentially, in using immunogold techniques, specific antibodies are bound with colloidal gold particles and then allowed to react with antigen sites in biological specimens where the gold particles allow electron microscopical visualization.

Fixation, dehydration, and embedding processes affect the retention of antigenicity. Hydrophilic resins such as Lowicryl K4M and LR gold are more reliable than epoxy resins for staining. Low temperature processing is of additional benefit. The degree of labelling is subject to many variables such as the concentration of the gold markers, their size, temperature, media viscosity, and staining duration. Efficiency of labelling is inversely related to the square root of size of the gold particles.

Numerous techniques abound though all share some common basis. *Staphylococcus aureus* protein A in combination with colloidal gold is used to interact with the Fc fragment of immunoglobulin G via tyrosine and histidine amino acids groups of the protein and is a widely used approach. Most gold is bound to protein A near its isoelectric point of approx. pH 5.1 but best results are obtained by binding at approx. pH 6.0 and the number of protein A molecules adsorbed per gold particle depends on particle size. The smaller the particle size, the fewer are the number of protein molecules adsorbed. Protein G isolated from *Streptococcus G* by enzymic digestion with papain can also be used as a colloidal gold binding agent. Each protein has an optimal concentration at which it stabilizes colloidal gold. The pH range at which the colloidal gold is stable has to be determined, as has the optimal pH for stability of the gold protein complex.

A general procedure for preparing a protein A–gold complex is provided in *Protocol 1* and *Protocol 2* gives details of sample handling.

Protocol 1. Preparation of a protein A–gold complex

Equipment and reagents

- Ultracentrifuge with fixed-angle (e.g. 10 × 10 ml) and swinging-bucket (e.g. (6 × 14 ml) rotors
- Colloidal gold
- 0.1 M NaOH
- Protein A solution
- Phosphate-buffered saline (PBS)
- 10–30% (w/v) glycerol gradient in PBS

Method

1. Dissolve the protein in distilled water at 1 mg/ml.
2. Adjust the pH of the colloidal gold solution to 6.0 using 0.1 M NaOH.
3. Prepare the protein–gold complex by adding protein A solution to the gold solution at pH 6.0.
4. Centrifuge the protein A–gold complex for 45 min at 125 000 *g* for 5 nm gold particles. Other sizes require different centrifugation conditions.
5. Remove the loose part of the pellet and resuspend in phosphate-buffered saline (PBS) at pH 7.3 for layering over a 10–30% (w/v) continuous glycerol gradient in PBS containing 1% (w/v) Carbowax 20M.
6. Centrifuge the gradient for 45 min at 125 000 *g* for 5 nm gold particle.
7. Identify the dark red band for collection which contains protein A–gold.

Whilst the complex is reasonably stable on storage in 20% (w/v) glycerol at −70 °C deterioration nevertheless occurs.

Protocol 2. Sample handling and labelling

Equipment and reagents

- Nickel grids
- 1% (w/v) ovalbumin in PBS pH 7.4
- Antibody solutions
- Protein A–gold solution
- 5% (w/v) aqueous uranyl acetate and lead citrate

Method

1. Thin sections are mounted on nickel grids.
2. Float grid sections on 1% (w/v) ovalbumin in PBS pH 7.4 to block non-specific attachment of antibodies to glutaraldehyde residues.
3. Transfer grids directly to antibody solution for several hours at room temperature or overnight at 4 °C.
4. Transfer to drops of protein A–gold solution for 30–60 min in a moist chamber.
5. Subsequent staining may be carried out with 5% (w/v) aqueous uranyl acetate and lead citrate if required.

As an alternative to antibody staining of antigen sites using colloidal gold complexes, the carbohydrate sequences in mucopolysaccharides can also be investigated using lectins to which colloidal gold has been bound. The reader is referred to a number of practical texts (21) for exact experimental details.

5.6 Enzyme cytochemistry

Enzyme histochemistry is a general term which describes the techniques used to localize the site of enzyme activity in biological specimens *in situ* (21). The fundamental aim of any preparative procedure designed to demonstrate enzymic activity *in situ* is the retention of both the activity and the topology of the enzyme. Fixation must bind the enzyme to its correct site (topology) but not destroy the enzyme activity. The fundamental procedure is to provide a specific substrate for the enzyme, the activity of which creates a primary reaction product (PRP). Subsequently, the site of the PRP (its topology) must then be demonstrated in a section either by its intrinsic electron density or by a subsequent procedure to convert the electron lucent PRP to an electron dense final reaction product (FRP).

There are obviously many opportunities for error in localizing sites of enzyme activity. The staining (localization) procedure must not inhibit enzyme activity. As with all enzyme and staining studies, the establishment of stoichiometric relation between staining and activity must be established before final interpretation can be attempted. The activity of the enzyme must survive experimental manipulation (including chemical fixation where appropriate), embedding in the appropriate medium, and allow a substrate to bind and be acted upon the active sites of the enzyme. The PRP formed at the enzyme site must remain localized at the enzyme site either by insoluble deposition or binding to structures local to the enzymic site. The PRP must not inhibit enzyme activity excessively and must always be capable of being either sufficiently electron dense itself to permit visualization or be converted, without removal, to a suitable electron dense FRP.

There are, or eventually will be, as many thousands of histochemical methods as there are enzymes which ultimately all will require topological localization. Therefore, only general principles and specimen technical details can be provided since each enzyme has its own unique method of demonstration. However, those factors which result in false localization of enzymic activity remain applicable to all cytochemical demonstrations.

As a specific example a simple phosphohydrolytic enzyme (acid phosphatase, E.C. 3.1.3.2) of lysosomes will be considered. Phosphohydrolytic activity can be demonstrated using a wide range of simultaneous azo-dye, azoindoxyl, tetrazolium, indigogenic, and metal salt techniques. Of these only metal salt techniques are relevant to electron microscopy. Essentially, the technique is to incubate the specimen in the presence of a suitable phosphate-containing substrate which on diffusing into the enzyme site is rapidly hydrolysed to release inorganic phosphate. The released phosphate is immediately captured to form an insoluble PRP by the presence of suitable ions incorporated in the incubating medium. Most recently soluble non-inhibitory trichlorides of lanthanum and other lanthanides such as gadolinium, didymium, praseodymium and neodymium have been used. The phosphate is precipitated as the PRP lanthanide phosphate. Subsequent incubation in sodium fluoride converts the lanthanide phosphate into almost insoluble lanthanide fluoride which is sufficiently electron dense to be visualized within the lysosomes by electron microscopy.

Original techniques involved the use of lead ions but often the lead ions were either inhibitory to enzyme activity or formed contaminating precipitates yielding false staining. More modern techniques introduced the use of cerium as a lead substitute to avoid enzyme inhibition. Care must be taken not only with careful choice of fixative but also of buffers used during fixation to avoid enzyme inhibition.

Staining of acid phosphatase in lysosomes is given as an example in *Protocol 3*.

Protocol 3. Acid phosphatase staining in lysosomes

Reagents

- 1.5% (w/v) glutaraldehyde/7.5% (w/v) sucrose in 0.05 M cacodylate buffer pH 7.2
- 0.5 mg/ml sodium borohydride, 0.1% (w/v) saponin in 0.05 M Tris–maleate pH 7.2
- 3 mM $CeCl_3$, $LaCl_3$, $DiCl_3$, or $GdCl_3$, 10 mM 3-amino-1,2,4-triazol in 0.05 M Tris–maleate pH 5.2
- 15 mM 2-glycerophosphate
- 7.5% (w/v) sucrose in 0.05 M cacodylate buffer pH 6.0
- 5–10 mM sodium fluoride in 0.05 M cacodylate buffer pH 7.2

Method

1. Fix specimens in 1.5% (w/v) glutaraldehyde/7.5% (w/v) sucrose contained in 0.05 M cacodylate buffer at pH 7.2.
2. Incubate in 0.5 mg/ml sodium borohydride and 0.1% (w/v) saponin contained in 0.05 M Tris–maleate buffer at pH 7.2 for 60 min at 20°C.
3. Incubate in substrate medium containing 3 mM $CeCl_3$, or $LaCl_3$, or $DiCl_3$, or $GdCl_3$, and 10 mM 3-amino-1,2,4-triazol contained in 0.05 M Tris–maleate buffer at pH 5.2 at 37°C for 1 h.
4. Post-incubate in a medium containing 15 mM 2-glycerophosphate for 2 h at 37°C.
5. Wash overnight at 4°C in 0.05 M cacodylate buffer containing 7.5% (w/v) sucrose at pH 6.0.
6. Incubate in 5–10 mM sodium fluoride in 0.05 M cacodylate buffer at pH 7.2 for 1 h at 20°C to convert PRP lanthanide phosphate to FRP lanthanide fluoride.

A further use of enzyme staining is that enzymes may be used as staining probes when conjugated with antibodies used to localize specific antigen sites. The underlying rationale is that whilst an individual antibody molecule may not be seen with an electron microscope, the presence of an enzyme (such as horse-radish peroxidase) attached to the probe may allow sufficient final reaction product to be formed for its identification.

5.7 Histochemistry

The simplest histochemical procedures are designed to localize the position of a specific chemical compound or macromolecular structure within cell components or organelles, for example, the carbohydrate and mucopolysaccharide complexes associated with membranes. The staining of 1,2-glycol groups within the Golgi apparatus using silver techniques is instructive. Periodate reactive carbohydrates are demonstrable using a periodic acid–thiosemicarbohydrazide–silver proteinate (PA–TCH–SP) technique. The method is described in *Protocol 4*.

Protocol 4. Carbohydrate staining using the PA–TCH–SP technique

Reagents

- 1% (w/v) periodic acid
- 1% (w/v) silver proteinate
- 1% (w/v) thiosemicarbohydrazide (TCH), in 10% (w/v) acetic acid

Method

1. Oxidize sections for 40 min using 1% aqueous periodic acid.
2. Rinse sections in distilled water.
3. Place sections on surface of 1% TCH in 10% acetic acid at room temperature for at least 1 h.
4. Rinse repeatedly and stain with 1% solution of silver proteinate for 30 min in the dark.

The raising of the pH of the silver proteinate from 6.4 in its native state to pH 9.2 significantly enhances staining of the glycocalyx.

For the staining of the carbohydrate associated with the Golgi apparatus a periodic acid-chromic acid-silver methenamine technique is available.

Protocol 5. Staining Golgi carbohydrate

Reagents

- 1% (w/v) periodic acid
- 10% (w/v) chromic acid
- 1% (w/v) sodium bisulfite
- Silver methenamine

Method

1. Oxidize sections for 20 min using 1% aqueous periodic acid.
2. Wash in distilled water for 30 min.
3. Oxidize in aqueous 10% chromic acid solution at room temperature for 5 min.
4. Wash with 1% sodium bisulfite solution for 1 min.
5. Wash in distilled water for 30 min.
6. Stain with silver methenamine for 30 min at 60°C.
7. Wash in distilled water for approx. 20 min.

Staining intensity may be varied according to the duration of staining time with silver methenamine.

5.8 Negative staining

The morphology and also the structure, though to a lesser extent, of simple particulate specimens such as isolated cell components, e.g. ribosomes, membranes, and isolated macromolecules can be studied using the techniques of negative staining (14,15). The techniques are also available for shape resolution of particulate structures (including viruses) using the high voltage electron microscope. The techniques of negative staining reduce electron radiation damage of specimens and assist thereby the preservation of macromolecular structure.

In principle, the specimen is embedded directly in a negative stain and on subsequent drying electron dense metal stain covers the specimen in three dimensions. With the electron microscope the specimen appears electron lucent surrounded by an electron dense stain. The boundary image between stain and specimen reveals surface morphology. Such structure as is revealed within the specimen is derived from penetration of the stain into the specimen through fissures or holes which might be artefactual.

The most commonly used negative stains are phosphotungstic acid (PTS) and uranyl acetate which are also used for positive staining. Unlike positive staining, the techniques of negative staining do not depend on chemical reaction or significant diffusion processes and binding is minimized by selection of an optimal pH at which minimal specimen–stain attraction occurs. Satisfactory staining still retains an element of artistry and practice.

The choice of stain depends on the aims of the investigation and the ability of the stain to yield adequate images. Negative staining techniques vary considerably between biological specimens and initially in an investigation multiple staining procedures should be undertaken. Image clarity depends on the nature of the stain, the degree to which it is affected by the drying process, and the amount of stain relative to the size of the particulate specimens. A suitable negative stain must be highly soluble to achieve high specimen loading, high melting and boiling points which assist in the stability under the electron beam and, as with positive stains, must naturally be sufficiently electron dense.

Staining may either be by simultaneous one step or sequential two step procedures. In the simple one step procedure, a drop of a 1:1 specimen–stain mixture is placed on a carbon coated grid and dried normally by evaporation or is vacuum dried. In the slightly more complicated two step method, the specimen is adsorbed on to the grid and the negative stain subsequently (i.e. sequentially) applied. This method is useful for low concentrations of particles and allows other procedures to be used as an intermediate stage between adsorption and negative staining. Alternative sequences may also be used to form negative stain–carbon films (NS–CF). Although the NS–CF techniques are well established, technical modifications are required according to the properties of the macromolecule under investigation. For example, a procedure given by Harris (1991) for studying human erythrocyte catalase (E.C. 1.11.1.6) is given in *Protocol 6*.

Protocol 6. NS–CF study of human erythrocyte catalase

Equipment and reagents

- Mica sheets
- Carbon rod
- Uncoated grids
- Ammonium molybdate, 0.1% (w/v) polyethylene glycol
- 2% (w/v) uranyl acetate

Method

1. Mix 10 μl of protein solution of 1.0–20 mg/ml with 10 μl of ammonium molybdate containing 0.1% (w/v) polyethelene glycol (use lower range of molecular weights of the glycol).
2. Adjust pH using sodium hydroxide or ammonia within pH range 5.5–9.0.
3. Place 10 μl of mixture on to a freshly cleaved sheet of mica and dry at room temperature.
4. Coat dried specimen with carbon film generated *in vacuo* by resistance heating of pointed carbon rod.
5. Float layer of carbon plus adsorbed protein on to the surface of 2% (w/v) aqueous uranyl acetate.
6. Mount small regions of carbon film from beneath the surface of uranyl acetate on uncoated grids.
7. Allow specimen to air dry and examine with an electron microscope.

Negative staining procedures can also be applied to freeze-dried material for some specimens, usually isolated virus particles, and in immunoelectron microscopy to visualize interactions between antibodies and macromolecular structures.

6. Further reading

No single chapter can possibly prepare the investigator for carrying out modern electron microscopy. I have indicated here some of the classical texts, notably those by Glauert and co-authors (1) and the encyclopaedic series of volumes by Hayat (2,3,8,11–14,19–21). In addition, several journals are invaluable sources of old and continuing new information. The *Journal of Microscopy* published by the Royal Microscopical Society, Oxford, *The Journal of Histochemistry and Cytochemistry* and the *Journal of Ultrastructure Research* (variously renamed after the late 1980s and now the *Journal of Structural Biology*). Other journals include *Ultramicroscopy*, the *Journal of*

Electron Microscopy Techniques, Scanning Electron Microscopy and *Micron and Microscopica Acta* recently reconstituted as *Micron: The International Research and Review Journal for Microscopy* in 1992.

References

1. Glauert, A. M. (1974–80). *Practical methods in electron microscopy*, Vols 1–8. North-Holland, Amsterdam.
2. Hayat, M. A. (1973–78). *Principles and techniques of electron microscopy: biological applications,* Vols 1–8. Van Nostrand Reinhold, New York.
3. Hayat, M. A. (1989). *Principles and techniques of electron microscopy: biological applications*. Macmillan Press, Scientific and Medical, London.
4. Weakley, B. S. (1981). *A beginner's handbook in biological electron microscopy*. Churchill Livingstone, London.
5. Goodhew, P. J. and Humphreys, F. J. (1988). *Electron microscopy and analysis*. Taylor and Francis, London.
6. Robards, A. W. (1986). *Botanical microscopy*. Oxford Scientific Publications, Oxford.
7. Kay, D. H. and Coslett, V. (1965). *Techniques for electron microscopy*. F. A. Davies, Philadelphia.
8. Hayat, M. A. (1981). *Fixation for electron microscopy*. Academic Press, London.
9. Steinbrecht, R. A. and Zierold, K. (1986). *Cryotechniques in biological electron microscopy*. Springer-Verlag, Berlin.
10. Rash, J. E. and Hudson, C. S. (1979). *Freeze-fracture: methods, artifacts, and interpretations*. Raven Press, New York.
11. Hayat, M. A. (1979). *Introduction to biological scanning microscopy*. University Park Press, Baltimore.
12. Hayat, M. A. (1986). *Basic techniques for transmission electron microscopy*. Academic Press, London.
13. Hayat, M. A. (1975). *Positive staining for electron microscopy*. Van Nostrand Reinhold, New York.
14. Hayat, M. A. and Miller, M. (1989). *Negative staining: methods and applications*. McGraw-Hill, New York.
15. Harris, J. R. and Horne, R. W. (1991). In *Electron microscopy in biology* (ed. J. R. Harris), pp. 203–28. IRL/OUP, Oxford.
16. Horne, R. W. and Cockayne, D. J. H. (1991). *Micron Microsc. Acta*, **22**, 321.
17. Bullock, G. R. and Petrisz, P. (1985). *Techniques in immunocytochemistry*. Academic Press, London.
18. Polak, J. M. and Varndell, I. M. (1984). *Immunolabelling for electron microscopy*. Elsevier, Amsterdam.
19. Hayat, M. A. (1989). *Colloidal gold: principles, methods, and applications*, Vols 1–2. Academic Press, London.
20. Hayat, M. A. (1992). *Micron Microsc. Acta*, **23**, 1.
21. Hayat, M. A. (1973–77). *Electron microscopy of enzymes*, Vols 1–5. Van Nostrand Reinhold, New York.

List of suppliers

Amersham

Amersham International plc., Lincoln Place, Green End, Aylesbury, Buckinghamshire HP20 2TP, UK.

Amersham Corporation, 2636 South Clearbrook Drive, Arlington Heights, IL 60005, USA.

Anderman

Anderman and Co. Ltd., 145 London Road, Kingston-Upon-Thames, Surrey KT17 7NH, UK.

Beckman Instruments

Beckman Instruments UK Ltd., Oakley Court, Kingsmead Business Park, London Road, High Wycombe, Bucks HP11 1J4, UK.

Beckman Instruments Inc., PO Box 3100, 2500 Harbor Boulevard, Fullerton, CA 92634, USA.

Becton Dickinson

Becton Dickinson and Co., Between Towns Road, Cowley, Oxford OX4 3LY, UK.

Becton Dickinson and Co., 2 Bridgewater Lane, Lincoln Park, NJ 07035, USA.

Bio

Bio 101 Inc., c/o Statech Scientific Ltd, 61–63 Dudley Street, Luton, Bedfordshire LU2 0HP, UK.

Bio 101 Inc., PO Box 2284, La Jolla, CA 92038–2284, USA.

Bio-Rad Laboratories

Bio-Rad Laboratories Ltd., Bio-Rad House, Maylands Avenue, Hemel Hempstead HP2 7TD, UK.

Bio-Rad Laboratories, Division Headquarters, 3300 Regatta Boulevard, Richmond, CA 94804, USA.

Boehringer Mannheim

Boehringer Mannheim UK (Diagnostics and Biochemicals) Ltd, Bell Lane, Lewes, East Sussex BN17 1LG, UK.

Boehringer Mannheim Corporation, Biochemical Products, 9115 Hague Road, P.O. Box 504 Indianapolis, IN 46250–0414, USA.

Boehringer Mannheim Biochemica, GmbH, Sandhofer Str. 116, Postfach 310120 D-6800 Ma 31, Germany.

British Drug Houses (BDH) Ltd, Poole, Dorset, UK.

Difco Laboratories

Difco Laboratories Ltd., P.O. Box 14B, Central Avenue, West Molesey, Surrey KT8 2SE, UK.

Difco Laboratories, P.O. Box 331058, Detroit, MI 48232–7058, USA.

Du Pont

Dupont (UK) Ltd., Industrial Products Division, Wedgwood Way, Stevenage, Herts, SG1 4Q, UK.

Du Pont Co. (Biotechnology Systems Division), P.O. Box 80024, Wilmington, DE 19880–002, USA.

European Collection of Animal Cell Culture, Division of Biologics, PHLS Centre for Applied Microbiology and Research, Porton Down, Salisbury, Wilts SP4 0JG, UK.

Falcon (Falcon is a registered trademark of Becton Dickinson and Co.).

Fisher Scientific Co., 711 Forbest Avenue, Pittsburgh, PA 15219–4785, USA.

Flow Laboratories, Woodcock Hill, Harefield Road, Rickmansworth, Herts. WD3 1PQ, UK.

Fluka

Fluka-Chemie AG, CH-9470, Buchs, Switzerland.

Fluka Chemicals Ltd., The Old Brickyard, New Road, Gillingham, Dorset SP8 4JL, UK.

Gibco BRL

Gibco BRL (Life Technologies Ltd.), Trident House, Renfrew Road, Paisley PA3 4EF, UK.

Gibco BRL (Life Technologies Inc.), 3175 Staler Road, Grand Island, NY 14072–0068, USA.

Arnold R. Horwell, 73 Maygrove Road, West Hampstead, London NW6 2BP, UK.

Hybaid

Hybaid Ltd., 111–113 Waldegrave Road, Teddington, Middlesex TW11 8LL, UK.

Hybaid, National Labnet Corporation, P.O. Box 841, Woodbridge, NJ 07095, USA.

HyClone Laboratories 1725 South HyClone Road, Logan, UT 84321, USA.

International Biotechnologies Inc., 25 Science Park, New Haven, Connecticut 06535, USA.

Invitrogen Corporation

Invitrogen Corporation 3985 B Sorrenton Valley Building, San Diego, CA 92121, USA.

Invitrogen Corporation c/o British Biotechnology Products Ltd., 4–10 The Quadrant, Barton Lane, Abingdon, OX14 3YS, UK.

Kodak: Eastman Fine Chemicals 343 State Street, Rochester, NY, USA.

Life Technologies Inc., 8451 Helgerman Court, Gaithersburg, MN 20877, USA.

Merck

Merck Industries Inc., 5 Skyline Drive, Nawthorne, NY 10532, USA.

Merck, Frankfurter Strasse, 250, Postfach 4119, D-64293, Germany.

Millipore

Millipore (UK) Ltd., The Boulevard, Blackmoor Lane, Watford, Herts WD1 8YW, UK.

Millipore Corp./Biosearch, P.O. Box 255, 80 Ashby Road, Bedford, MA 01730, USA.

New England Biolabs (NBL)

New England Biolabs (NBL), 32 Tozer Road, Beverley, MA 01915–5510, USA.

New England Biolabs (NBL), c/o CP Labs Ltd., P.O. Box 22, Bishops Stortford, Herts CM23 3DH, UK.

Nikon Corporation, Fuji Building, 2–3 Marunouchi 3-chome, Chiyoda-ku, Tokyo, Japan.

Perkin-Elmer

Perkin-Elmer Ltd., Maxwell Road, Beaconsfield, Bucks. HP9 1QA, UK.

Perkin-Elmer Ltd., Post Office Lane, Beaconsfield, Bucks, HP9 1QA, UK.

Perkin-Elmer-Cetus (The Perkin-Elmer Corporation), 761 Main Avenue, Norwalk, CT 0689, USA.

Pharmacia Biotech Europe Procordia EuroCentre, Rue de la Fuse-e 62, B-1130 Brussels, Belgium.

Pharmacia Biosystems

Pharmacia Biosystems Ltd. (Biotechnology Division), Davy Avenue, Knowlhill, Milton Keynes MK5 8PH, UK.

Pharmacia LKB Biotechnology AB, Björngatan 30, S-75182 Uppsala, Sweden.

Promega

Promega Ltd., Delta House, Enterprise Road, Chilworth Research Centre, Southampton, UK.

Promega Corporation, 2800 Woods Hollow Road, Madison, WI 53711–5399, USA.

Qiagen

Qiagen Inc., c/o Hybaid, 111–113 Waldegrave Road, Teddington, Middlesex, TW11 8LL, UK.

Qiagen Inc., 9259 Eton Avenue, Chatsworth, CA 91311, USA.

Schleicher and Schuell

Schleicher and Schuell Inc., Keene, NH 03431A, USA.

Schleicher and Schuell Inc., D-3354 Dassel, Germany. Schleicher and Schuell Inc., c/o Andermann and Company Ltd.

Shandon Scientific Ltd., Chadwick Road, Astmoor, Runcorn, Cheshire WA7 1PR, UK.

Sigma Chemical Company

Sigma Chemical Company (UK), Fancy Road, Poole, Dorset BH17 7NH, UK.

Sigma Chemical Company, 3050 Spruce Street, P.O. Box 14508, St. Louis, MO 63178–9916.

Sorvall DuPont Company, Biotechnology Division, P.O. Box 80022, Wilmington, DE 19880–0022, USA.

Stratagene

Stratagene Ltd., Unit 140, Cambridge Innovation Centre, Milton Road, Cambridge CB4 4FG, UK.

Strategene Inc., 11011 North Torrey Pines Road, La Jolla, CA 92037, USA.

United States Biochemical, P.O. Box 22400, Cleveland, OH 44122, USA.

Wellcome Reagents, Langley Court, Beckenham, Kent BR3 3BS, UK.

Index